D1516598

CERAMIC AND GRAPHITE
FIBERS AND WHISKERS

REFRACTORY MATERIALS

A SERIES OF MONOGRAPHS

John L. Margrave, *Editor*
DEPARTMENT OF CHEMISTRY
RICE UNIVERSITY, HOUSTON, TEXAS

VOLUME 1. L. R. McCreight, H. W. Rauch, Sr., W. H. Sutton
Ceramic and Graphite Fibers and Whiskers
A Survey of the Technology

CERAMIC AND GRAPHITE
FIBERS AND WHISKERS
A Survey of the Technology

L. R. McCreight, H. W. Rauch, Sr., and W. H. Sutton

SPACE SCIENCES LABORATORY, GENERAL ELECTRIC COMPANY, VALLEY FORGE, PA.

1965

ACADEMIC PRESS   New York and London

ACADEMIC PRESS INC.
111 Fifth Avenue, New York, New York 10003

United Kingdom Edition published by
ACADEMIC PRESS INC. (LONDON) LTD.
Berkeley Square House, London W.1

Library of Congress Catalog Card Number: 65-27784

PRINTED IN THE UNITED STATES OF AMERICA

PREFACE

Continuing advances toward the maturation of space technology have emphasized the marginal response of many commercially available synthetic fibers and whiskers in the rigorous environment of aerospace. Research efforts in the field of fiber technology are aimed at development of efficient, high-temperature structural materials to fulfil future requirements.

A number of research programs are being carried out by various organizations, but the wealth of text data are not readily available to the engineer for his evaluation and analysis.

As a result, the Air Force Systems Command, through its Ceramics and Graphite Information Center at Wright-Patterson Air Force Base, Ohio, initiated a contract* with the Space Sciences Laboratory of the General Electric Company to conduct a survey of the state-of-the-art of whisker and fiber technology.

The authors especially wish to acknowledge and express their appreciation for the contributions and cooperation of:

Messrs. S. W. Bradstreet and B. R. Emrich for their assistance and support in obtaining reports, samples and contracts in this field.

Dr. Louis Navias, Consultant and retired "Dean of Ceramics" in the General Electric Company, for his abstracting of patents and as general consultant.

Messrs. A. C. Harrison and J. H. Wood, and Mrs. Laurel Kuttner, for their editing.

Each of the many personnel in over sixty organizations who contributed information to the survey.

Valley Forge, Penna.

July, 1965

L. R. McCreight
H. W. Rauch, Sr.
W. H. Sutton

*USAF Contract No. AF 33(615)-1618

v

TABLE OF CONTENTS

LIST OF ILLUSTRATIONS

LIST OF TABLES

Page

I. INTRODUCTION

Ceramics and graphite offer potentially greater strengths than any other materials. They possess the desirable combination of high modulus of elasticity, high strength retention at high temperatures, and relatively low density that makes them prime candidates for many structural applications. They are inherently brittle, though, which necessitates carefully designed applications in order to avoid excessive shear and tensile loadings. However, in filamentary form they contain fewer gross flaws than in the bulk forms and, consequently, have very high tensile strengths and even show flexibility.

Ceramic and graphite filaments are currently being used as insulation, fillers, textiles, and as reinforcement for plastics. While the applications for single crystal filaments (whiskers) are still in the development stage, the ability to make them into tape and paper forms as well as the feasibility for reinforcing both metal and plastic matrices have been demonstrated.

A review of fiber technology reveals that most commercially available ceramic and graphite filaments have already become marginal in the rigorous and hostile environments of aerospace applications. Deficiencies, principally in high temperature strength, have resulted in accelerated efforts to develop stiffer, more refractory filaments having high strength-to-density ratios. As a result, the number of companies developing filamentary materials is greatly increasing and the volume of related literature (including patents) is rapidly expanding. The current results of the research and development programs are so widely disseminated throughout the numerous reports and publications that many engineers and scientists have neither the time nor the familiarity with data retrieval to pursue this information. Therefore, a prime objective of this book is to make available the results of an Air Force-sponsored survey of the technology of

ceramic and graphite fibers and whiskers to research, development, engineering, and manufacturing personnel. Thus, this book
presents a unified, comprehensive and authoritative source of the
unclassified and non-proprietary literature, patents, and current
activities of many organizations.

In order to cover the vast amount of information available,
an extensive literature search, including patents, was undertaken.
Questionnaires were also sent to numerous educational, industrial,
and government institutions. Visits to several laboratories,
including 23 in Europe, were made in order to gain a comprehensive insight into the progress, trends, and anticipated future
developments in fibers and fibrous composites.

The main subjects discussed in the book are the theoretical
strength of materials, the classification and applications of fibers,
and the factors affecting fiber strength. In addition, the information presented is supplemented by a 550-entry bibliography, cross-
indexed according to subject and author, abstracts of over 200
patents, and over 60 contact reports based on visits to both
domestic and foreign organizations currently working with filamentary materials.

The type of materials considered in this book primarily
includes those compositions useful above $1200^{\circ}F$, such as non-
conventional glasses (E-glass is used as a reference point), glass-
ceramics, ceramic oxides, graphite, carbides, and other similar
materials. Both single crystal whiskers and polycrystalline fibers
are discussed.

The compilation and evaluation of data include such properties as tensile strength, modulus of elasticity, temperature
resistance, flexibility, abrasion resistance, chemical stability,
and other properties which may characterize a certain material.
When possible, other characterizing information such as fiber
diameter, micro-structure, composition, surface conditions,
forming method, test method, and other defining parameters is
incorporated into the data. The data are summarized and presented in a format for easy retrieval. Various types of forming
and processing techniques are discussed. Since most of the high-
temperature fibers pose problems in forming them into woven
structures, and in the ability to be converted into flexible and
resilient woven forms, some methods for overcoming forming
problems are mentioned.

II. THE POTENTIAL STRENGTH OF MATERIALS

The achievement of near-theoretical strengths by single crystal whiskers of numerous materials has prompted new impetus to the development of new high-temperature structural composites. The high temperature performance of sapphire single crystal whiskers, for example, is extremely noteworthy. They retain considerable strength at temperatures far above the softening point of the best commercially available glass fibers, as shown in Figure 1.

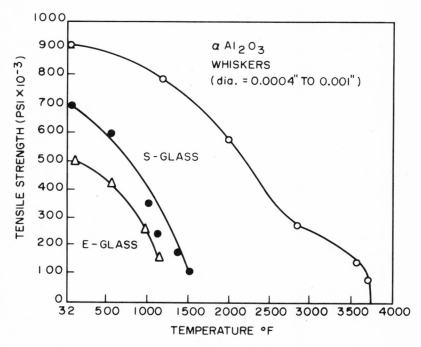

Figure 1. *Tensile Strength of αAl_2O_3 Whiskers (Brenner, Ref. 61) Compared with E-and S-Glass Filaments (Contact Report No. 45) at Various Temperatures.*

One of the prime advantages in using fibrous materials for reinforcements is their high inherent strength and modulus. Nearly all materials are far stronger in the fiber form than in the bulk. This arises from the fact that the best available fiber-forming processes and the shape and size of the fiber all tend to favor a higher degree of surface and internal (crystallographic) perfection which in turn leads to higher strengths. These effects are discussed in greater detail in Chapter IV. However, it is helpful to consider the theoretical strength of materials in order to establish a basis for predicting the maximum potential strength and for evaluating the extent to which current fibers have achieved this potential.

The cohesive energy, which is a measure of the strength of a solid, is the energy required to separate the constituent atoms (ions) to an infinite distance. Polanyi[370]*, in 1921, made a simple estimate of what the theoretical cohesive force might be for solids. Basing his theory on a single ionic model (NaCl) he derived the tensile strength, σ_{max}, in terms of the surface energy, γ, the elastic modulus, E, and the 'elongation', L_o, necessary to cause fracture. This relationship is shown as follows:

$$\sigma_{max.} \cong \sqrt{\frac{4E\gamma}{L_o}} \qquad\qquad (1)$$

Brenner[59] reports that several other investigators have since modified this equation to yield:

$$\sigma_{max} = K \sqrt{\frac{E\gamma}{a}} \qquad\qquad (2)$$

where

    K = a coefficient of the order of unit

    a = the crystal lattice parameter

---

*Superscript numbers refer to bibliography in Section IX-B.

Charles[70], using equation (2) to determine a, and estimating the fracture energy for Si-O-Si linkages in fused silica, calculated a theoretical value for $\sigma_{max} \cong 4 \times 10^6$ psi. Brenner[59], applying equation (1) to the data for Al_2O_3, has estimated the upper limit for the strength of Al_2O_3 as $0.17E$. By comparison, the measured strength of Al_2O_3 single crystal rods is about $0.001E$.

Brenner[59] has also discussed the theoretical shear strength, τ_{max}, necessary to initiate plastic flow in a perfect crystal. Based on Frenkel's analysis, Brenner has assumed that

$$\frac{\tau_{max}}{G} \cong \frac{\sigma_{max}}{E} \qquad (3)$$

where:

$$\tau_{max} = \left(\frac{b}{a'}\right)\left(\frac{G}{2\pi}\right) \qquad (4)$$

b = the distance between atoms in the slip plane,

a' = the distance between slip planes, so that $b/a' \approx 1$,

G = the shear modulus.

Based on the assumption in equation (3), Brenner[59] has shown that $\sigma_{max} \approx 0.16E$. However, a lower limit would be $\sigma \approx 0.03E$, because plastic deformation in a perfect crystal occurs more easily by the nucleation of a patch of slipped region bounded by a dislocation loop. Thus Brenner[59] concludes that the range for the theoretical strength of crystals is $0.03E - 0.17E$. Generally, an average value of about $0.1E$ is used.

Table I shows the properties, both theoretical (S_f in terms of E) and measured, of several reinforcing filaments. In general, the glass filaments and whisker materials exhibit strengths closest to the theoretical value. The values for glass filaments range between $0.024E$ and $0.081E$, for ceramic whiskers between $0.012E$ and $0.083E$, and for metal whiskers between $0.018E$ and $0.066E$. It is noteworthy that the whiskers which have been studied the most, viz., Al_2O_3 and Fe, have also shown the highest strength values, $\sigma_{max} = 0.1E$ and $\sigma_{max} = 0.066E$, respectively. This is probably due to 1) improved methods for controlling the growth of whiskers, and 2) testing of a larger population. Similarly, the technology for producing high strength steel wires is more advanced, and shows higher values in terms of E, for example, than molybdenum-wires ($\sigma_{max} = 0.021E$ and $0.006E$, for steel and molybdenum, respectively).

TABLE I. PROPERTIES OF REINFORCING FILAMENTS

Filament Type	Melt. Pt. or Soft. Pt. (^{o}C)	Density (lb/in^3)	Experimental Tensile Str. * ($\times 10^{-3}$ psi)	Young's Modulus ($\times 10^{-6}$ psi)	S_f in terms of E	References
I. Cont.						
A. Glass						
E-glass	840	0.092	250	10.5	0.024E	378,391,509
E/HTS	840	0.092	500	10.5	0.018E	378,391,509
YM 31A	840	0.103	500	16.0	0.031E	378,391,509
S-994	840	0.090	650	12.6	0.052E	378,391,509
29-A	900	0.096	800	14.5	0.055E	378
SiO_2	1660	0.079	850	10.5	0.081E	320
B. Metal						
B	2100	0.09	300-500	55-60	0.007E	CR/51
W	3400	0.697	580	52	0.011E	245
Mo	2622	0.369	320	52	0.006E	245,393
Rene 41	1350	0.298	290	24.2	0.012E	200,245
Steel	1400+	0.280	600	29.0	0.021E	391,393
Be	1284	0.066	185	35	0.005E	393
II. Discont. (whiskers)						
A. Ceramic						
$Al_2 O_3$	2040	0.143	6200	62-67	0.099E	61,388
BeO	2570	0.103	2800**	58-60	0.047E	436
$B_4 C$	2490d***	0.091	934	70	0.013E	184
SiC	2690d	0.115	1650	70	0.024E	547
Graphite	3650s	0.060	2845	102	0.036E	30
B. Metal						
Cu	1083	0.322	427	18	0.002E	63
Ni	1455	0.324	560	31	0.002E	63
Fe	1540	0.283	1900	29	0.066E	63
Cr	1890	0.260	1290	35	0.037E	174,281

* Specific Strength = S_f/D.
** Flexure Test.
*** d = decomposes; s = sublimes.

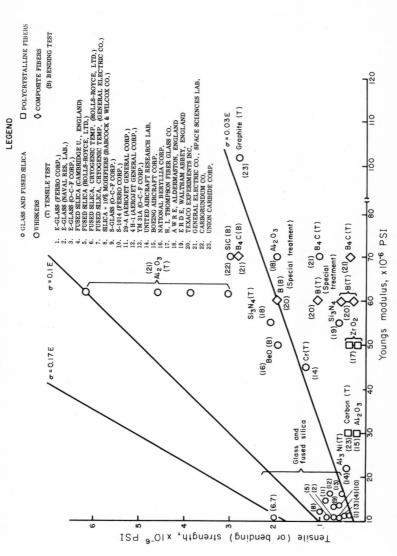

Figure 2. Selected Values of Tensile and/or Bending Strength at Accepted Values of Young's Modulus for Various Filaments. (All Data Obtained at Room Temperature Except Nos. 6 and 7.)

The comparison of tensile strength, theoretical tensile strength ($\sigma_{max} = 0.1E$), and elastic modulus is further illustrated in Figure 2. Of all the glass filaments, fused silica approaches $0.1E$ at room temperature and shows higher values at cryogenic temperatures ($\sigma_{max} \cong 0.2E$). On the other hand, most of the whisker (crystal fibers) have exhibited strengths which fall in the range $\sigma_{max} = 0.03E$ to $0.1E$. In one case the strength of an Al_2O_3 whisker approaches $E/10$ based on a value of 62×10^6 psi for E. However, the measured value of E for this whisker was unusually high (198×10^6 psi), which suggests the possibility of an error in determining the whisker's cross-sectional area (see section on testing).

In terms of potential reinforcements, Figure 3 presents values of the density divided by the third root of the modulus ($\rho/E^{1/3}$) as a function of the melting point or dissociation temperature. The best compositions from a structural viewpoint would be those having the lowest $\rho/E^{1/3}$ values and highest temperature capabilities. Materials such as Be, B, BN, B_4C, BeO, SiO_2, Al_2O_3, MgO, TiC, and diamond show the most desired characteristics on this basis. However, many other properties have to be considered to determine the useful or engineering value of filaments, for example, chemical durability, fabricability, ease of forming, and economical aspects are also most important. Therefore, it is the actually attainable properties and behavior, as well as the applications, that define their practical usefulness. The full merits of a given filament are based on several parameters, which are described in greater detail in this report.

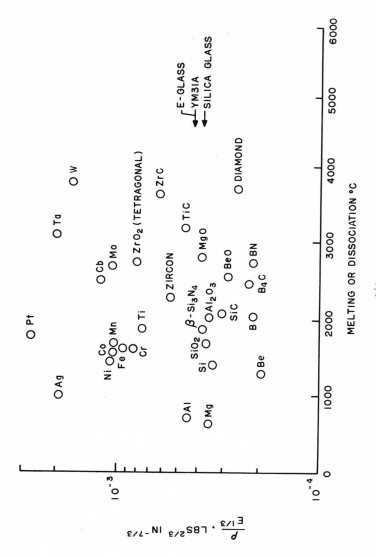

Figure 3. Relative Structural Efficiency, $\rho/E^{1/3}$, for Various Materials in Terms of Their Melting or Dissociation Temperatures.

III. APPLICATIONS OF FIBERS

A. Introduction

Ceramic and graphite fibers and whiskers are used primarily to reinforce composites; other applications include use as insulators, fillers, and filters. They are produced, with varying degrees of difficulty, as either long, continuous, filaments or short fibers, and formed into yarns, rovings, felts, papers, batts, etc. In general, the methods of handling, fabrication, and application of ceramic and graphite fibers are similar to those used for glass, mineral wool, asbestos, and related fibers although in some cases, these newer fibers may necessitate modification or further development of the present conventional techniques.

There are, however, some major dissimilarities between the ceramic and graphite fibers under consideration here and the asbestos, mineral wool, and (the majority of) glass fibers currently in use. Ceramic and graphite fibers with specific application to reinforcement have higher melting points, a higher modulus of elasticity, and are carefully made to provide high tensile strengths. In addition, the great hardness and relatively high degree of chemical inertness of the ceramic and graphite fibers help them to resist mechanical and chemical degradation more effectively than most other fibers.

The fibers of interest fall into three composition classes: (1) special glasses and silica fibers, (2) carbon, and (3) ceramic and intermetallic whiskers and fibers of materials such as alumina, beryllia, boron, boron carbide, silicon carbide, zircon, and zirconia. The reinforcing potential of some of these fibers is such that the tensile strength of the reinforced material may be increased by as much as an order of magnitude. This is first reviewed and then some of the applications of the composites are considered.

B. Reinforcement

The primary area of application for high strength fibers is that of composite-reinforcement. While ceramic, metallic, and organic materials are also of interest for use as matrix materials in composites, the latter two offer the widest range of applicability as structural matrices. To reinforce plastics, however, the fibers should be either long or continuous within the geometrical shape of the structure because the loading to the embedded fibers is accomplished via a shear transfer process at the matrix-fiber interface. Since the shear strength of polymers is low, a greater transfer length (i. e., long fibers) is necessary if the fibers are to carry the major portion of the load. Short fiber-reinforced plastics generally show tensile strengths of about 50,000 psi, whereas -all things being equal- the same polymers reinforced with continuous fibers may exhibit strengths of 250,000 psi or greater. Sutton, Rosen, and Flom[493] have shown recently that an epoxy resin reinforced with 14 volume percent alumina whiskers had a tensile strength of 113,000 psi, and a Young's modulus of 6×10^6 psi (as compared to 0.4×10^6 psi for the unreinforced specimen). Their results are shown in Figure 4. In this case, the aspect ratio (fiber length to diameter) was large, viz., ~ 1000. Similar work is being continued by G. A. Schmidt under a Bureau of Naval Weapons contract (NOw 64-0330-d).

While the newer, continuous, ceramic fibers under development are only 10% to 20% as strong in tension as the best whiskers, they offer usable strengths about equal to glass fibers and, in addition, offer improvements of 4 to 6 times in the Young's modulus. In many applications, the low modulus of fiber-glass reinforced plastics has been a limitation; therefore, the development of high modulus filaments (e.g., boron, boron carbide, silicon carbide, etc.) is indeed an important advance. The economic factor in the use of these new fibers must also be considered. Due to the high costs of both whiskers and continuous high-modulus fibers, thought is being given to utilizing them in conjunction with glass fibers. In such cases, the glass fibers would be used to provide tensile strength, and the high modulus filaments and whiskers to provide the improved modulus.

In fiber-reinforcement of metals, both continuous and discontinuous fibers can be used effectively. However, the higher shear strength available in metals permits the efficient use of short fibers or whiskers, provided that they are well bonded to the matrix. The greatest gains achieved by using continuous filaments for reinforcing metals have been silica-fiber-reinforced aluminum, which is apparently at an advanced stage of development at Rolls-Royce Ltd., England. The successful development

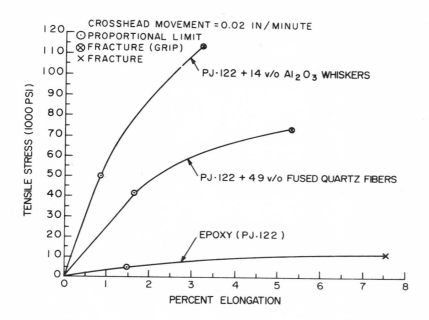

Figure 4. Stress—Elongation Curves for Epoxy (PJ—122) and for Epoxy Reinforced with Continuous Fused Quartz Fibers and with Discontinuous Al_2O_3 Whiskers (Sutton et al, Ref. 493). Permission Soc. of Plastic Eng., Inc.

of a proprietary coating*, which minimizes the reaction between the silica and the molten aluminum, has removed one of the major limitations to the use of this material.   With a better under-standing of the long-term behavior, such as creep, fatigue, etc., of silica-fiber-reinforced metals and an economical means of production of silica fibers, an immediate application for fused-silica reinforced aluminum may be in compressor blades for jet engines.   Data from Rolls-Royce Ltd.[114] as shown in Figure 5 compares the tensile behavior of silica-fiber-reinforced alum-inum with that of other aluminum alloys.

---

*Applications for patents have been filed.

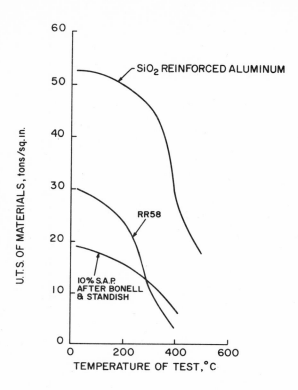

Figure 5. Comparison of Tensile Strengths of Silica Reinforced Aluminum, an Aluminum Alloy RR58, and 10% SAP Alloy; After Bonell and Standish, R61 CAP 69, June 1962, AECL 1532 (Cratchley and Baker, Ref. 114). Permission Metallurgia.

Other combinations of metals reinforced with continuous ceramic fibers have shown improved properties, but the materials are in the early stage of development. Continuous boron fibers in aluminum, magnesium, titanium, and nickel are examples. Figure 6 is a photomicrograph of a boron filament in a nickel matrix. Metals reinforced with refractory metal wire have shown some promise, although the number of chemically compatible combinations for high temperature applications is quite limited.

The research and development of whisker-reinforced metals has generally proceeded along two routes. First, the in-situ precipitation method for producing a composite material has been extensively investigated at several laboratories, especially at Cambridge University, Rolls-Royce Ltd., and United Aircraft

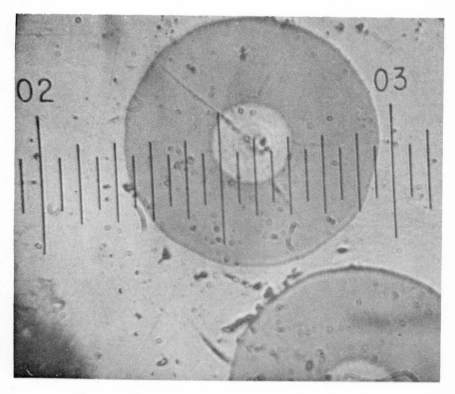

Figure 6. Photomicrograph of Boron Filaments in a Nickel Matrix. Courtesy General Technologies Corp.

Research Laboratory[174] (see contact report Nos. 17, 46,
and 56). This technique results in a geometrical separation and
alignment of a second phase which precipitates as whiskers that
grow parallel to the heat flow during cooling. Some of the
alloys produced by in-situ precipitation are three times stronger
than those produced by conventional methods; among the systems
studied are aluminum-nickel, aluminum-copper, chromium-
copper, nickel-beryllium, nickel-boron and tantalum-carbon.
In-situ whiskers have been chemically extracted from their parent
alloys for tensile measurements. For example, an Al_3Ni whisker
from the Al-Al_3Ni system exhibited a 400,000 psi tensile strength,
while a similar measurement on a Cr whisker, from the Cu-Cr
eutectic composition, has a strength in excess of 1.2×10^6 psi.
The objective has been to precipitate an intermetallic composition
of whiskers in a metal matrix to yield a product in the near-final
shape. Meanwhile Owens-Corning has used similar techniques
as a production method to prepare TiO_2 and ZrO_2 whiskers or
filamentary crystals (some of which are twins) in a glass batch
from which the fibers are later separated. This is discussed
in the contact reports and cited in the patent section (see contact
report no. 45).

In general, the in-situ growth of whiskers has not yielded the
best results although the technique appears promising. This
method may be significant as an economical method of making a
product of a certain specified composition in near-final, but
simple, shapes.

The other approach has been to grow whiskers separately and
then incorporate them into a suitable matrix. While this process
requires more steps, it offers a greater performance latitude and
should be sufficiently amenable to automation to make production
of such composites economically feasible.

Sutton and Chorne[484] have shown that silver reinforced with
sapphire whiskers at $1400^\circ F$ is 20 times stronger in tension than
pure silver tested under similar conditions. Figure 7 shows the
cross-section of a sapphire whisker-silver composite in which
the whiskers constitute about 45% of the volume. Later studies[494]
have shown that major strengthening has resulted from the
whiskers at temperatures to within 98% of the melting point of
silver ($1760^\circ F$). They point out, however, that this work does
not constitute an endorsement for sapphire-whisker-reinforced
silver as a structural material, but serves primarily to demonstrate the feasibility and potential of this strengthening mechanism. Other studies[495] at General Electric's Space Sciences
Laboratory with whisker-reinforced aluminum and nickel may
demonstrate comparable strengthening at elevated temperatures.

Silver, silver -1% silicon, and AZ-18 epoxy resin were used as matrices in the study of the reinforcement capabilities of silicon nitride whiskers at the Explosives Research and Development Est. (ERDE), Waltham Abbey, England (see contact report No. 25). Hot pressing was used to fabricate the metal composites, but damage to the whiskers limited the amount of whisker reinforcement to 15 volume percent for the metals and 20 volume percent for the epoxy. Despite these limitations, convincing reinforcement was obtained.

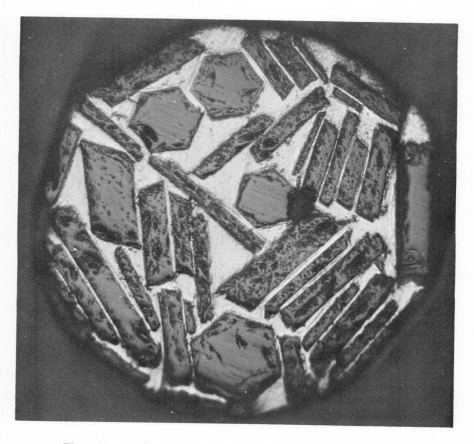

Figure 7. 45 v/o Al₂O₃ Whiskers in Ag Matrix. Magnification ca. 340X. (Sutton and Chorne, Ref. 495).

C. Insulation

Fibrous wools and batts of silica, chromium oxide-modified silica, high silica glasses, alumino-silicates, potassium titanate, stabilized zirconia, and very fine, wool-type whiskers have three properties which make them highly desirable as insulation materials: low bulk density, small diameters, and thermal properties that make them applicable as insulation from cryogenic to elevated temperatures. In addition these materials are, or are becoming, commercially available from more than one manufacturer.

The useful temperature limit of a given insulation is determined by the composition of the fiber and by the resistance of the wool or batt to sintering or densification. In the case of bonded-shapes, the thermal resistance of the bond may be the limiting factor. Bonded insulations are used primarily at temperatures below $600^{\circ}F$, and thus are generally excluded from this survey.

A recent evaluation[163] of various fibrous insulations revealed that materials specified by the manufacturer for intermediate temperature service (near-continuous operation at $2300^{\circ}F$) performed reasonably well, while those specified for more refractory service showed rather severe degradation after exposure to $3000^{\circ}F$. This study and several others[5,6,530] point out the need for further development work before the problems of shrinkage, embrittlement, and fiber sintering at service temperatures near $3000^{\circ}F$ are solved.

The nominal characteristics of some commercially available fibrous insulations are given in Table II; some of the properties of various experimental low density fibrous insulations are shown in Table III.

D. Fillers

It is common practice in the manufacture of plastics to add a filler to the resin to extend it and to control its viscosity and thermomechanical properties. Except in the case of extending the resin for economic reasons, selection of a filler of proper size, shape, and composition may be very important. Minerals such as clay, talc, calcium carbonate, and calcium silicate are often used as fillers because they are readily obtained, economical, and relatively inert. However, chopped, milled, or short fibers of glass, silica, carbon, graphite, or polycrystalline oxides can make another contribution as fillers, in that they can also serve as reinforcements.

E. Fabrics, tapes and paper

Glass, fused silica, carbon, and graphite fabrics are commercially available and have been successfully used in the aerospace industry. Although the major users have been the reinforced plastics manufacturers, several developmental applications appear promising[104] . Among these are retardation and recovery devices for re-entry vehicles, expandable rocket nozzles, and expulsion bladders for cryogenic fluids in liquid propellant rockets.

Most glass and fused silica filaments, as well as some ceramic fibers and whiskers, can be twisted into strands and woven into cloth by available textile fabricating methods. However, the relatively poor abrasion resistance and brittle nature of glass require special handling. In addition, short fibers or whiskers may require mixing with other (usually organic) fibers to permit handling on textile machinery, these carrier fibers to be removed later by chemical or thermal treatments. The manufacture of carbon and graphite fabrics by ordinary textile producing methods has been limited by the low strength of the fibers. Therefore, the principal manufacturing technique has been the precursor conversion by suitable thermal treatment[417-419] of selected organic fabrics containing stronger fibers.

The strength of a filament is usually much greater than that of a fabric made from the same filaments. This discrepancy can be largely explained by the poor abrasion resistance and/or lack of flexibility due to the high modulus of some of these fibers. Thus the fibers are not stressed uniformly in the fabric, resulting in low fabric strengths. A fruitful area for study would be the application of sizes, lubricants, and finishes to the fibers to produce a significant improvement in the fabric strength.

Another approach to improved fabric strength is in the selection of the fabric weave[152] . This is especially important in the production of carbon and graphite cloth by precursor conversion. For example, a satin weave has fewer crimps in both the warp and fill directions than a square weave, and thus is stronger[416].

The use of fiber glass tapes, ribbons, and webbings for decelerator applications was aided by the development of a silicone finish for these products[157] . A partial heat-cleaning technique used in conjunction with silicone finish improved the flexure and abrasion resistance for such articles at room temperature. After a two-hour exposure at 750°F, strength retention was approximately 75% of that prior to the finish treatment.

A tape, woven four inches wide from a combination silica/cellulose fiber, was one of several fabrics examined as a candidate material for use in retardation and recovery devices for re-entry vehicles[104] . Air exposure at temperatures up to

TABLE II. NOMINAL CHARACTERISTICS OF SOME COMMERCIAL FIBROUS INSULATIONS [a]

Manufacturer Generic Code	Principal Component(s)	Rated Max T. (°F)[b]	Fiber Length (in.)[c]	Fiber Dia. (μ)[c]	Nom. Mt. Pt. (°F)	Fiber Sp. Gr.	Natural Bulk Density (lb/ft³)[d]	Bulk Density Range (lb/ft³)[e]	Thick. Range (in.)	Binder	Other Textile Forms[f]
Babcock and Wilcox Co.											
Kaowool Blanket	45% Al_2O_3 42% SiO_2	2300°	to 10	(2.8)	3200°			3-8	$1/4$ - 2	none	bulk fibers, strip
Carborundum Company											
Fiberfrax Blankets											
Short Staple Fiber	51% Al_2O_3 47% SiO_2	2300°	to 1.5	>1-10 (2.5)	3200°	2.7	6	6	$1/2$	varies	various
Long Staple Fiber (Fine Quality)	51% Al_2O_3 45% SiO_2		2-3	2-30 (7)			2	4-6	$1/4$ - $1/2$	varies	various
duPont											
Tipersul Sheet	K-titanate ($K_2Ti_6O_{13}$)	2200 (air)	0.008-0.02	to 1	2500°	3.6	12-15	14-16	$1/16$ - $1/4$	varies	various
Johns-Manville											
Micro-Quartz Felt	98-99% SiO_2	2000		to 1.3	3000°		5	3-3.5	$3/16$ - $1/2$	none (or trace)	bulk fibers, paper
Dyna-Quartz Board	99% SiO_2	2750		to 1.3				4.5-10	$1/4$ - 3	none	none
Thermoflex Felt	Al-silicate	2000	to $1/2$	1.5		2.7	4.5	3-12	$1/4$ - 2	organic	bulk fibers
Cerafelt Felt[g]	Al-Silicate	2000		(3)				3-24	$1/8$ - $1/2$	organic	bulk fibers

Manufacturer Generic Code	Principal Component(s)	Rated Max T. (°F)[b]	Fiber Length (in.)[c]	Fiber Dia. (μ)[c]	Nom. Mt. Pt. (°F)	Fiber Sp. Gr.	Natural Bulk Density (lb/ft³)[d]	Bulk Density Range (lb/ft³)[e]	Thick. Range (in.)	Binder	Other Textile Forms[f]
H.I. Thompson Fiber Glass Co. (HITCO)											
A-100 Refrasil Batt	97%+ SiO_2	2000	to 0.12	to 1.7		2.3		3-7	0.18	none	various
B-100 Refrasil Batt	96%+ SiO_2	2000	to 1	to 12.2		2.3		2-6	0.19	none (or trace)	various
Zirconia "A" Batt	ZrO_2 4-6% CaO	none	1-3	2-8	4500°	4.5-5.2	3-5	3-20	to 1	none	
Zirconia "C" Batt	ZrO_2 9-11% SiO_2	none	1-3	2-8	4500°	4.5-5.2	3-5	3-20	to 1	none	
Zirconia "E" Batt	ZrO_2 10-15%Nd_2O_3	none	1-3	2-8	4500°	4.5-5.2	3-5	3-20	to 1	none	

[a]Nominal property values extracted from manufacturers' brochures.
[b]For near continuous service; higher values usually prescribed for "short time" or "transient" conditions.
[c]Prefix "to" indicated no specified minimums. Values in parenthesis specified as being "average" or "nominal."
[d]As made, prior to textile processing.
[e]Typical limitations for stocked generic code insulations; other dimensions usually available on special order.
[f]Various available forms usually include block, batting, board, cordage, felt, sleeving, yarn, etc.
[g]Unreported values similar to those for Thermoflex Felt.

TABLE III.   EXPERIMENTAL LOW DENSITY FIBROUS INSULATIONS [a]

Manufacturer Generic Code	Principal Component(s)	Rated Max T. (°F)[b]	Fiber Length (in.)[c]	Fiber Dia. (μ)[c]	Nom. Mt. Pt. (°F)	Fiber Sp. Gr.	Nominal Density (lb/ft³)	Nominal Thick. (in.)	Binder	Nominal Relative Cost ($/ft²)	Cost ($/lb)
Carborundum Company											
Fiberfrax Blankets Type XSW Type XV–Felt	{51% Al$_2$O$_3$ 47% SiO$_2$}	2300°	to 1.5	>1–10 (2.5)	3200°	2.7	6	$\frac{1}{2}$	none organic	0.60	2.40
DuPont											
Tipersul Sheet	{K-titanate (K$_2$Ti$_6$O$_{13}$)}	2200° (air)	0.008– 0.02	to 1	2500°	3.6	14–16	$\frac{1}{4}$	carrier fibers	1.56	––
Johns-Manville											
Micro–Quartz Felt Dyna–Quartz Board	98–99% SiO$_2$ 99% SiO$_2$	2000° 2750°	–– ––	to 1.3 to 1.3	3000° ––	–– ––	3.5 6.2	$\frac{1}{2}$ $\frac{1}{4}$	traces none	1.92 ––	1.32 ––
Thermoflex Felt Type CRF 600	Al–silicate	2000°	1.5 (3)	1.5	––	2.7	6.	$\frac{1}{2}$	organic	1.20	4.80
Cerafelt Felt[d] Type RF 600	Al–silicate	2000°	––	––	––	––	6	$\frac{1}{2}$	organic	0.74	2.96
H.I. Thompson Fiber Glass Co. (HITCO)											
A–100 Refrasil Batt B–100 Refrasil Batt	97% SiO$_2$ 96% SiO$_2$	2000° 2000°	to 0.12 to 1	to 1.7 to 12.2	–– ––	2.3 2.3	3–7[e] 2–6[e]	0.18 0.19	none none	1.55 0.78	–– ––

Manufacturer Generic Code	Principal Component(s)	Rated Max T. (°F)[b]	Fiber Length (in.)[c]	Fiber Dia. (μ)[c]	Nom. Fiber Mt. Pt. (°F)	Sp. Gr.	Nominal Density (lb/ft³)	Nominal Thick. (in.)	Binder	Nominal Relative Cost ($/ft²)	($/lb)
H.I. Thompson Fiber Glass Co. (HITCO)											
"Irish" Refrasils	{Cr₂O₃ mod. / SiO₂}	2600°	--	--	--	--	--	--	--	--	--
A-1573 Batt[f]		--	--	--	--	--	--	0.18	none	--	--
B-1576 Batt[g]		--	--	--	--	--	--	0.19	none	--	--
B-1575 Fab-Bat	{2-4% Cr₂O₃ / 97% SiO₂}	--	--	--	--	--	14-16[e]	0.23	none	--	--
Zirconia-Base											
Zirconia "A"	Zr₂+stab. 4-6% CaO		1-3	2.8	4500°	4.5-5.2	3-5	1/4-1/2	none	--	--
Zirconia "E"	10-15%Nd₂O₃										
Thermokinetic Fibers, Inc.											
Sapphire Paper	99.5% Al₂O₃	--	0.02-1.0	1-3	3780°	3.96	--	0.015 / 0.050	none	--	--

^{a}Nominal property values extracted from manufacturers' brochures.
^{b}For near-continuous service; higher values usually prescribed for "short time" or "transient" conditions.
^{c}Prefix "to" indicates no specified minimums. Values in parenthesis specified as being "average" or "nominal".
^{d}Unreported values similar to those for Thermoflex Felt.
^{e}Min-max values for specifications. Variation typically much less for a given lot.
^{f}Unreported values similar to those for A-100 Batt.
^{g}Unreported values similar to those for B-100 Batt.

1800°F proved that slow pyrolysis is necessary to prevent the cellulose from flaming, which disrupts the silica residue.

Papers made of ceramic fibers have been available for several years. The ease with which they can be cut, trimmed or shaped, their low thermal conductivity, and excellent thermal shock resistance has made them useful in missiles, rockets, and jet engines[24]. The upper temperature limit of usefulness is about 2300°F for most of these materials. The recent appearance of a sapphire wool paper* on the market may open up an entirely new temperature range of applications, viz. 3000°F and higher.

---

*Thermokinetic Fibers, Inc.

IV. FACTORS AFFECTING FIBER STRENGTH

A. Introduction

The numerous factors that contribute to the measured strength of a fiber will be considered in some detail. Usually, the apparent strength of most fibers falls considerably below the theoretical (potential) value discussed in Section II. Thus, in order to discuss the strength of a fiber, the factors which determine the potential (theoretical) strength and those which contribute to the weakening or lowering of the ideally high values must be considered. The major factors which influence the measured (or useable) strength can be categorized into four areas:

1. Compositional factors

2. Processing factors

3. Structural factors

4. Testing factors

Compositional factors determine the (inherent) potential strength of the fiber as well as the resistance of the fiber to degradation. Such factors include: a) the resistance to corrosion, chemical degradation or oxidation, b) the resistance to abrasion (hardness), c) the type and number of bonds (and the atomic structure) which govern the physical and thermal properties, and d) the time-dependent properties such as creep, stress-rupture, and stress-corrosion.

Processing factors play an important role in determining the internal atomic and microscopic structure of the fiber, the surface perfection, and whether or not the fiber will be a single crystal, a polycrystalline agglomeration, or an amorphous material.

Structural factors are dependent upon both the composition and processing parameters, but their influence can be correlated with the observed mechanical behavior of the fiber. These factors include the microscopic and macroscopic structure of the fibers; the type, number, and location of defects; the inhomogenieties in composition; the state of internal stresses; and the size and shape of the fiber.

Testing factors affect the measured strength of a fiber, since the test method chosen determines how the load is transmitted to the fiber. The test method can also affect the state of stress within the fiber prior to fracture. It is the interaction of the stresses and the imperfections (or stress-concentrators) that determine the value of the measured strength.

Since all of the above-mentioned factors have an important bearing on the strength of a fiber, they will be included in the following discussion on the strength of specific types of ceramic and graphite fibers and whiskers now available either commercially or in the laboratory.

B. Glass and fused silica fibers

The large difference generally observed between the measured strength of glass fibers and the theoretically predicted value illustrates the sensitivity of this material to degradation by mechanical abrasion and chemical corrosion. Based on theoretical estimates, the ultimate tensile strength of glass should be about E/5, where E is Young's modulus. Thus, some known high modulus glasses[84, 294, 378] should have ultimate tensile strengths approaching 3×10^6 psi. However, glass in bulk form subjected to normal handling is probably no stronger than 10,000 psi and glass in fiber form (without a protective coating) after handling is rarely stronger than 200,000 psi. Because of the relatively poor abrasion and corrosion resistance of glass with regard to strength, the filament surface is usually protected by applying sizings, lubricants, and finishes after the drawing process. As a result, treated glasses with ultimate tensile strengths of 500,000 to 700,000 psi are commonly produced today.[285, 312]

Since there is still room for significant improvement in the strength of commercial glass fibers, factors affecting fiber strength are briefly discussed to emphasize some of the critical problem areas in which further research and development are needed.

1. Size Effect

The classic work of Griffith[202] has led to a plausible explanation of effect of increasing strength with decreasing fiber diameter (and length). Since that time, numerous investigators have studied this relationship. Until a decade ago, the general belief that the increase in strength of the smaller fibers was an intrinsic property, was supported by the work of Anderegg[12], Murgatroyd[326] and several others. However, more recent studies[342, 502] indicate that strength can be independent of fiber diameter so long as fibers are formed under carefully controlled conditions that result in relatively flaw-free surfaces. These results were obtained with fibers having dimensions in the range from 5μ - 15μ, the same range in which Griffith found strength to be most dependent on diameter. Further evidence against the "strength/dependence on diameter" theory was presented by Bartenev[37] who claims to have attained approximately the theoretical strength of glass. The following is quoted directly from the abstract of his paper:

"Recently, the author, jointly with Mrs. L. K. Ismailova, reported about 'flawless' glass fibres possessing a strength independent of their length, fracturing into 'dust', and showing no scatter of experimental results from sample to sample. The strength of these fibres practically does not depend on their diameter, as the reduction in diameter from 20 down to 2μ m increases their strength from 280 to 310 kfg/cm^2, i. e., a total increase by 10%. Such a small increase in strength may be caused by changes in molecular orientation in the process of drawing of glass fibres. The high strength achieved (300 kgf/cm^2) is rather close to the theoretical strength of glass (350 kgf/cm) calculated recently by G. K. Demishev and I. V. Rasumovskaya (Steklo, 15, No. 1 (1962); Steklo, 15, No. 4 (1962); Institute of Glass, Moscow)."

A summary of the relationship between fiber diameter and strength reported by Griffith, Anderegg, Otto, and Thomas is shown in Figures 8 and 9.

The importance of gage length in determining the strength of E- and S-glass fibers is emphasized by Metcalfe and Schmitz[311]. Their results indicate that fiber length has a marked effect on strength; data obtained from their study are shown in Figures 10 and 11. The general relationship of strength, length, and surface damage is shown in Figure 12.

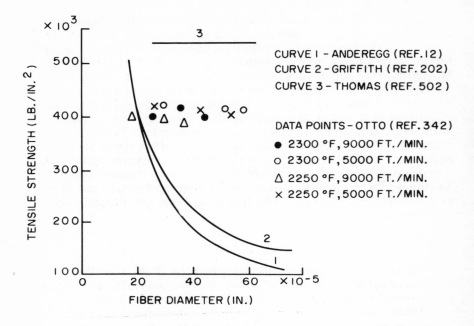

Figure 8. *Strength—Diameter Relationships for Glass Fibers.*

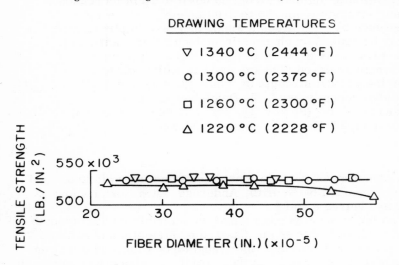

Figure 9. *Variation of Tensile Strength with Fiber Diameter (Thomas, Ref. 502). Permission J. Soc. Glass Tech.*

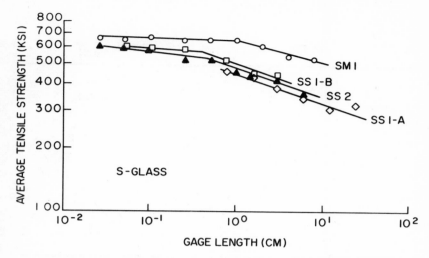

Figure 10. Length Effect on Strength of S-Glass Fibers (Metcalfe and Schmitz, Ref. 311). Permission Am. Soc. for Testing and Materials.

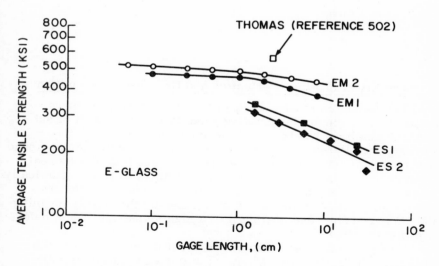

Figure 11. Length Effect on Strength of E-Glass Fibers (Metcalfe and Schmitz, Ref. 311). Permission Am. Soc. for Testing and Materials.

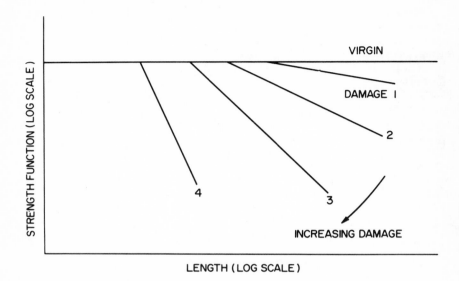

Figure 12. General Strength-Length Relationship for Glass Fibers (Metcalfe and Schmitz, Ref. 311). Permission Am. Soc. for Testing and Materials.

2.   The Effect of Fiber Perfection and Chemical Composition

The reduction of fiber strength by the presence of flaws, formed either during the drawing process or as a result of subsequent handling, has been well documented. Holloway[230], working with Pyrex, E-glass, and a soda-lime-silica glass, found that chemical contamination of a glass surface followed by heat treating at temperatures at which glass can be worked, drastically reduces the strength of the fibers. He also found that the frequency and size of flaws can be minimized by thoroughly cleaning the glass rod first and heating it to a high temperature just before drawing the fibers.

Using an improved technique to decorate glass fibers with sodium, Gordon and co-workers[197] reported a correlation between an elaborate system of surface cracks and the mechanical strength of the glasses studied. They postulated that fine-scale devitrification on the glass surface during drawing might be an alternate mechanism to abrasion for the origin of surface cracks.

Metcalfe and Schmitz[311] identified two types of flaws in virgin fibers of both E- and S-glass. One type is severe and governs
failure of long filaments; the other, which is less severe, determines failure at short gage length.

The role of chemical composition in determining the strength
of glass may be one of secondary nature[375]. Alkali-free or very
low alkali-content glasses are usually more resistant to abrasion
and therefore less likely to incur surface damage than glasses
containing normal amounts of alkali oxides. The composition of a
glass also influences its viscosity and its tendency to devitrify,
which in turn can affect ease of fiber drawing[12]. Weisbart[534]
demonstrated that there is no fundamental difference in the tensile
strength between glass fibers of alkali-containing glass and glasses
having very low alkali content. The results of this study are
shown in Figure 13. The compositions of some currently used
glass fiber are shown in Table IV.

3.   Strength Dependence on Surface Treatment

The wide usage of glass fibers as structural materials is
due largely to the development of lubricants and sizings that pro-

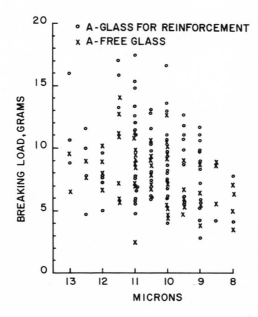

*Figure 13. Breaking Load of Single Glass Fibers (Weisbart, Ref.
534). Permission Soc. of Plastic Eng., Inc.*

TABLE IV. COMPOSITIONS OF VARIOUS FIBER GLASSES

Oxide	Fused Silica[1]	E-Glass[1]	HM 905*	S-994[2]
SiO_2	99.97	54.5	48.0	64.0
Al_2O_3	0.015	14.5	0.0	26.0
CaO	0.0032	17.0	11.85	
B_2O_3	0.00	8.5	8.0	
BeO	0.00	0.0	8.0	
Li_2O	0.00	0.0	6.0	
V_2O_5	0.00	0.0	4.0	
MoO_2	0.00	0.0	2.0	
MgO	0.00	4.5	8.5	10.0
TiO_2	0.00		4.0	
Na_2O	0.0022	1.0	0.0	
K_2O	0.0007		0.0	
Fe_2O_3	0.0010		0.0	

[1] Eakins (Ref. 141)

[2] Tomashot (Ref. 504)

* Experimental high modulus glass. This composition, Owens-Corning Fiberglas Corporation's YM 31-A, and several other BeO-containing glasses are not currently being used because of possible beryllia toxicity problems.

tect the as-drawn fiber surface and the evolution of coupling agents to promote the bonding with a suitable matrix. Both the glass fibers and the reinforced plastics industries have devoted considerable effort toward improving the methods for utilizing the potential strength of glass fibers. Unless the pristine surface of the glass fiber is protected from abrasion and other environmental influences, the strength is sharply reduced. Yet paradoxically, the application of surface protectants results in some loss of strength. Morrison and co-workers [322] state that the best method of protection is still in doubt. Whether chemical bonding to the glass fiber surface or a mechanical gripping of the surface by a resin is more desirable is still a debated subject. Furthermore, the reactions between the fiber surface and the matrix and/or coupling agents vary with the specific kind of resins and coupling agents used. Several review articles [143,142,147,403] have appeared recently which thoroughly cover this subject. The differences in flexural strength of an epoxy-E-glass laminate, in which the silane coupling agent (A-1100) has been applied by different methods, is illustrated in Figure 14. The effect of the type of resin system used on the strength of an epoxy-E-glass laminate is shown in Table V.

The need to optimize the reactions between a fiber surface and the fiber agents and/or the matrix has resulted in the development of several methods for studying interfacial reactions. The use of electron microscopy [458] and the applications of classical surface chemistry (e.g., contact angle measurement, x-ray diffraction, infrared analysis, etc) are widespread [360]

A review [162] of metal coatings on glass filaments indicates that insufficient information is available to determine protective strength-enhancing capabilities of such coatings in a marine environment. However, Figures 15 and 16 illustrate the importance of metal coatings for preventing fiber-to-fiber contact in composites.

In an effort to eliminate the influence of surface flaws on fiber strength, Morley [319] applied a Pyrex coating to a fiber of the composition: CaO - 23.3%, Al_2O_3 - 14.7%, and SiO_2 - 62.0%. Tensile tests at 700°C suggest higher strengths for the coated fibers, as seen in Figure 17.

More recent work by Morley et al [318], at Rolls-Royce Ltd., has produced aluminum coated-fused silica fibers having average tensile strengths in excess of 800,000 psi.

The deposition of carbon (see contact report No. 28-D) on freshly drawn fused silica fibers is another method for protecting their surface. Fibers about 10μ in diameter, treated this way, have average tensile strengths of 500,000 psi.

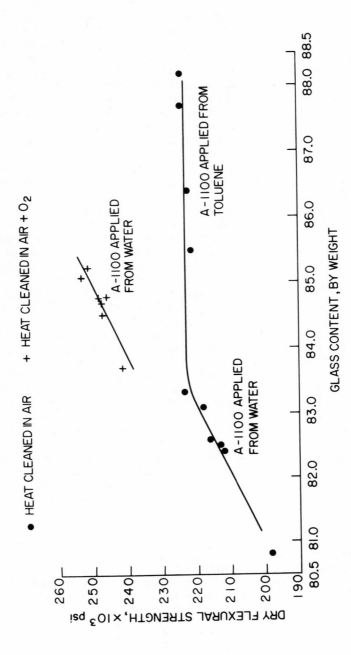

Figure 14. Flexural Strength vs Glass Content for Samples with A-1100, Which Were Heat Cleaned in Air or Oxygen-Enriched Air (Eakins, Ref. 146). Permission Soc. of Plastics Eng., Inc.

TABLE V. EFFECT OF TYPE OF RESIN SYSTEMS USING E-GLASS

Resin	Curing Agent	Time / Temp. (hrs.) (°C)		Ultimate Tensile Strength in Glass ($\times 10^6$ psi)	Number of tests
828	MPD	15 @	20°		52
		2 @	100°	0.3494 ± 0.0076	
828	MPD +1% A1100	15 @	20°		10
		2 @	100°	0.3601 ± 0.0135	
828	DDM	15 @	20°		12
		1 @	100°	0.3497 ± 0.0153	
828	Aliphatic Amine*	48 @	20°	0.22575 ± 0.0117	12
828/X71	MPD	15 @	20°		11
		2 @	100°	0.2868 ± 0.0206	
828/X71	NAEP	15 @	20°		5
		2 @	100°	0.2104 ± 0.032	
828/X71	Aliphatic Amine	48 @	20°	0.35825 ± 0.0163	4

*Hardener 951, Ciba (A.R.L.) Limited, England.   Brookfield and Pickthall (Ref. 64)

Figure 15.   Cross-Section of a Glass-Fiber-Reinforced Aluminum
Specimen. The Arrows Indicate the Cracks Which Were Initiated
at Points of Fiber-Fiber Contact. Magnification ca. 400X.
(Sutton and Chorne , Ref. 495).

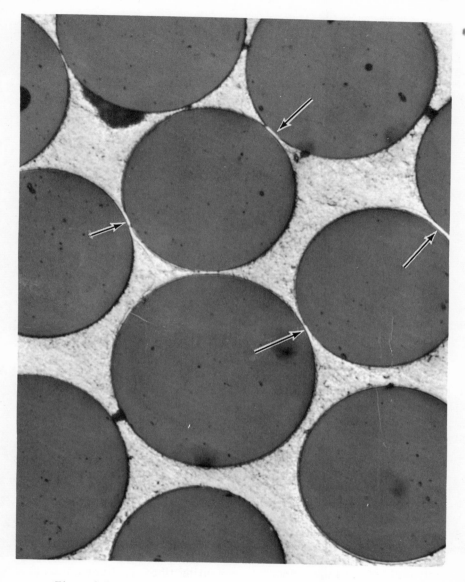

Figure 16. Cross-Section of a Glass-Fiber-Reinforced Aluminum Specimen, Where the Fibers were Pre-Coated with a Thin Metallic Coating Prior to Fabrication. The Arrows Indicate Absence of Direct Fiber-Fiber Contact Due to This Coating. Magnification ca. 400X (Sutton and Chorne, Ref. 495).

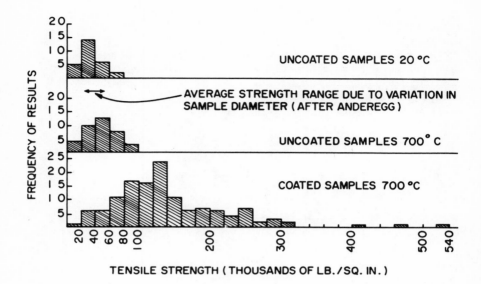

4.   The Effect of Forming Methods

Morley and co-workers[320] investigated the drawing of fused
silica fibers about 25μ in diameter under a variety of controlled
conditions.   Tensile tests on various fibers showed an average
strength of 850,000 psi.   Their results indicate that the variations
in the drawing process did not affect the mean fiber strength.
Thomas[502], working with E-glass, found that, as long as the
temperature of the molten glass was sufficiently high to draw a
fiber of uniform diameter, variations in composition, nozzle
diameter, drawing speed, and several other factors had no significant effect on fiber strength.   Otto[342] reports similar findings
in studying the effect of variation in fiber diameter and forming
method on the strength of the fibers.   Each of these investigators
emphasizes the importance of preventing fiber-to-fiber contact,
or any other surface damage that might reduce the fiber strength.
The rate of cooling experienced by a glass fiber causes profound changes in the structure and has been the subject of numerous investigations[325,341,461].   The great cooling rate of the
fibers is generally believed to the primary cause of property
differences between bulk glass and fibers.   It is postulated that

rapid cooling favors an orientation of the weak bonds normal to the axis of drawing and results in higher strength of the fibers[325].

Murgatroyd[325] studied the delayed elastic effect in both massive and fiber glass of two different compositions in an effort to confirm the orientation effect resulting from rapid cooling. He measured the residual strain at intervals from 1 to 10^4 minutes after an imposed stress was released. The data he obtained for a sheet glass and a borosilicate glass are shown in Figure 18.

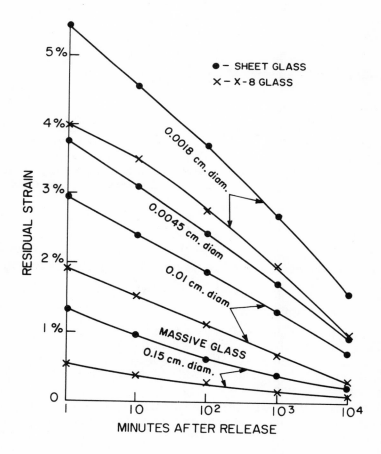

Figure 18. *The Delayed Elastic Effect in Sheet Glass and X-8 Glass in Massive and Fibrous Form. The Effect is Shown as Residual Strain Remaining After Release of a Specimen Strained for 10 Days (Murgatroyd, Ref. 325). Permission J. Soc. Glass Tech.*

5.   Environmental Effects

a.   Moisture

The effect of moisture on the strength of glass has been the subject of considerable study. Holland[224] and others have extensively documented this subject. In general, experimental evidence is conclusive that glass strength decreases upon exposure to moisture. Thomas[502] studied the effect of 0% and 100% relative humidities on the strength of E-glass fibers over a period of 120 days. The results of his study are shown in Figure 19.

The composition of a glass has an indirect effect on the strength of the fiber. For example, it is common knowledge that the surface of a high-alkali glass will be more readily attacked by moisture than that of a low-alkali glass. These effects on the strength of high- and low-alkali fiber glass by immersion in water at room temperature are shown in Figure 20.

b.   Temperature

Commercially produced glass fibers have physical and mechanical properties that are considerably different from the properties of bulk glass[341]. These properties tend to approach those of the bulk glass if glass fibers are heated to elevated temperatures. In general, fiber glass differs from bulk glass in that:

a.   the tensile strength is higher,

b.   the density is lower,

c.   the Young's modulus is lower,

d.   the index of refraction is lower,

e.   the thermal conductivity is lower, and

f .   the specific heat is lower.

In addition, the chemical reactivity is increased (e.g., as as-formed fiber can be leached more rapidly than one which has been heat-treated), and changes in these fiber glass properties can be brought about at temperatures well below the specified temperature required to anneal the bulk glass.

Thomas [502] investigated the variation in tensile strength with temperature and length of heat treatment, at temperatures ranging from 150° to 600°C. Fibers were heated at various temperatures (up to 600° C) for a period of four hours. The room temperature breaking stress of the treated fibers decreased steadily up to about 450° C, then remained fairly constant. These results are shown graphically in Figure 21. Thomas also heated fibers for 5, 15, 30, 60, 120, and 240 minutes at temperatures from

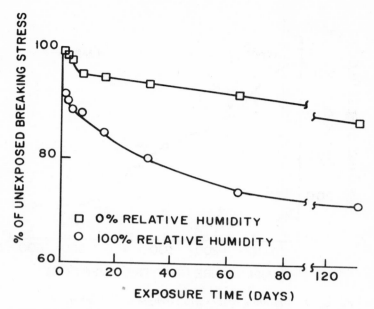

Figure 19. Variation of Percentage Tensile Strength with Exposure Time at 0% and 100% Humidities (Thomas, Ref. 502). Permission J. Soc. Glass Tech.

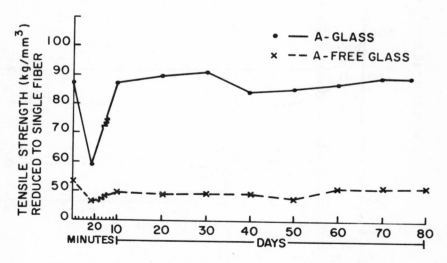

Figure 20. Tensile Strength of Glass Fibers; Immersion in Water of 20°C (Weisbart, Ref. 534). Permission Soc. of Plastic Eng., Inc.

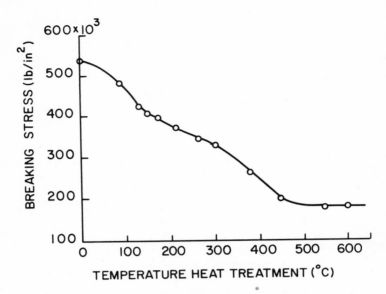

Figure 21. Variation of Tensile Strength with Temperature of Heat Treatment (Thomas, Ref. 502). Permission J. Soc. Glass Tech.

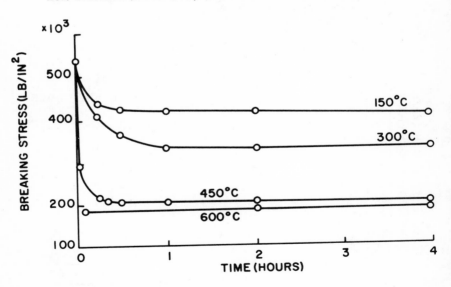

Figure 22. Variation of Tensile Strength with Period of Heat Treatment at Four Temperatures (Thomas, Ref. 502). Permission J. Soc. Glass Tech.

150° to 600° C and found that the time necessary to reduce the breaking stress to a constant value is a function of the temperature of the heat treatment. These values are shown in Figure 22.

The effect of heating on the room temperature strength of E-glass rods (0.080" to 0.117" dia.) was determined by previously subjecting the rods to temperatures ranging from 180° to 1100°F[78].

It was found that the strength decreased with increasing temperature of heat treatment. This is true for E-glass fibers that have been similarly treated. Ordering or devitrification at the rod surface during the heating period is suggested as the most likely reason for the strength decrease. The strength decrease of E-glass rods and fibers after heating appears in Figure 23.

Fused silica fibers, manufactured under standard conditions, were heated in air for periods of up to one hour, then loaded to fracture at the same temperature[320]. In general, the fiber strength decreased with both the time and temperature of the heat treatment prior to testing. This effect is illustrated in Figure 24.

Eakins[143] describes some work conducted on fused silica fibers at Rolls-Royce, Ltd., in which tensile tests were performed in air and in vacuum over the temperature range -273° to 700°C. The results, Figure 25, show that, at temperatures between -200° and 0°C, the strength decreases sharply; it then in-

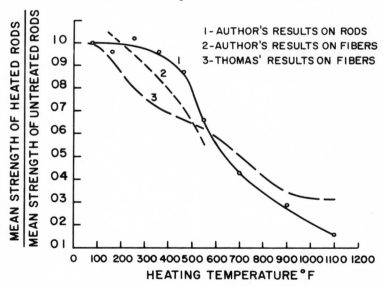

Figure 23. Dimensionless Strength vs Heating Temperature Comparing the E-Glass Fiber Results of Thomas and the E-Glass Fiber and Rod Results of the Author (Cameron, Ref. 78).

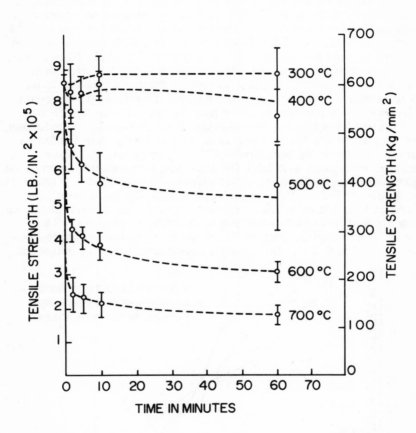

*Figure 24. Effects on Fused Silica Fibers (Morley et al, Ref. 320).
Permission J. Soc. Glass Tech.*

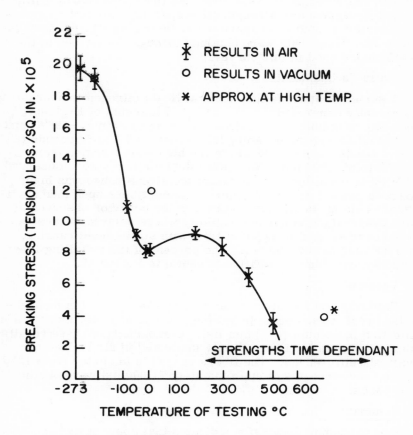

Figure 25. Strength of Silica Fibers vs Temperature. Courtesy Rolls-Royce, Ltd.

creases slightly from 0° to $200^{\circ}C$, only to fall off sharply again
after 200°C. The high strength (2. 0 $\times$ 10^6 psi) of fused silica at
-273°C shown in Eakins' report compares favorably with the 1.96
1.96$\times$10^{6}psi strength of fused silica at -196°C reported by Hillig[218].

The dependence of strength on temperature for two commercially available, and two experimental fiber glasses, is shown in
Figure 26. These data were obtained through the replies to the
questionnaire used in this survey.

c.   Static Fatigue (Static Loading)

A series of static fatigue tests at liquid nitrogen temperatures
was recently completed[225-227, 229].   Filaments of E-glass,
about 11μ in diameter were dead-loaded in tension at -196°C and
maintained at that temperature for at least 1. 7 $\times$ 10^5 seconds (2
days). Loads were varied within the high stress region from
400, 000 to 650, 000 psi. No static fatigue failures were observed
under these conditions, even though the stress range was high
enough to cause immediate failure of some fibers upon load application. This is in distinct contrast to the behavior observed at
room temperature in normal humidity where delayed failures occurred over several decades of time with stress level ranging
from 200, 000 to 400, 000 psi. The percent of fibers failing immediately upon load application is shown in Figure 27.

d.   Vacuum

Holloway and Schlapp[231] developed a technique for measuring the breaking strength, in bending, of Pyrex glass fibers drawn
and tested in vacuum (10^{-6} mm Hg). Comparison of these results
with similar data for fibers drawn and tested in air yielded no
significant differences. The fibers ranged in diameter from 125μ
to 130μ, and strengths of approximately 71,000 psi were obtained
in all cases.

6.   Summary

The preceding discussion indicates that some of the well established explanations for the strength behavior of glass fibers have
been somewhat modified. While the dominant factors governing
the strength of glass have been generally identified, the detailed
mechanisms controlling this property are still rather obscure.
The strength of glass fiber is closely related to the fiber surface
perfection (freedom from flaws), which, in turn, is dependent
on composition, forming and drawing variables, and test environment. Therefore, a more complete understanding of the mechanisms influencing strength depends on the development of better
techniques for characterizing the nature of the surface more fully.

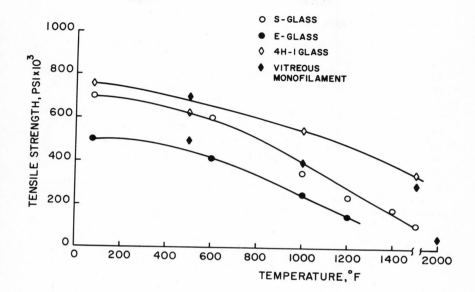

Figure 26. Strength/Temperature Relationship for Various Glass Fibers: S-Glass and E-Glass, Contact Report No. 45; 4H-1 Glass, Contact Report No. 2; Vitreous Monofilament, Contact Report No. 9.

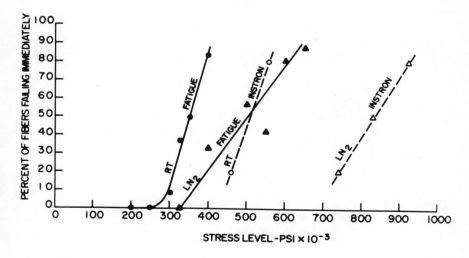

Figure 27. Immediate Failures Upon Load Application (Hollinger et al, Ref. 229).

C. Polycrystalline and composite fibers

The strength of a material is highly dependent on its micro-structure. Pores, for example, not only decrease the cross-sectional area on which the load is applied, but can also serve as stress concentrators. Even smaller defects, such as inclusions at grain boundaries, as well as crystallite size, markedly affect the microstructure and properties. The microstructure in turn is directly related to the forming process, and the crystalline nature of the fibers now being discussed warrants a review of the methods whereby polycrystalline and composite fibers are produced. The strength dependency on microstructure becomes very significant if we consider the differences in tensile strength between polycrystalline and composite fibers, amorphous (glass and silica) filaments, and single crystal whiskers. Strength varies in the first group from 200,000 to 340,000 psi (see contact report No. 54) with occasional values to about 500,000 psi. Glass filaments are now commercially available with strengths up to 700,000 psi (see contact report No. 45) and Rolls-Royce, Ltd.[318] reports aluminum coated-fused silica fibers with an average strength of 850,000 psi. Tensile strengths over 1,000,000 psi are commonly reported for various ceramic whiskers, and a preliminary value of 6,200,000 psi for an Al_2O_3 whisker was recently reported[308].

The large variations in the strengths of these three classes of filaments are, of course, related to their structural differences. As previously mentioned, the relationship between structure and forming methods is extremely important to any consideration of factors affecting strength, thus the following sections describe and compare the various methods used to produce polycrystalline and composite fibers:

1. Evaporation of a colloidal suspension
2. Extrusion
3. Rayon spinnerette technique
4. Vapor deposition onto a heated substrate
5. Core-sheath method
6. Taylor wire method
7. Other processes
   a) flame deposition
   b) drawing or extruding from the melt
   c) dip forming
   d) fused salt electrolysis
8. Comparison of processes

1. Colloidal Suspensions

Horizons, Inc., as well as other organizations subsequently, has studied this approach by spreading a thin film of a colloidal suspension containing the desired oxide (or oxides) on a flat surface which is then rapidly heated[88]*. As evaporation takes place, fibers are formed by cracking of the film along radii of the deposit. The fibers are then collected and heat treated to render them both thermally and mechanically stable. Fibers produced this way have a rectangular cross-section and vary from 0.25 to 2 or more inches in length with widths about 0.125 inch or less. Generally the ratio of width to thickness ranges from 3:1 to 10:1. Fibers of ZrO_2, SiO_2, mixtures of these oxides, Al_2O_3, and ThO_2 have also been prepared by this method. The tensile strengths have been relatively low, but are sufficient to permit these fibers to be used as thermal insulation.

2. Extrusion

Fibers of several materials such as zircon, zirconia, alumina and magnesium oxide[172] have been formed by this method. In the last case, finely divided magnesium oxide was added to a dispersing medium of $Mg(OAc)_2 \cdot 4H_2O$ and anhydrous MeOH. This slurry was then pressure extruded through platinum orifices, of the type used for synthetic fiber production, to obtain filaments about 0.005 inch in diameter. Thermal treatment was used to convert the filament to a moderately strong but brittle condition, and no strength data were reported. The principal use for this kind of material would probably be as an insulation.

The H. I. Thompson Fiber Glass Company (see contact report No. 54) has developed a process for producing a very fine-grained polycrystalline material in a continuous form by a conventional ceramic extrusion technique in which finely divided oxide particles mixed with an organic binder are extruded. Fibers are regularly made this way having tensile strengths varying from 200,000 to 340,000 psi with a maximum value of 500,000 psi being obtained. However, extremely careful control of the thermal treatment during processing is necessary to prevent excessive grain growth of the originally small oxide particles. In addition to close thermal control, grain growth inhibitors are added in an attempt to prevent the grain size from becoming larger than 10%-20% of the fired diameter. Lachman and Sterry[274] have demonstrated that grains having diameters in the size range from 0.25 to 0.75 microns will result in fiber strengths lower than 100,000 psi.

---

*See also Section VIII.

3.   Rayon Spinnerette Method

A method for producing continuous polycrystalline fibers is under development at Horizons, Inc. (see contact report No. 31) and is understood to be under consideration by several other organizations for a variety of metallic and ceramic fibers. Fibers of Al_2O_3, $ZrSiO_4$, $3Al_2O_3 \cdot 2SiO_2$, ZrO_2, and TiO_2 have been studied but current interest is centered on γ-alumina-spinel. The process employed is similar to the techniques used to produce rayon thread. Sixty fibers are made at a time and one pound is produced in about 4-5 hours. The fibers are then very carefully fired at 1550° C to yield polycrystalline, transparent to opalescent, fibers of about $25\,\mu$ diameter. Further work on reducing the diameter and improving other aspects of the process are expected to increase the strength, which has been measured up to 0.5×10^6 psi for an 8-inch gage length sample.

Studies so far indicate that the initial work with unstabilized γ-alumina was unsatisfactory because γ converts to α at about 750° C and crystals grow to the full diameter of the fiber. Such fibers are relatively weak, easily corroded, and can be rehydrated. By stabilizing the γ-alumina until it can transform (at least partially) to alumina spinel, the strength is retained through the 750° C temperature region and is further enhanced by the spinel formed during firing to 1550° C.

4.   Vapor Deposition

The technique for producing filaments by vapor deposition consists of reducing or decomposing a volatile compound of the desired coating material onto a heated surface usually a fine wire or other conductive filament. The volatile compound formed must decompose at a temperature well below the melting point of the deposit or the substrate, yet must be sufficiently stable to resist decomposition until it contacts the substrate. Among the many organizations investigating this method for continuous filament production are: Air Force Materials Laboratory, General Electric Company, General Technologies Corporation, and Texaco Experiment, Inc. (see contact report Nos. 2, 28-A, 28-D, 28-F, 29, 51).

The feasibility of making composite polycrystalline filaments by this method has been demonstrated with the following materials: boron, boron carbide, silicon carbide, titanium boride, titanium carbide, aluminum oxide, beryllium oxide and beryllium. A typical reaction for vapor depositing boron carbide is:

$$BCl_{3\,(g)} + CH_{4\,(g)} \xrightarrow[\text{at the substrate}]{\sim 1200^\circ C} B_4C_{(s)} + HCl_{(g)} + H_{2(g)}.$$

The substrate material must be electrically conducting so that it can be resistance-heated. Tungsten filaments, of varying diameter from 0.5 to 5.0 mils, are generally used for this purpose, although other materials such as aluminum, nickel, titanium, molybdenum[124], and pyrolytic-carbon coated fused silica have been used (see contact report Nos. 17 and 28-D).

Filaments produced by this technique are, by the nature of the process, composite materials containing a core surrounded by a polycrystalline (or amorphous) coating. Although such a structure is likely to contain inherent flaws, close process control has resulted in some rather strong filaments. A study to determine the feasibility of vapor depositing a variety of materials is being conducted by the General Technologies Corporation[124]. The tensile strengths, Young's moduli and densities of the filaments they produced are shown in Table VI. Measurements on boron carbide filaments at the General Electric Company Space Sciences Laboratory[188] show tensile strengths as high as 420,000 psi, bending strengths up to 3×10^6 psi, Young's modulus of 69×10^6 psi, and a density of 0.091 lbs/in^3. The tensile strength of boron filaments produced by Texaco Experiment, Inc. ranges from 300,000 to 500,000 psi with modulus values from $55-60 \times 10^6$ psi (see contact report No. 51).

5. Core-Sheath Method

In an attempt to combine the formability of vitreous fibers with the high modulus of polycrystalline fibers, a composite core-sheath configuration was studied by Narmco Research and Development Division of Whittaker Corp.[344, 345] Fused silica was selected as one of the sheath candidates because it has an inherent high tensile strength and is resistant to high temperatures. It also has a temperature-viscosity relationship which permits fiber drawing over a wide temperature range. Alumina, or a system in which alumina was a major constituent, was chosen for a core material because of its high modulus and high modulus-to-density ratio. Other core materials considered were ZrO_2, TiO_2, polycrystalline glasses, and pyrophosphates.

In this process, a tube of fused silica is filled with one of the core materials; the end of the tube is flame melted, and test samples are drawn by suitably modified fiber drawing techniques. In addition to fused silica, Vycor and Pyrex have been used as sheath materials.

The results thus far indicate that, in principle, Al_2O_3, ZrO_2, and many other materials can be maintained as a core in a vitreous sheath using the tube fill technique. Fibers formed were generally short and of varying diameters (1-8 mils). With Al_2O_3 as a core, modulus of elasticity values in excess of 20×10^6 psi were ob-

TABLE VI. SUMMARY OF TEST RESULTS

Material	Tensile Strength ($\times 10^3$ psi)	Modulus of Elasticity ($\times 10^6$ psi)	Density of Coating (g/cm^3)
Boron Carbide	28–113	34–47	2.0
Silicon Carbide	76–200	6–20 (SiCl$_4$ Route) 30–80 (Silane Route)	1.3–1.5 2.5–2.7
Boron	20–190	---	1.8–2.0
Titanium Carbide	8.7–15	2	3.2
Titanium Boride	3.3–15	63–74	3.3
Aluminum Oxide	24	8.5–35	3.1
Beryllium Oxide*	63–96	50	---
Beryllium*	116–246	---	---

Davies and Withers, (Ref. 123)

*Coatings were thin.

tained and subsequent heat treating of the composite increased both the density and the modulus values.

Another program at Narmco[511] has as its objective the exploration of various core-sheath systems with a dual melt vitreous fiber-forming system. Compatibility, fiberizing feasibility, and effectiveness of the concept as a means to a high-modulus ceramic vitreous fiber are being investigated.

E-glass and Ferro Corporation RKl014 glass were used as sheath materials, while an iron-free E-glass, a nucleating glass, and two calcium alumina-silicate glasses were selected as core materials.

Initial results established that a) the ratio of core-to-sheath is virtually unlimited for materials having somewhat similar temperature-viscosity relations, b) there is no significant difference in the tensile strength or elastic modulus between E-glass composite fibers and solid E-glass fibers, c) of the three candidate core materials, a nucleating glass exhibited the best physical characteristics, and d) at higher temperatures, during recrystallization, the candidate core materials reacted with the sheath glass, causing partial conversion of the sheath to crystalline material.

6. Taylor Wires

Perhaps the oldest method used for making core-sheath or composite fibers is the technique of drawing Taylor wires. Briefly, this approach utilizes a glass tube filled with molten metal which is drawn into a metal filled glass filament. The principle of this method is currently being studied by several organizations. Among them are the Jet Propulsion Laboratory and the United Aircraft Research Laboratory[110-112]; the latter is engaged in an effort to produce continuous lengths of beryllium wire 1 micron in diameter. Preliminary work in this program indicates that a reaction at the interface between the molten beryllium and the glass is a limiting factor. To reduce this reaction, protective coatings (1 mil thick) of chromium, rhodium, and palladium were placed on the wire and protection tubes of boron nitride, beryllium oxide, aluminum oxide, and water cooled stainless steel inside the glass tube. Another major problem is dissolved gases in the molten metal which causes column separation during the drawing process. While only a few short lengths of beryllium-glass composite fibers were obtained, the general technique warrants further study.

Initial work with copper-glass composite fibers has been more successful. For copper core diameters in the range of 20-25 microns, continuity has been attained in lengths up to about four feet. Copper core diameters as small as 3 microns have been drawn, but, for larger diameters (up to about 16-20 microns), continuity has not been obtained over any appreciable length.

The production of fine wire (1 micron diameter) by this technique is in the early development stage and preliminary studies reveal many technical difficulties to be overcome.

7. Other Processes

Numerous other processes, and variations of them, as well as those processes already discussed include: a) flame deposition, which has been used to apply thin silica coatings to metals and to make pyrolytic graphite; b) drawing or extruding polycrystalline materials from a melt; c) dip-forming, in which a wire or substrate fiber is pulled through a melt to pick up a coating of the melt material; and d) fused salt electrolysis, electro- or cataphoretic deposition processes, in which an electrical potential is applied to a bath containing the desired material in order to deposit it on a substrate (conductive) that serves as the cathode. In these methods, a necessary next step in most cases would be to consolidate the deposit.

Each of these processes by which composite polycrystalline fibers may be made have inherent difficulties. All are in relatively early stages of development, and the scarcity of details regarding process parameters and/or data obtained prevents further discussion.

8. Comparison of Processes

Colloid evaporation, extrusion, and the "rayon spinnerette" method for making polycrystalline fibers yield a product which is porous, has a high shrinkage rate during firing, and is subject to grain growth during firing. Any one of these conditions is detrimental to high strength.

The vapor deposition of certain materials onto a heated filamentary substrate appears to be feasible for the continuous production of composite fibers. A large number of materials systems which are not otherwise amenable to formation into fibers can be filamentized by this technique. Such filaments have displayed the best mechanical properties, but at their present state of development are the least economical. Among the problem areas of immediate concern to the current level of technology are the following: gas flow dynamics and its effect on mass transport mechanisms, kinetics and temperature control, residual stress effects, core/sheath interactions, and process effects (on growth rate, bulk and surface structure, and electrical and mechanical properties). All of these factors must be understood and manipulated to produce high quality materials rapidly and economically. Much useful background information pertinent to these problems can be found by reviewing the literature describing the deposition of pyrolytic graphite.

The similarities in the Taylor wire and core-sheath techniques lead to mutual problems in the preparation of filaments by these methods. Reaction at the interface between the outer and the inner materials can seriously affect the strength of the filaments produced.

Undoubtedly there are inherent strength-influencing conditions associated with the remaining processes, but, as noted above, there is a lack of information about these methods.

A review of the forming methods for these materials points out how much the strength of the product depends upon the method of processing. In order to produce high strength filaments, careful consideration must be given to this factor.

D. Carbon and graphite fibers

1. Introduction

Carbon base fibers, while they may be thought of as a class of polycrystalline fibers, are of sufficient importance to warrant separate discussion in this section, because of (a) the special nature of carbon, graphite, and diamond in the materials spectrum, and (b) the properties and processes associated with carbon and graphite fibers. There is also a nomenclature problem associated with this class of materials. While they are all carbon, chemically, they represent the widest possible range of crystallographic features. For example, carbon may exist as (1) "amorphous" or "noncrystalline" lampblack, (2) graphite with a hexagonal layered structure, or (3) diamond with a highly ordered structure.

The following describes carbon filaments with respect to classification, properties, and processes applicable to their manufacture.

2. Classification

Synthetic filaments or fabrics which have been thermally converted to a carbonaceous condition are commonly referred to as either partially carbonized, carbonized, or graphitized. In fact, all three terms are often used to describe the same material. To minimize the confusion and misunderstandings attendant with such closely related materials, several systems of classification have been proposed. Schmidt and Jones[417] described the following two systems:

1. Fibers prepared at moderate temperatures, ranging from 1300° to 1700°F, are referred to as partially carbonized and carbonized. Graphite fibers, although synthesized

from the same precursory material, are processed at
temperatures ranging from 4900° to 5400° F.

2.   Fibers having a carbon content up to 90 weight percent
     are described as "partially carbonized". Fibers classi-
     fied as "carbonized" have an elemental carbon content in
     the range 91-98 weight percent. Filaments are described
     as "graphite" fibers when their elemental carbon content
     exceeds 98 weight percent.

However, these methods fail to take into account the actual crystal
structure of the fiber. A "graphitized" filament, for example,
could be one processed at a very high temperature or one having a
very high elemental carbon content. That the term "carbon" encom-
passes a family of materials rather than a single material, almost
dictates that an arbitrary classification method be used. Any
method that reliably and reproducibly relates the properties of the
filament with each phase of the classifying technique, will be a
valuable guide to the engineer and the scientist in using these
materials.

3.   Strength

Although carbon and graphite fibers have valuable thermal
and chemical properties, their mechanical strengths have not yet
been optimized. The potential for high strength exists in these
materials; it is the current level of technology that restricts their
use primarily to reinforcements for ablating thermal protection
systems.

The potential strength of a highly perfect carbon fiber could
be as large as 14.5 to 20.0 $\times 10^6$ psi. This prediction is based on
the fibers displaying 10%-16% of the Young's modulus of 145×10^6
psi which has been estimated[30] for graphite whiskers. Values of
tensile strength up to 2.80 $\times 10^6$ psi and Young's modulus of
102 $\times 10^6$ psi have been reported by Bacon[30] for graphite whiskers. Gilman[175] has estimated the Young's modulus for three
orientations of diamond as follows: $E_{100} = 152 \times 10^6$, $E_{110} =$
168.5 $\times 10^6$ and $E_{111} = 174 \times 10^6$ psi. These values of E infer
that the potential tensile strength of carbon (diamond structure)
fibers could approach 28 $\times 10^6$ psi.

Present commercially available carbon-graphite fibers have
tensile strengths of about 0.1- 0.25 $\times 10^6$ psi and a modulus of
6- 9 $\times 10^6$ psi. Some new laboratory values show considerable improvement. At the Parma Laboratories of Union Carbide Co. (see
contact report No. 55), for example, values of tensile strength to
0.35 $\times 10^6$ psi and modulus of elasticity of 30 $\times 10^6$ psi have
been obtained on small samples. Higher values were inferred
and are expected to be announced in technical papers in the near

future. These fibers have a density of 1.4 g/cc while other sources report densities up to 2 g/cc which gives carbon fibers a very attractive strength-to-density ratio. Other work in the field includes that of the H. I. Thompson Fiber Glass Co., Carbon Products Division of Carborundum Co., the Minnesota Mining and Manufacturing Co., and the Japanese Government Industrial Research Institute.

The gap between the theoretical potential strength and the actual strength of carbon and graphite fibers can be looked at from a structural viewpoint. Comparison of the microstructure of graphite whiskers with that of presently available carbon fibers (as shown by x-ray and microscopy) shows large differences. These differences rather clearly explain the variations in strength; at the same time they indicate what appears to be the inherent limitation for making graphite fibers that approach graphite whiskers in strength; i.e., the basic crystal shape and extremely anisotropic properties of graphite. The thin hexagonal plate-shaped crystals do not conform to circular fibers and, since the faces of these platelets are weakly bonded to each other, the whiskers and most graphite fibers tend to show concentric layers (see figures in contact report No. 57). This microstructural feature results in a peeling action when the whiskers are loaded in tension.

In spite of the above limitations for making carbon fibers having multi-million psi strengths, recent progress in technology has resulted in considerable strength improvements.

4. Processing

The preparation of carbon and graphite fibers (with the exception of whiskers) is generally based on the carefully controlled thermal degradation of synthetic organic fibers, yarns, or textiles; the choice dependent upon the desired product form. Rayon is probably the most widely used precursor, but numerous other starting materials have also been utilized.

A detailed discussion of carbon and graphite fibers made at the Japanese Government Industrial Research Institute (Osaka) by heating a polyacrylonitrile fiber in a controlled atmosphere is available[438]. Analysis of the evolved gases, measurements of electrical resistance, and x-ray examinations were made on fibers which were heated at various temperatures up to 1000° C. The effect of pre-oxidation on these properties and on the tensile strength of the fibers was determined. Variations in maximum load and tensile strength for both pre-oxidized fibers and fibers that were not pre-oxidized are shown in Figure 28.

A partially carbonized fabric known as "Pluton" is available from the Minnesota Mining and Manufacturing Company. The

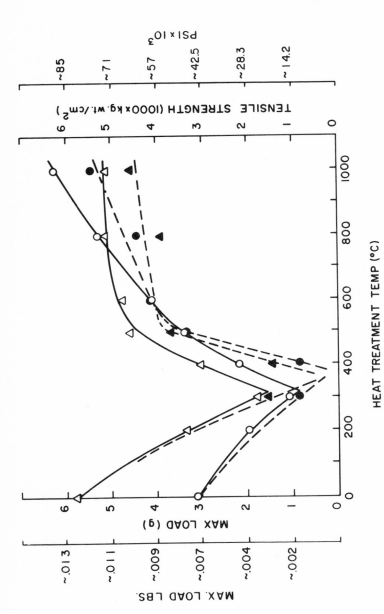

Figure 28. Variation in Maximum Load and Tensile Strength of the Polyacrylonitrile Fibers with Increasing Heat-Treatment Temperature. The Preoxidized Fiber: △ *Maximum Load,* ○ *Tensile Strength; The Unpreoxidized Fiber:* ▲ *Maximum Load,* ● *Tensile Strength. (Shindo, Ref. 438). Permission Jap. Govt. Ind. Res. Inst. Osaka*

tensile strength of "Pluton" filaments is about 38,400 psi in the warp and 47,400 psi in the fill direction. In contrast to conventional partially carbonized, carbonized or graphitized materials, "Pluton" is an excellent thermal and electrical insulator. The manufacturer does not recommend it as a filler for reinforced plastics, however, due, apparently, to excessive shrinkage of the fibers during the resin curing process which results in a very brittle laminate.

Other variations of carbon-graphite fibers as well as some graphite whiskers are generally prepared by vapor deposition or more specifically pyrolytic graphite techniques. One of these carbon coated (100-500 Å thick) 0.4 mil diameter fused silica fibers is discussed in a preceding section and in contact reports on Cambridge University and the General Electric Co., Missile and Space Division, Manufacturing Research Operation. The carbon coating is applied to preserve the virgin surface of the fused silica fiber. Another variation is obtained by applying a pyrolytic graphite coating to carbon fibers to improve the modulus of elasticity of the fibers. The coating also provides a dense, impervious, surface layer over the relatively porous carbon fiber, which may be advantageous in some applications.

Hough's work[232] is another interesting method for making "carbon" filaments. He deposited pyrolytic graphite on a resistance-heated filamentary tungsten substrate by pyrolyzing certain alkane hydrocarbons in a chamber surrounding the substrate. Filaments up to 2500 feet in length were produced by this technique. Although preliminary strength values were not particularly attractive (50,000 to 145,000 psi), further study should result in considerable improvement.

Graphite whiskers have been prepared by at least two techniques. One method is to crack methane at temperatures in the range of 900°-1200° C (see contact report No. 57) in a simple tube furnace at low (5-8 mm) to atmospheric pressures. This is also how massive pyrolytic graphite articles are generally made and whiskers often result as a by-product of this operation. The other method utilizes a high pressure arc technique described by Bacon[30].

In addition to the main applications of these fibers discussed in Section III, carbon fibers are also used as conducting substrates for preparing some of the composite polycrystalline composite fibers discussed previously, and as sources of carbon to make carbide fibers by diffusing and reacting metals into them.

In summary, while the conventional carbon and graphite fibers have inherent limitations such as oxidation susceptibility, high thermal conductivity, poor abration resistance and low strength, they are nevertheless being used in many applications. Carbon-

base fiber reinforced plastic ablation materials for thermal
protection in re-entry and rocket propulsion systems, carbon
fiber resistance elements for small furnaces, and substrates for
the production of composite fiber are some of the used for these
materials.

The considerable effort that is being devoted to these materials should result in greatly improved carbon and graphite
fibers. The low density of carbon coupled with predicted improvements in strength make this material extremely attractive
as a structural reinforcement.

E. Single crystal (whisker) fibers

A theoretical estimate[202] of the strength of an ideal crystal
(see Section II) was proposed several years prior to the actual
demonstration that near-perfect single crystals were indeed extremely strong. Herring and Galt[216] were the first to report, in
1952, that tin whiskers exhibited strengths which approached the
estimated theoretical value. Their work, based on measured
elastic strains, led to further efforts by others to learn why some
whiskers from the same growth population were strong while
others were quite weak. The following attempts to delineate various parameters which may help to explain and illustrate the
scatter observed in whisker strength.

1. Effect of Size

The dependence of whisker strength on size has been well
documented[63,388]. However, in establishing the size dependency
on strength, the test method used to obtain the data is of great
importance. A comparison of the work by Strelkov and Shpunt[465]
with that of Fridman and Shpunt[180] emphasizes the importance of
test methods. The former made bend tests on LiF cleavage whiskers, the latter tested these fibers in tension. The results of both
studies are shown in Figures 29 and 30. While a size dependency
is indicated by both studies, as shown by the sharp rise in whisker
strength as size decreases below 2 μ, the bending strengths are
about 4 to 10 times greater than those obtained by tensile tests.
Furthermore, both studies show that the 20μ to 30μ fragments
have strengths ranging from 7,000 to 14,000 psi, which is about
6.5 to 10 times greater than the flow stress of the parent crystals
from which these cleavage whiskers are formed.

It appears that the strength dependency on the size of the
whiskers is due primarily to the greater probability of internal or
surface flaws on the larger whiskers, resulting in lower strengths.

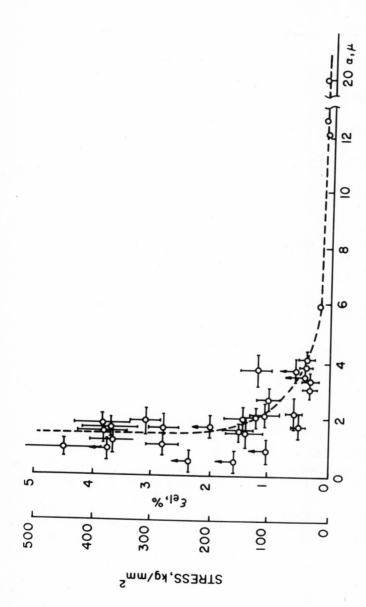

Figure 29. Dependence of the Elastic Deformation Limit (ϵ_{el}, %) of Lithium Fluoride Cleavage Whiskers on Their Transverse Dimensions (a, μ) (Strelkov and Shpunt, Ref. 465). Permission Am. Inst. of Physics.

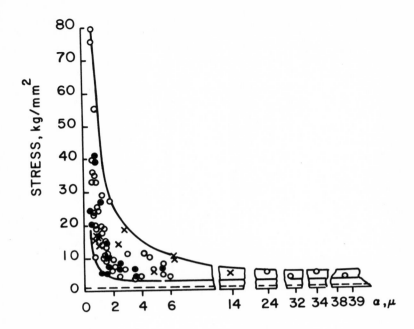

Figure 30. Dependence of the Tensile Strength of Lithium Fluor-ide Crystal Fragments on Their Transverse Dimension a (Fridman and Shpunt, Ref. 180). Permission Am. Inst. of Physics.

2.   Effect of Crystal Perfection

Although whiskers have exhibited strengths approaching the-oretical values, a full characterization of the crystalline perfec-tion of the tested segments is rarely performed. The very small cross sections (usually 5μ) with which one must work, makes the determination of structural perfection extremely tedious and dif-ficult. However, such methods as etch pitting[97], dislocation dec-orations[9], x-ray diffraction[187], and electron microscopy[309] have been successfully used to study whisker structure.

The observations of Webb et al[525, 527] with Al_2O_3 whiskers and Hamilton[209] with SiC whiskers rather well substantiate the theory that high-strength whiskers are nearly structurally perfect. Coleman[96] concludes that the low (volume) density of dislocations and the homogeneity of the crystal lattice contribute to the observed high strengths of whiskers. Further support of the high-strength/low-dislocation-density theory is shown in Table VII which summarizes Regester's work on the relationship between whisker size and number of dislocations in alumina whiskers[388]. In addition, the fact that dislocations are quite immobile in the ceramic whiskers helps to make them very useful as reinforcements.

Overgrowths which produce stress concentrations can seriously decrease whisker strength. For example, work at ERDE, Ministry of Aviation, Waltham Abbey, England (see contact report No. 25), indicates that a 90° step on the face of a Si_3N_4 whisker is essentially as detrimental to whisker strength as are Griffith flaws in glass fibers.

TABLE VII. VARIATION OF ETCH PITS WITH WHISKER
CROSS-SECTIONAL AREA

Area - μ^2	Distribution Classification			
	None	Individual	Cluster	Continuous
0-100*	↔			
100-200	↔ ↔ ↔			
200-300		↔ ↔		
300-400	↔	↔	↔ ↔	
400-500			↔	
500-750			↔	
750-1000				↔ ↔
1000-2000				↔
2000-3000				↔

*Many dissolved   Regester (Ref. 388)

3. Effect of Surface Layers

In view of the above discussion on the effects of overgrowths on the strength of whiskers, it is also important to consider the effect of thin (atomic) layers on the surface of whiskers. For example Weik[532] noted that some of the larger whiskers of iron grown by the reduction of ferrous chloride, exhibited a fracture during testing of the type where an outer sheath was removed from an inner core. A careful analysis of these whiskers revealed that the sheath or skin was of the same composition as the core and that no oxide or corrosion layer was present. $_0$Further studies of very fine whiskers, which were coated with 100Å of nickel, showed a pronounced increase in strength over that of the uncoated whiskers even though the fine uncoated whiskers exhibited no sheath effect. However, the larger whiskers exhibiting the 'sheath' effect were as strong as the nickel-coated whiskers having no sheath (see Figure 31). Thus, both the coated whiskers and those possessing an outer 'sheath' appeared to be strengthened by the outer layer structure. A possible conclusion drawn from these studies was that whiskers can grow by a mechanism different from that of an axial screw dislocation, viz., that outer layers may grow by another mechanism.

Metallic whiskers often may have an outer oxide layer. For example, Cabrera and Price[76] have stated that the thickness of the oxide layer on zinc whiskers is beneficial to whisker strength, if the thickness of the oxide layer is uniformly increased. However, they point out that, if oxidation is carried out at slightly above room temperature, the oxide film ceases to be uniform and strength decreases. Figure 32 shows the relationship between yield point and oxide thickness. On the other hand, Webb and Stern[529] found no difference in the strength (at 0.4% strain) of copper and silver whiskers having tarnished surfaces from those that had been electrochemically cleaned.

Removal of the outer surface of a whisker may influence its strength. Brenner[59] performed tensile tests on iron whiskers immersed in an alcohol-picric acid solution. The strain was monitored while a constant force was applied. As the whisker cross-section decreased due to etching, the elastic strain and strength increased significantly. Figure 33 schematically shows the arrangement of the whisker under load during the experiment and the increase in elastic extension with decrease in the whisker diameter. A typical result for the stress-strain behavior of the whiskers can be seen in Figure 34.

The fabrication of high strength whisker-reinforced composites depends on the strength of whiskers, individually and collectively. Thus the relationship between surface treatment and whisk-

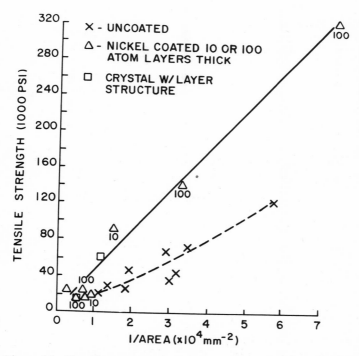

Figure 31. Tensile Strength vs Reciprocal Cross-Section Area of Iron Whiskers Grown Under the Same Conditions. (Weik, Ref. 532). Permission Am. Inst. of Physics.

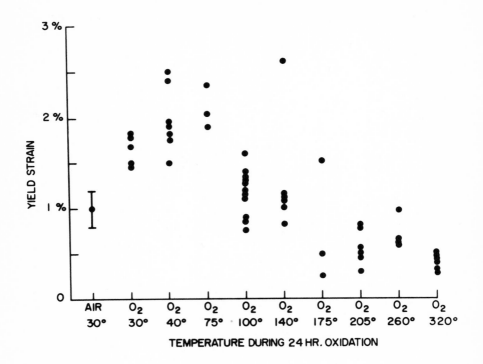

Figure 32. Yield Point as a Function of Oxide Thickness (Cabrera and Price, Ref. 76). Permission J. Wiley & Sons, Inc.

er strength is of vital importance, particularly if interfacial reactions roughen the whisker surface to the extent that stress concentrators are formed. Sutton* found indications that Ni-Pt and Ti-coated sapphire whiskers heated to 1000°C retain their high strength at room temperature; heated to 1500°C, the room temperature strength of Ti-coated whiskers was seriously reduced. The effectiveness of metal coatings in promoting matrix-to-whisker bonding can be seen in Figure 35.

This survey did not reveal any programs in which the effect of surface films on <u>high temperature strength</u> has been, or is being, studied specifically.

─────────

*Unpublished data.

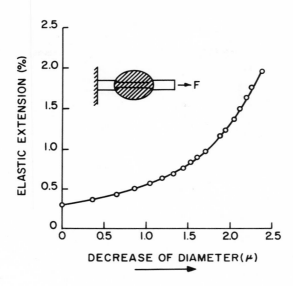

Figure 33. Relationship of Elastic Extension to Whisker Diameter and (in insert) Arrangement of Whisker Under Load (Brenner, Ref. 59). Reproduced from Chapter 2, "Factors Influencing the Strength of Whiskers", S.S. Brenner, in Fiber Composite Materials Copyright 1965 by American Society for Metals, Metals Park, Ohio.

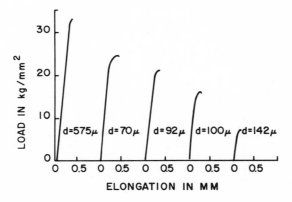

Figure 34. Stress-Strain Behavior of Whiskers (Brenner, Ref. 59). Reproduced from Chapter 2, "Factors Influencing the Strength of Whiskers", S.S. Brenner, in Fiber Composite Materials, Copyright 1965 by American Society for Metals, Metals Park, Ohio.

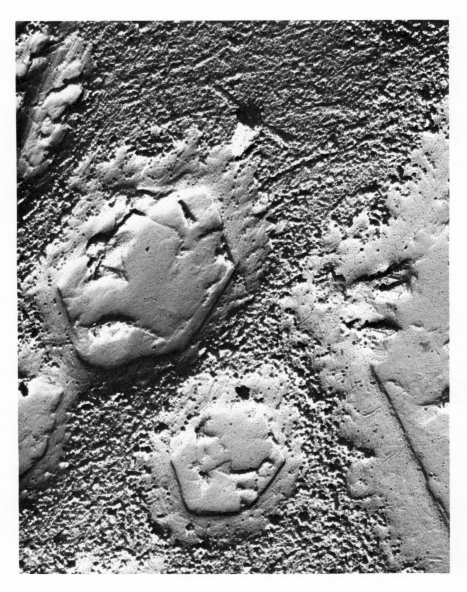

Figure 35.   Metal-Coated Sapphire Whiskers in a Silver Matrix.
Magnification 5000X.

4. Effect of Growth Techniques

Whiskers of at least 30 elements and more than 50 compounds have been grown[333] thus far. A variety of techniques has been employed for growing whiskers, especially for the metallic varieties. For example, metal whiskers have been grown spontaneously from plated metals[25] or in eutectic alloys[174, 217]. They have also been grown by condensation from the vapor phase, by electrolysis, by sudden cooling in aqueous solutions, from a solution with a cleaved surface of a bulk crystal as the substrate[331], from porous substrates[246], by changes in pressure[368], by a vapor-liquid-solid process[521, 522] and by the most widely used method of all — the high temperature reduction of metallic salts in a hydrogen atmosphere.

Within the limited variety of techniques used to grow oxide whiskers, the number of compounds is quite large[96]. The most commonly used technique is by the slow oxidation of a metal in the presence of a refractory substrate within a controlled atmosphere.

Other techniques have been used to grow whiskers of graphite[30], pyrolytic graphite[73, 347, 463], boron carbide[187], silicon carbide[258], aluminum nitride[125] and silicon nitride[247].

Some growth techniques result in a greater number of structural imperfections than others, and in this way are directly related to whisker strength (or weakness). Nadgornyi et al[334] have summarized various growth techniques, test methods and strengths as shown in Table VIII. The importance of growth technique on strength can not be overemphasized, since the growth technique markedly influences both the surface and bulk perfection of the whiskers. Careful control of process variables permits rather wide latitude in the types of whiskers grown, as illustrated in Figures 36 and 37. Of the various methods for growing whiskers and fibers, the vapor deposition technique generally produces the best results. However, it can also produce poor results if not properly used.

5. Effect of Environments

a. Temperature

The yield stress of silicon whiskers 16μ to 28μ in diameter decreased rapidly above $600^{\circ}C$ [362]. This is above the temperature at which silicon becomes plastic and the results are consistent for large whiskers of ductile materials.

Brenner[61] reports that sapphire whiskers also show a strength dependence on temperature. He explains this behavior by assuming that the whiskers are essentially dislocation free and that failure occurs by thermally activated crack propagation, which is preceded at high temperatures by localized dislocation nucleation.

TABLE VIII. FILAMENTARY CRYSTALS

No.	Mat'l	Characteristics of Growth		Mechanical Properties			
		Growth Technique	Growth Direction	Type of Deformation	ϵ_{max} (%)	σ_{max} (kg/mm²)	d (μ)
1	Sn	Spontaneous	[100], [110], [111]	Bending	1.0–2.0	--	1.8
2	Pb	"	[110]	--	--	--	--
3	Zn	"	[112̄0]	--	--	--	1.0–5.0
4	Zn	Vapor deposition	[112̄3]	Elongation	1.0	50	
5	Cd	Spontaneous	[112̄0]	--	--	--	--
6	Cd	Vapor deposition	[112̄0]	Elongation	1.0	52	3.0
7	Cu	Reduction of halide	[100], [110], [111]	"	2.8	300	1.25
8	Ag	"	[100], [110]	"	4.0	176	3.8
9	Fe	"	[100], [110], [111]	"	4.9	1340	1.6
10	Ni	"	--	"	1.8	--	--
11	Co	"	--	--	--	--	--
12	Pt	"	--	--	--	--	--
13	Au	"	--	--	--	--	--
14	Hg	Vapor deposition	[100]	In electric field gradient	--	100	0.01
15	Mn	Reduction of halide	--	Bending	1.5	310[a]	1.7
16	Cr	Precip. from solid phase	--	"	3.8	800[a]	1.6
17	Si	Reduction of halide	[111]	Elongation	2.0	390	--
18	Si	Vapor deposition	[111]	Bending	2.6	500	16–28

| No. | Mat'l | Characteristics of Growth | | Type of Deformation | Mechanical Properties | | |
		Growth Technique	Growth Direction		ϵ_{max} (%)	σ_{max} (kg/mm²)	d (μ)
19	Ge	Reduction of halide	[11̄1], [1̄12̄]	Bending	--	30	20
20	ZnO	Synthesis of Zn, O, and H	[0001]	"	1.5	150[a]	21
21	Zn	Vapor deposition	[0001]	"	1.5	105[a]	up to 50
22	CdS	" "	--	"	2.4	--	--
23	Mn$_5$Si$_3$	Reduction of halide	--	"	2.7	--	--
24	Si$_3$N$_4$	Nitriding of silicon	--	"	4.0	320[a]	2.8
25	Al$_2$O$_3$	Oxidation in gas phase	[0001]	"	2.3	1200[a]	3
26	Cr$_3$O$_4$	Precip. from solid phase	--	"	0.3	67[b]	4.1
27	Cr$_2$N	" " " "	--	"	5.8	--	2.2
28	Fe$_3$C	" " " "	--	"	4.1	800[a]	1.9
29	C	In high-pressure arc	--	--	2.0	2100	5.0
30	NaCl	Evap. thru porous mat'l.	[100], [110]	Elongation	2.3[a]	110	1.0
31	KCl	Evap. thru porous mat'l.	[110]	--	--	--	--
32	LiF	From supersat. soln.	[100]	Bending	1.2	--	4.5
33	Hydro-quinone	From supersat. soln.	--	Bending	6.0–8.0	--	up to 5

[a] Estimated from elastic constants of ordinary single crystals.
[b] Not the maximum; still under investigation.

(Nadgornyi et al, Ref. 334, Permission Am. Inst. of Physics).

*Figure 36.   Needle-Like Sapphire Whiskers Grown in Alumina
Boat.   (Sutton et al, Ref. 486).*

Figure 37. Wool-Like Sapphire Whiskers Grown in Alumina Boat. (Sutton et al, Ref. 486).

Although the average and maximum fracture stresses of these
whiskers decreases significantly with increasing temperature,
their fracture stress remains at least 20 times greater than the
observed flow stress of the bulk crystals at comparable temperatures. Since no plastic flow was observed in the whiskers, the
fracture stress was assumed to be equal to the yield stress.
Brenner reported that the average strength of the sapphire whiskers at 20° C below the melting point was greater than the fracture
strength of bulk crystals at room temperature. The data of Pearson et al[362] and Brenner[61] are shown in Figures 38 and 39 respectively.

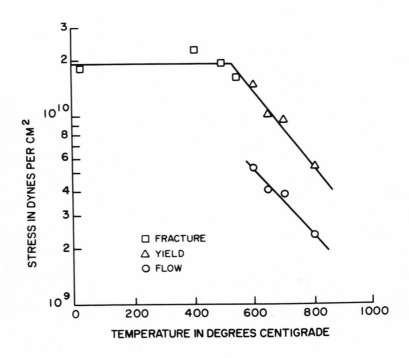

*Figure 38. Fracture, Yield, and Flow Stress vs Temperature for
Silicon Whiskers (Pearson et al, Ref. 362). Permission Acta
Metallurgica.*

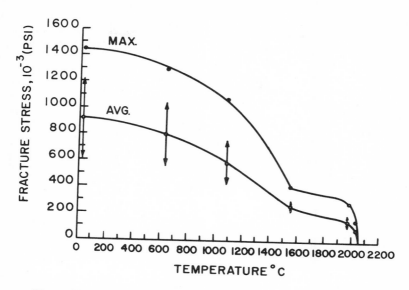

Figure 39. Fracture Stress vs Temperature for α - Al₂O₃ Whiskers (Brenner, Ref. 61). Permission Am. Inst. of Physics.

b. Impurities

The growth of sapphire whiskers in refractories containing silica or in systems where silica is present in the growth chamber results in whiskers that contain 2%-3% silica as an impurity (see Section VIII, patent abstract 3,077,380). Brenner[59] states that iron and copper whiskers grown by the reduction of commercial halides contain about 100 ppm of impurities and that sapphire whiskers contain from 1% to 2% silicon. Lynch et al[292] grew Al_2O_3 whiskers in a carbon resistance furnace and in a simple tungsten wound furnace. Analysis of the whiskers thus produced showed SiO_2 and Fe_2O_3 to be the major impurities, with MgO present in considerably lesser amounts. Although three different grades of alumina were used as the parent material, whiskers were grown only from the two less pure grades, which helps to confirm the theory that impurities should be present for whisker growth. For complete results of their analyses, see Table IX.

6. Summary

The foregoing discussion has attempted to emphasize the fact that the strength of ideal whiskers should approach the theoretical cohesive strength of perfect single crystals. This, in fact, has been demonstrated. However, experimental tests have also shown

TABLE IX. RESULTS OF WHISKER ANALYSIS

	A. AVERAGE			B. HIGH IRON CONTENT		
Al_2O_3	90 %	±	2.0%	85 %	±	2.0%
SiO_2	9 %	±	1.0%	10 %	±	1.0%
Fe_2O_3	1.3%	±	0.2%	5 %	±	0.5%
MgO	0.4%	±	0.1%	< 0.1%	±	---
	100.7%	±	3.3%	100 %	±	3.5%

Lynch et al (Ref. 292)

a wide scatter in the strength data due to a wide variety of reasons discussed above; viz. , that not all whiskers have an ideal structure because of bulk and surface imperfections and because of oxide or corrosion surface layers. Thus, it cannot be over-emphasized that careful control over all the growth parameters is mandatory in order to assure a supply of uniformly high-strength whiskers. The influence of these various parameters is reflected in the data scatter of Figure 40, which shows the relationship between tensile strength and cross-sectional area of Al_2O_3 whiskers.

Many problems remain to be solved before the high strength of certain whisker materials can be fully utilized. However, a comparison of bulk material strength with the strength of the same material in whisker form, as shown in Table X, presents a convincing argument for the potential of whisker materials.

F. Test and evaluation

It is evident from the above discussion that testing methods and procedures are most important, since they provide one of the best means for evaluating the fiber's potential as a reinforcing material. However, there can be a wide scatter in the data due to (a) inhomogenieties in the fibers, and (b) the test method itself. Therefore, it becomes important to distinguish between the two sources of scatter in order to fully characterize the useful strength of the fiber.

Three methods commonly used for determining the strength of a fiber are the tensile test, the bend test, and the loop test.

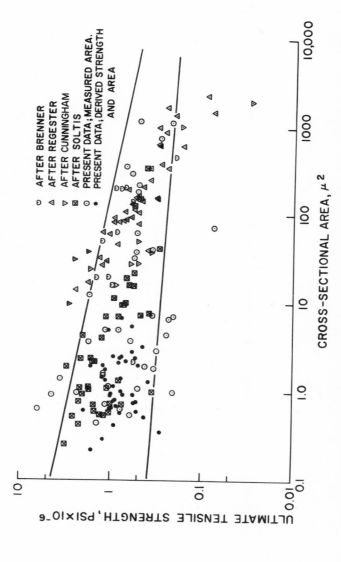

Figure 40. Tensile Strength of Alumina Whiskers as a Function of Area (Meban et al, Ref. 309).

TABLE X. DIFFERENCES BETWEEN BULK AND
WHISKER STRENGTH OF VARIOUS MATERIALS

Material	Tensile Strength ($\times 10^6$ psi)	
	Bulk	Whisker
Iron	0.004	1.9
Copper	0.0002	0.4
Silicon	0.005	0.55
Graphite	0.04	3.0
Boron Carbide	0.0225	0.965
Alumina	0.08	6.2
Silicon Carbide	0.03	3.0*

*Bend test, all others tensile data.

Each method has its limitations, but the tensile test usually yields the most precise values since the stress is distributed evenly over the entire fiber volume. The other two methods are usually simpler to perform, but only the outer perimeter of the fiber at the point of maximum curvature is stressed to the maximum value. Thus, the volume element under maximum stress is considerably smaller than that of a fiber tested in tension. Since the probability of finding a critical flaw increases with the volume under stress, fibers tested in tension usually appear to be weaker than those tested in bending. In addition to these differences in test methods, other factors are important and will be considered for each test method.

1. Tensile Test

This test is based essentially on pulling a fiber apart by gripping and applying a load to its ends. The tensile stress, σ_{max}, is determined simply by dividing the breaking load by the fiber's cross-sectional area:

$$\sigma_{max} = \frac{P}{A} \tag{1}$$

where P is the load, and A the cross-sectional area.

The prime difficulties arise in gripping the fibers (since they are brittle) and ensuring proper alignment of the tensile load along the fiber's axis. Other problems arise when there is taper or other geometrical changes along the fiber's axis. Many fibers are not circular in cross-section and, as the dimensions of the fiber diminish to a few microns (e.g., in the case of whiskers) measurement of the true area becomes the greatest source of error in determining σ_{max}.

The bend and loop tests depend on imparting a symmetrical curvature to a fiber, which places the outer surface of the fiber at the point of minimum curvature under maximum stress. In three point bending, for example, the maximum (fracture) stress is given by:

$$\sigma_{max.} = \frac{8P\ell}{\pi d^3} \quad \text{(for circular fibers)} \tag{2}$$

and

$$\sigma_{max.} = \frac{3P\ell}{2bh^2} \quad \text{(for rectangular fibers)} \tag{3}$$

where ℓ is the length of fiber between the two outer anvils, and d, b, h refer to fiber diameter, breadth, and height (thickness), respectively. It becomes immediately evident that σ_{max} depends on $1/d^2$ for the tension test, and on $1/d$ for the bend test of circular fibers. Thus the errors introduced in the measurement of d are more critical in the bend tests, especially for the very smallest of fibers and whiskers. Also of great importance is the ℓ/d of the test specimen. For example, when ℓ is relatively large, the fiber tends to slip past the anvils and thus cause deviations in the measured strength value. In general, the ℓ/d ratio should be less than 100. Hence, for a $1\,\mu$ diameter fiber, the ideal length would be only $100\,\mu$ or 0.005 inch. The maximum (outer) stress in a circular fiber that has been formed into a loop is determined by the following relationship:

$$\sigma_{max} = \frac{Ed}{2r}$$

where E is the Young's modulus of the fiber, d the fiber diameter, and r the radius of loop curvature. The reliability of this test also depends on a precise measurement of the fiber dimensions, on an accurate knowledge of the Young's modulus (which may vary considerably for single crystal fibers of a given growth direction), and on the radius of curvature, which becomes increasingly difficult to measure for high-strength, very fine filaments. Also, the plane of the loop (under the microscope) may not be normal to the line of sight and thus a spurious value of r is obtained.

The bend and loop tests are therefore inherently not as precise as the tensile tests because the values of σ_{max} are based on the following numerous assumptions: the fiber has a uniform cross-section throughout, plane stresses remain plane within the fiber, Hooke's law is obeyed, the fibers are homogeneous and isotropic, and the elastic modulus in tension is equal to that in compression. Most of these conditions are satisfied in the case of glass fibers, but when whiskers and inhomogeneous fibers are tested, some of these assumptions are not possible. Therefore, when the strength data is of critical importance to the development of improved or new fibers or is critical to the design application, information other than the strength values should be considered, viz.,

1. the type of test

2. the specimen type size, shape, cross-sectional area and gage length

3. the ambient atmosphere and temperature

4. the handling of the fibers (and the possible introduction of surface flaws)

5. the specimen history (how formed, exposure to moisture, temperature, etc.)

6. rate of load application or duration

7. the statistical data; strength scatter, standard deviation, coefficient of variation, confidence limits, etc.

V. RESULTS OF VISITS AND QUESTIONNAIRES

A. Introduction

A search of the patent reviews and literature on ceramics and graphite fibers provides an extensive and solid base of relevant information. However, to ascertain the present state of the art, a survey of recent and current activities was necessary. During this survey, about 200 organizations (in the United States, Canada, Australia, Japan, England, and Western Europe) were contacted personally, or by telephone or letter, and invited to describe their work as fully as possible within the limits of classified or proprietary information. Although some organizations undoubtedly have been missed, we feel that the primary result of further investigation at this time would be duplicative, except where classified or proprietary work is in progress.

Of those contacted, more than 60 supplied data on their activities in the field of fibers and whiskers. The resulting reports, summarized in this section, present a brief review of the work in each organization and are a comprehensive representation of the ceramic and graphite fiber field. They are intended to acquaint the reader with current developments and enable him to contact directly those organizations of interest for additional information.

For more convenient reference, each contact report has been numbered and two tables of contents are provided. The first lists the organizations alphabetically with their areas of principal interest; the second groups the organizations under each of the various subjects. In addition, a tabular summation of the reported properties of the fibers and whiskers discussed in the contact reports is presented at the end of this section in Tables XX through XXII.

1. Alphabetical Listing of Organizations

Organization	Principal Field of Interest
1. Aerojet-General Corporation	Glass fibers
2. Air Force Materials Laboratory	Glass, polycrystalline fibers, and whiskers
3. Army Materials Research Agency	Whiskers and composites
4. Astro Research Corporation	Boron fibers and fiber-reinforced composites
5. Atomic Energy of Canada, Ltd.	Fused silica and aluminum on silica fibers
6. Atomic Weapons Research Establishment Aldermaston, England	Silicon nitride whiskers and composites
7. Atomics International	Beryllia whiskers
8. Avco Corporation	Magnesia, tungsten oxide, et al, whiskers; filament-reinforced plastics
9. Babcock and Wilcox Co.	Refractory glass; e.g., aluminium silicate fibers
10. Battelle Institute e.V. Frankfort, Germany	Iron whiskers, silica fibers
11. Battelle Memorial Institute	Polycrystalline alumina fibers
12. Bell Telephone Laboratories	Whisker growth mechanisms and properties
13. Bjorksten Research Laboratories	Silica fibers
14. Boeing Company	Polycrystalline Al_2O_3 and ZrO_2
15. Brunswick Corporation	Alumina silicate glass and metal fibers
16. Bureau of Naval Weapons	Glass fibers, alumina whiskers

Organization	Principal Field of Interest
17. Cambridge University Cambridge, England	In situ deposition of whiskers and fiber-composites
18. Carborundum Company	SiC whiskers and aluminum-silicate glass fibers
19. Central Institute for Industrial Research, Olso, Norway	Oxide whiskers; e.g., Nb_2O_5
20. Cincinnati Testing Laboratory	Fiber-reinforced ablative plastics for missiles
21. Compagnie Francaise Thomson-Houston, Paris, France	Al_2O_3 whiskers
22. Cornell University	Whisker and fiber-reinforced composites
23. DeBell and Richardson, Inc.	Special cross-section shapes of glass fibers
24. E.I. duPont de Nemours and Company	Potassium titanate fibers
25. Explosives Research and Development Establishment, Waltham Abbey, Essex, England	Silicon nitride and carbide whiskers
26. Ferro Corporation	Glass fibers
27. Fritz-Haber Institute, Max Plank Gesellschaft Berlin, Germany	SnO coated silica fibers
28. General Electric Company	
A. Advanced Engine and Technology Department	Boron and other fiber composites
B. Lamp Glass Department	Fused silica fibers
C. Manufacturing Engineering Services Laboratory	Glass fibers
D. Manufacturing Research Operation	Carbon coated fused silica
E. Research Laboratory	Whiskers, glass, silica, and intermetallics
F. Space Sciences Laboratory	Whiskers and fibers; organic and metallic composites

Organizations	Principal Field of Interest
29. General Technologies Corporation	Oxide, carbide, silicide, nitride continuous fibers, and composites
30. Haveg Industries, Inc.	High silica fibers and composites
31. Horizons, Inc.	Single and poly-crystalline fibers; e.g., Al_2O_3, ZrO_2
32. Hughes Aircraft Company	Refractory glass fibers
33. Illinois Institute of Technology, Research Institute	Polycrystalline alumina
34. Johns-Manville	Fibrous ceramic insulations
35. Lexington Laboratories	Alumina whiskers
36. P. R. Mallory and Co.	Ceramic whiskers
37. Marquardt Corporation	Silicon carbide filaments
38. Narmco Division, Whittaker Corporation	Silica, glass, and composite filaments
39. National Aeronautics and Space Administration	Whiskers, and fiber composites
40. NASA-Lewis	Whiskers, and metal fiber composites
41. National Beryllia Corporation	Beryllia whiskers
42. National Bureau of Standards	Single crystal properties
43. Naval Ordnance Laboratory	Glass fiber and whisker composites
44. Naval Research Laboratory	Mechanics of composites
45. Owens-Corning Fiberglas Company	Glass fibers, TiO_2 and ZrO_2 crystal fibers
46. Rolls-Royce, Ltd. Derby, England	Aluminum on silica fibers, Al_2O_3 whiskers

Organizations	Principal Field of Interest
47. Royal Aircraft Establishment Farnborough, England	Carbon and carbide fibers, metal wire-metal composites
48. Rutgers University	Glass fibers
49. Solar	Glass fiber strength
50. Swedish Institute for Metals Research Stockholm, Sweden	CuO, etc., metal whiskers, growth mechanisms
51. Texaco Experiment, Inc.	Boron fibers
52. Thermokinetic Fibers, Inc.	Al_2O_3, SiC whiskers
53. Thiokol Chemical Corporation	Metal whiskers
54. H. I. Thompson Fiber Glass Company (HITCO)	Glass, polycrystalline ceramic fibers
55. Union Carbide Corporation	Carbon fibers
56. United Aircraft Corporation	In situ whiskers, composite fibers
57. University of Grenoble Grenoble, France	Graphite whiskers
58. Watervliet Arsenal	Whiskers and composites

2. Organization Listing by Principal Fiber and Whisker Types

Special Glass Fibers

Aerojet-General Corporation
Air Force Materials Laboratory
Babcock and Wilcox Company
Bjorksten Research Laborator-
   ies
Brunswick Corporation
Bureau of Naval Weapons
Carborundum Company
Cincinnati Testing Laboratories
DeBell and Richardson, Inc.
Ferro Corporation
Haveg Industries, Inc.

Hughes Aircraft Company
Johns-Manville
Narmco
National Aeronautics and Space
   Administration
Naval Ordnance Laboratory
Naval Research Laboratory
Owens-Corning Fiberglas Co.
Rutgers University
Solar
H. I. Thompson Fiberglass Co.
   (HITCO)

Polycrystalline Continuous Fibers

Air Force Materials Laboratory
Astro Research Corporation
Avco Corporation
Battelle Memorial Institute
Brunswick Corporation
Bureau of Naval Weapons
Cincinnati Testing Laboratories
General Electric Company
General Technologies Corpora-
   tion

Horizons, Inc.
Illinois Institute of Technology,
   Research Institute
Marquardt Corporation
Narmco
National Aeronautics and Space
   Administration
Texaco Experiment, Inc.
H. I. Thompson Fiber Glass
   Company (HITCO)
United Aircraft Corporation

Fused Silica Fibers

Atomic Energy of Canada, Ltd.
Battelle Institute e.V.
Fritz-Haber Institute
General Electric Company
Haveg Industries, Inc.

Johns-Manville
Narmco
Rolls-Royce, Ltd.
H. I. Thompson Fiber Glass
   Company (HITCO)

Carbon Fibers

Royal Aircraft Establishment
   (England)

Union Carbide Corporation

Discontinuous Fibers (excluding single crystal whiskers)

Babcock and Wilcox Company
Boeing Company
E. I. duPont de Nemours and
Company
Owens-Corning Fiberglas
Company
Thiokol Chemical Corporation

Single Crystal Whiskers

Air Force Materials Laboratory
Army Materials Research
Agency
Atomic Weapons Research Establishment (England)
Atomics International
Avco Corporation
Battelle Institute e. V.
(Germany)
Bell Telephone Laboratories
Bureau of Naval Weapons
Cambridge University (England)
Carborundum Company
Central Institute for Industrial
Research (Norway)
Compagnie Francoise Thomson-
Houston (France)
Cornell University
E. I. duPont de Nemours and
Company
Explosives Research and Development Establishment
(England)

General Electric Company
Horizons, Inc.
Lexington Laboratories
P. R. Mallory and Co.
National Aeronautics and Space
Administration
National Beryllia Corporation
National Bureau of Standards
Naval Ordnance Laboratory
Owens-Corning Fiberglas
Corporation
Rolls-Royce, Ltd. (England)
Swedish Institute for Metals
Research
Thermokinetic Fibers, Inc.
United Aircraft Corporation
University of Grenoble (France)
Watervliet Arsenal

B. Contact reports

Contact Report No. 1

Aerojet-General Corporation   January 1965
Azusa, California

Contact: Al Lewis

By: L. R. McCreight

As demonstrated by their response both to a questionnaire and personal contacts, Aerojet-General is engaged in work on both refractory vitreous (aluminum-silicate glass) fibers and continuous "amorphous" (or very fine polycrystalline) fibers for use in reinforcing resins in such applications as rocket motors.

The experimental aluminum-silicate glass fibers (designated 4H-1) are 0.0004" to 0.0005" dia. and in continuous lengths. Strength and some other properties are listed as follows:

Ult. Tensile Strength Rm. Temp.	6.5 to 7.9 $\times 10^5$ psi (1" gage length at 0.2"/min)
Rm. Temp. (median)	7.3 $\times 10^5$ psi
500°F Median	6.3 $\times 10^5$ psi
1000°F Median	5.5 $\times 10^5$ psi
1500°F Median	3.5 $\times 10^5$ psi
Young's Modulus (range)	12.4-15.7 $\times 10^6$ psi (1" gage length at 0.2"/min)
Median	14.0 $\times 10^6$
Elongation	4.5 - 5.0%
Density	2.55 g/cc
Melting/Softening Point	2750°F ($\sim$1650°C)

Some of this work is sponsored by the U. S. Air Force under contract AF 33(657)-8904 and has been reported in papers to the American Ceramic Society.

Contact Report No. 2

Air Force Materials Laboratory   April 7, 1965
Research and Technology Division
U. S. Air Force Systems Command
Wright-Patterson Air Force Base, Ohio

Contacts: Colonel L. R. Standifer
          Dr. A. M. Lovelace
          Dr. Harris Burte
          Colonel Melvin Fields
          Mr. George Peterson
          Mr. R. T. Schwartz

By: L. R. McCreight

This organization sponsors the majority of the DOD programs
on filaments and filament reinforced composites. In addition,
several in-house programs are conducted on both the filaments
and composites. The activities are in four of the six divisions of
the Air Force Materials Laboratory: Metals and Ceramics, Non-
Metallic Materials, Advanced Filaments and Composites, and
Manufacturing Technology. The principal AFML activities were
initiated several years ago in the Non-Metallic Division so their
activities are more fully described below. While the other divisions have more recently initiated programs, both in-house and
on contract, the results are not as extensive nor as available at
this time, but will be reported and published in the future.

                    *   *   *

The Nonmetallic Materials Division, Air Force Materials
Laboratory, conducts both in-house and contractual research
leading to new and improved fibers for structural composites, ablative composites, and flexible fibrous structures, (such as decelerators and expandable structures). For structural composites,
the emphasis is on high modulus, high strength, low density continuous filaments in the areas of glass, carbon and graphite, and
materials such as boron carbide, silicon carbide, and beryllium.
For ablative composites, fibers based on refractory materials and
carbon and graphite are being emphasized. For flexible structures,
glass, silica, carbon, graphite, and high temperature alloy fibers
are being investigated. Typical technical reports available from
DDC to qualified requestors are as follows:

   1. Refractory Reinforcements for Ablative Plastics. Part I-
       Synthesis and Reaction Mechanisms of Fibrous Zirconium
       Nitride. ASD-TDR-62-260 Part I.

2. Refractory Reinforcements for Ablative Plastics. Part II - Synthesis of Zirconium Nitride and Zirconium Oxide Flakes. ASD-TDR-62-260 Part II.

3. Refractory Reinforcements for Ablative Plastics. Part III - Pyrolytic Boride Reinforcing Agents. ASD-TDR-62-260 Part III.

4. Refractory Reinforcements for Ablative Plastics. Part IV - Synthesis Apparatus for Continuous Filamentous Reinforcements. ASD-TDR-62-260 Part IV.

5. Continuous Pyrolytic Graphite Filaments. AFML TR 64-336.

6. Research to Obtain High-Strength Continuous Filaments from Type 29-A Glass Formulations. RTD-TDR-63-4241.

7. AF-994 - A Superior Glass Fiber Reinforcement for Structural Composites. ASD-TDR-63-81.

8. Pyrolyzed Rayon Fiber Reinforced Plastics. AFML TDR 64-47.

9. Carbon Fiber Reinforced Plastics. ASD-TDR-62-365.

10. Investigation of the Thermal Behavior of Graphite and Carbon-Based Fibrous Materials. ASD-TDR-62-782.

Contact Report No. 3

Army Materials Research   February 26, 1965
    Agency (AMRA)
Watertown Arsenal
Watertown, Mass.

Contacts: A. P. Levitt,
                Chief High Temp. Branch
            A. Tarpinian

By: W. H. Sutton

    The U. S. Army Materials Research Agency is conducting in-house and contracted research with Lexington Laboratories aimed at developing improved methods for growing whiskers, and for determining the mechanical properties of whiskers. In addition, the wetting and bonding characteristics of whiskers to possible metals matrices, such as Al_2O_3 to Ni and Ni-alloys, are being investigated at the General Electric Space Sciences Laboratory. AMRA is also supporting research with the United Aircraft Research Laboratories in the development of whisker reinforced high temperature metals by the unidirectional cooling of binary eutectic alloys. Systems under study include $Ta-Ta_2C$, $Ni-NiBe$, $Ni-Ni_3B$, and $Ti-TiB$. The in-house studies include an investigation of the wetting and bonding of Al to Al_2O_3 and the determination of the mechanical properties of whiskers.

Contact Report No. 4

Astro Research Corporation   January 25, 1965
1309 Cacique Street
Santa Barbara, California

Contacts: Dr. Hans Schurech,
                President
           Dr. Robert Witucki

By: L. R. McCreight

     Astro Research Corporation is primarily concerned with the
design and development of composites for space structures and
retardation devices. In this work they usually buy the fibers of
interest which include carbon, glass and silica, but have also
made their own boron fibers. Some boron fiber-metal matrix
(Cu, Al, Mg) composites have been made and tested under Contract NASw-652. The results of this exploratory work have been
published in a report, NASA CR-202, titled "Boron Filament
Composite Materials for Space Structures". They are currently
studying the micromechanics of boron fibers, under Air Force
Contract AF 33(615)-2177. Progress reports describing their
initial efforts in this program have been issued.

Contact Report No. 5

Atomic Energy of Canada, Ltd.   July 1964
Research Metallurgy Branch
Chalk River, Canada

Contact: M. R. Piggott

By: W. H. Sutton

     $10-100\mu$ silica fibers are being drawn on a recently initiated
program at this laboratory. The ultimate interest, as indicated
by a response to a questionnaire, is in fiber reinforced aluminum.
     A few strength values of the silica fibers have been measured
with the results falling in the 0.1 to 1.0×10^6 range with a
median of 0.3×10^6 psi. The tests were run with 1-1/2" gage
lengths at a loading rate of 10^4 lbs/sq. in. /second.

Contact Report No. 6

Atomic Weapons Research Establishment   November 1964
   (AWRE)
Aldermaston, Berkshire, England

Contact: Dr. R. J. Wakelin
             Visited G. E. on
             November 2, 1964

By: W. H. Sutton

   One area of interest to AWRE is the development of high
strength-to-weight ratio materials for elevated temperature applications. The mechanical properties of whiskers are being investigated, and the growth of Al_2O_3 whiskers is being studied.
Tensile strengths for Al_2O_3 and Si_3N_4 whiskers have been
measured and fall in a range of 0.4 to 2.0 $\times$ 10^6 psi, with corresponding elongations varying from 2% to 5%. The elastic mod-
ulus of the Si_3N_4 whiskers is 55 $\times$ 10^6 psi while that for the Al_2O_3
whiskers is 70 $\times$ 10^6 psi. Metallic coatings of Ni, Cu, and Ag
are being applied by an electroless plating technique. Also, Ni
coatings are being applied by a nickel carbonyl process. The purpose of these coatings is to 1) protect the whisker surfaces,
2) facilitate the fabrication process, and 3) serve as the matrix.
Techniques are being developed for sorting and orienting the
whiskers prior to composite fabrication. By using a slurry of
finely divided silver powder and Si_3N_4 whiskers and applying
techniques used in the paper industry, a tape about 3" wide of
Ag-Si_3N_4 was prepared, and in the green (unsintered) state was
sufficiently strong to withstand handling.

Contact Report No. 7

Atomics International   February 17, 1965
Division of North America Aviation, Inc.
P. O. Box 309
Canoga Park, California 91304

Contacts: Dr. S. C. Carniglia,
              Section Chief
              Materials & Process
              Research

By: L. R. McCreight

   Atomics International personnel have both participated in and
reviewed the ceramic whisker and crystal field with particular

emphasis on beryllium oxide. Both massive and filamentary
crystals have been grown and studied by microscopy and x-ray
techniques.

Modulus and tensile strength from bend test data for BeO were
discussed, as obtained from the work of Atomics International and
other organizations; notably the University of California, Lawrence Radiation Laboratory work reports in UCRL-2476 among
the latter sources. The highest tensile strength, as determined by
bending, is reported as 2.7×10^6 psi on a 7 mil whisker. The
modulus of elasticity values were found to vary from about 54.5
to about 59.5×10^6 psi as a function of angle with the c axis as is
shown in Figure 41.

Other divisions of North American Aviation were reported to
be working on company funded efforts in refractory high modulus
non-metallic fibers, but information on this work is not available at this time.

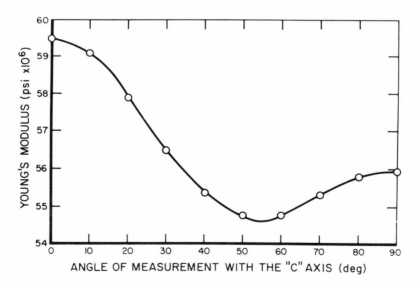

*Figure 41. The Young's Modulus of BeO as Calculated from the
Single Crystal Elastic Constants. Reference: "Room Temperature
Single Crystal Elastic Constants of BeO" (to be published) by
G. G. Bentle.*

Contact Report No. 8

Avco Corporation   October 1964
Research and Development Division
Wilmington, Mass.

Contact: Dr. E. G. Wolff

By: W. H. Sutton

Magnesia and γ-tungsten oxide whiskers have been studied at Avco as reported by a questionnaire.

The majority of magnesia whiskers grown in this study have a square cross section measuring 2-10μ on the diagonal by 1-8 mm long.   The ultimate tensile strength, determined on whiskers with diagonal lengths ranging from 9-24μ, gave values between 11,800 psi and 81,300 psi with some indication of a dependence upon whisker size.   Tests were performed on 2mm gage lengths loaded initially at 0.364 gm/min, then at 0.879 gm/min.   The density of the magnesia whiskers is 3.57-3.58 g/cc but only .06 g/cc as a felt.

Crystallographic, x-ray and related studies have been the prime emphasis in the case of the tungsten oxide whisker work. They are grown from the $W-H_2O$ reactions in sizes of 0.1-1μ diameter by 3-10 mm long and have a composition corresponding to $W_{18}O_{49}$.   These whiskers usually grow in bundles up to 0.01-0.1 mm in diameter.   Their density is 7.7 g/cc; they melt at > 1500°C.

Contact Report No. 9

Babcock and Wilcox Company   September 1964
Old Savannah Road
Atlanta, Georgia

Contacts: C. L. Norton

By: W. H. Sutton

Alumina and zirconia polycrystalline staple fibers as well as aluminum-silicate and high silica monofilaments are under development at Babcock and Wilcox Company, as indicated by the results of a questionnaire.

The alumina fibers have a density of 2.5 g/cc, an average diameter of 3μ, 90$^+$% purity, and show room temperature tensile strengths in the range of 0.02 to 0.30 $\times 10^6$ psi with a median value of 0.125 $\times$ 10^6 as tested at a loading rate of 0.050"/min. and with a 3/8" gage length.

The zirconia fibers have a density of 5 g/cc, an average diameter of 6μ, 90^+% purity, and show room temperature tensile strengths of 0.02 to 0.50 $\times 10^6$ psi under the same test conditions as used for the alumina.

Alumina-silicate glass fibers have been made for many years and sold under the name, Kaowool. It is made for use as an insulating fiber and has not been extensively evaluated for other applications for which mechanical properties would be important.

The high silica (90%) vitreous continuous monofilaments are a new development showing strengths of 1.0 to 10.32 $\times 10^5$ psi with a median of 5-7 $\times 10^5$ psi at room temperature and 500°F, then decreasing to 4 $\times 10^5$ psi at 1000°F, 3 $\times 10^5$ psi at 1500°F, and 0.5 $\times 10^5$ psi at 2000°F. These were all on 3" gage lengths loaded at 0.050 in/min. The Young's modulus falls in the range of 10.8 to 12.2 $\times 10^6$ psi, with a median value of 11.6 $\times 10^6$ psi. The composition includes 10% of modifiers which include Al_2O_3, Cr_2O_3, CaO, MgO, TiO_2 and ZrO_2.

Contact Report No. 10

Battelle Institute e. V.   May 29, 1964
Frankfort am Main, Germany

Contacts: Prof. Dr. Max Barnick,
                     Director

By: L. R. McCreight

Glass and fused silica fibers, as well as iron whiskers are subjects of recent or current efforts at Battelle. Other activities included various studies on special ceramics (e. g., nitrides, borides, carbides). The potential capability for work on ceramic fibers and whiskers was obvious and indeed confirmed verbally. However, no current specific technical activity in these fields was presented.

Of the activities discussed that were most closely related to this survey, the iron whisker work was particularly extensive and occupied the time of several persons. The work includes growing the whiskers by means of gas phase, then conducting extensive mechanical properties investigations.

Contact Report No. 11

Battelle Memorial Institute   January 7, 1965
505 King Avenue
Columbus, Ohio

Contacts: L. S. O'Bannon
            D. Bowers
            H. Wagner

By: L. R. McCreight

Extruded polycrystalline alumina fibers have been studied in an in-house program aimed at preparing fibers by an economical process. These were made about 10 mils in diameter and, after firing, were reduced to about 7 mils. Preliminary flexure tests on 5- and 10-mil fibers gave strength values of 150,000 and 125,000 psi, respectively.

Contact Report No. 12

Bell Telephone Laboratories   January 5, 1965
Murray Hill, New Jersey

Contacts: J. H. Scaff,
              Metallurgical Director
            R. S. Wagner
              Member of Technical Staff
            W. C. Ellis
              Member of Technical Staff

By: W. H. Sutton
      H. W. Rauch

The purpose of this visit was to discuss the VLS (vapor-liquid-solid) method for growing whisker crystals – a concept which was developed recently by Drs. R. S. Wagner and W. C. Ellis*.

The point that Wagner and Ellis emphasize is that the screw dislocation growth mechanism, which has gained wide acceptance

---

*See R. S. Wagner and W. C. Ellis, "Vapor-Liquid-Solid Mechanism of Single Crystal Growth", Applied Physics Letters, 4(5), pp. 89-90, 1 March 1964; and R. S. Wagner, W. C. Ellis, K. A. Jackson, and S. M. Arnold, "Study of the Filamentary Growth of Silicon Crystals from the Vapor", Jour. Appl. Physics, 35 (10), pp. 2993-3000, 1964.

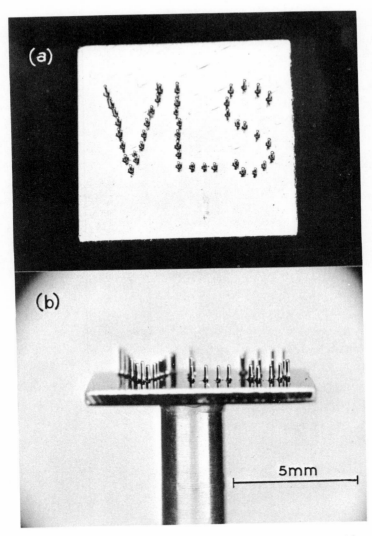

Figure 42. Reprinted with Permission from "The Vapor-Liquid-Solid Mechanism of Crystal Growth and its Application to Silicon" by R. S. Wagner and W. C. Ellis, Transactions of the Metallurgical Society of AIME, 233, 1965 (Copyright AIME).

as the mechanism for filamentary growth (whiskers), is not the mechanism responsible for the VLS whiskers. For example, silicon whiskers grow when a small particle of the selected impurity (Au, Pt, Ag, Pd, Cu or Ni) is placed on a (111) surface of a silicon wafer and heated to 950°C. A small droplet of Au-Si (in the case of Au as the impurity) alloy then forms and, in the presence of H_2 and $SiCl_4$, the liquid surface of the alloy becomes a preferred sink for Si atoms. As the reaction continues, the Si entering the liquid freezes out at the solid-liquid interface. The liquid alloy "rides" the top of the growing whisker, thus becoming more and more displaced from the original silicon substrate. The diameter of the whisker is determined by the original diameter of the globule, and values ranged from a few microns to hundreds of microns. It was demonstrated that whiskers could be nucleated at specific sites on the substrate by placing fine particles of Au at desired locations. Thus, it was possible to grow silicon whiskers arranged as a "VLS" pattern. (see Figure 42). The method is applicable to the growth of crystals of many substances. Numerous advantages, for example, growth at a relatively low temperature, accrue from the VLS mechanism. No mechanical property data were available, although the whiskers appear to have highly perfect structures.

Contact Report No. 13

Bjorksten Research Laboratories   August 1964
P. O. Box 265
Madison, Wisconsin

Contacts: Dr. J. A. Bjorksten,
                 President
             Irvin J. Leichtle,
                 Administrative Assistant

By: W. H. Sutton

    Bjorksten has worked on the production and characterization of glass and silica fiber, and in related areas of coatings, sizes, and finishes, for over 20 years, in programs sponsored by governmental agencies as well as industrial clients. Of most interest to this survey are the silica fibers which were drawn at speeds up to 6000 ft/min, with average tensile strength of 0.4×10^6 psi and occasional strengths to 0.6×10^6 psi. These fibers have been coated with aluminum, silver, and other metals. No indication of the use of a diffusion barrier was given in the data supplied by Bjorksten.

Contact Report No. 14

The Boeing Company   July 1964
Aerospace Division
P.O. Box 3707
Seattle, Washington

Contact: E. E. Bauer,
                Chief, Research
                Materials & Processes

By: W. H. Sutton

Stabilized ZrO_2 and aluminum oxide in polycrystalline fiber form have been studied in the past at Boeing, according to a questionnaire response.

The ZrO_2 fibers, stabilized with 1% Y_2O_3, have a $40\mu \times 80\mu$ cross-sectional dimension and gave strength values in the range of 0.05 to 0.19 $\times 10^6$ with a median at 0.1 $\times 10^6$ psi. A potential application is as thermal insulation to 4000°F for which they show a bulk density of 40 lb/ft^3 and a thermal conductance of 1.25 Btu-in/hr-ft^2/°F at a mean temperature of 1600°F.

The alumina fibers were of similar dimensions but gave strength values of 0.002 to 0.14 $\times 10^6$ psi, also with a median value of 0.110 $\times 10^6$. The Young's modulus was determined to be in the 14 - 30 $\times 10^6$ psi range with a median at 20 $\times 10^6$. At a bulk density of 14 lb/ft^3, and a mean temperature of 1900°F, the conductance was 2.0 Btu-in/in-ft^2/°F.

Both types of fibers were made by a colloid evaporation method in lengths of about 3" for the ZrO_2 and 4-8" for the alumina.

Contact Report No. 15

Brunswick Corporation   June 1964
Defense Division
Marion, Virginia

Contact: J. B. Patton

By: W. H. Sutton

Alumina-silica fibers are being developed at Brunswick Corporation, as indicated by a questionnaire response. Fibers, containing up to 69% Al_2O_3 and the balance silica, having di-

ameters ranging from 0.25 to 2 mils, have been made in 10 foot
lengths. The room temperature tensile strength ranges from 0.15
to 1.0 $\times$ 10^6 psi with a median value of 0.35 $\times$ 10^6 psi which is
retained to 1200°F. The fibers devitrify in about 2 hours at
2000°F and melt or soften above 3000°F.

Continuous metallic fibers also have recently (January 1965)
been announced by Brunswick Corporation.

Contact Report No. 16

Bureau of Naval Weapons   December 30, 1964
U. S. Navy
Munitions Building
Washington 25, D.C.

Contacts: N. E. Promisel
             R. Schmidt
             J. Lee
             J. Wright

By: L. R. McCreight

Various branches of the Bureau of Naval Weapons have
sponsored work in glass, ceramic, and graphite fiber reinforced
composites, ranging from exploratory to advanced development,
and into the usage of such materials in naval weapons and
vehicles. Among these have been: glass fiber reinforced plastics,
glass fiber reinforced aluminum at Owens-Corning, and alumina
whisker reinforced metals at Horizons and at the General Electric
Company, Space Sciences Laboratory.

The composite materials work at the Naval Ordnance Laboratory, White Oak (widely noted by virture of the NOL ring test),
investigation of alumina whisker reinforced resin composites at
the General Electric Co., study of the effects of moisture on the
strength of glass fibers and the development of composite glass-
ceramic fibers at the Narmco Division of the Whittaker Corp.,
are also being supported by the Bureau of Naval Weapons.

In addition, Paul J. Soltis, at the Aeronautical Materials
Laboratory of the Naval Air Engineering Center located at the
Philadelphia Navy Yard, has performed many tests of fibers and
whiskers for BuWeps. He has kindly supplied Figure 43 and
Table XI, showing test data for alumina whiskers, from his re-
port "Anisotropic Mechanical Behavior in Sapphire (Al_2O_3)
Whiskers", Report NAEC-AML-1831.

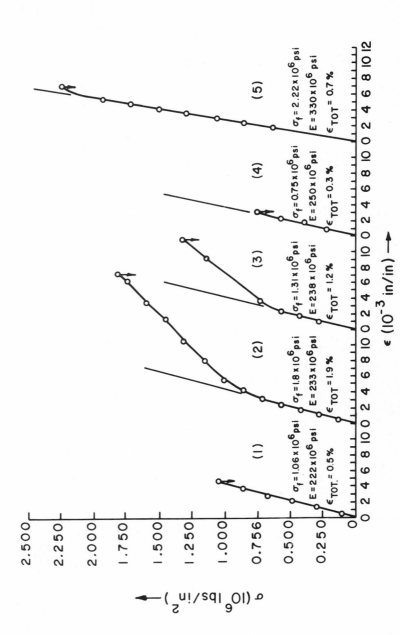

Figure 43. Range of Tensile Stress-Strain Curves for $AQ_2O_3(a_{11}$- axis) Whiskers

TABLE XI. ANISOTROPIC TENSILE PROPERTIES
OF Al_2O_3 WHISKERS AT 25°C

Orientation	c - axis	a_I - axis	a_{II} - axis
Measured Elastic Modulus* ($\times 10^6$ psi)	67.0	180	330
Initial Yield Stress* ($\times 10^6$ psi)	1.47	1.70	2.0
Fracture Stress* ($\times 10^6$ psi)	1.58	3.23	2.22
Total Elastic Strain* (%)	0.95	0.95	0.60
Initial Yield Stress** ($\times 10^6$ psi)	0.275	0.750	0.570

*The values recorded were obtained from whiskers believed to be of high purity and near structural perfection.

**The values recorded were obtained from whiskers believed to be of high purity but containing potential slip dislocations.

NOTE: a_I - axis is $<11\bar{2}0>$ and a_{II}-axis is $<1\bar{1}00>$

Contact Report No. 17

Cambridge University   August 1964
Dept. of Metallurgy
Cambridge, England

Contacts: Dr. A. Kelly
                   Visited G. E. , July 17-22, 1964
              Dr. G. Davies
                   Visited G. E. , July 2, 1964

By: W. H. Sutton

    The Department of Metallurgy at Cambridge University under
A. H. Cottrell has been investigating the fundamentals of fiber-
reinforced metals. Using a model system of copper reinforced
with fine tungsten filaments, Kelly and Tyson conducted critical
experiments which demonstrated the role of fiber aspect ratio,
orientation, and packing density on the strength of the reinforced
metal. Much of this information has been published recently.
    Other studies have been extended to examine the properties
of in-situ fibers (whiskers) grown from the melt [118] and to determine the shear stresses associated with a fiber-reinforced
matrix by a photoelastic analysis[506] .
    Fibers of fused silica are being coated with a thin film of
carbon by thermal decomposition of a hydrocarbon gas. The following data was taken from a questionnaire returned to us by Dr.
Kelly:

Fiber Code:	S/EP/c
Young's Modulus:	10.1×10^6 psi
Density:	2. 21 gm/cc
Diameter:	$\sim 35\mu$
Shear Modulus:	4.3×10^6 psi
Softening Pt.:	$\sim 3000^{\circ}F$
Tensile Strength:	400, 000 - 600, 000 psi
Elongation:	$\sim 4.5\%$
Coefficient & Thermal Expansion:	3.0×10^{-7} in/in/$^{\circ}F$

Contact Report No. 18

Carborundum Co.   February 11, 1965
Niagara Falls, N. Y.

Contacts: Dr. H. Dean Batha,
               Manager, Research Branch
            Dr. P. T. B. Schafer,
               Senior Research Associate

By: W. H. Sutton

The Carborundum Company has recently announced (January 28, 1965) that they can produce SiC whiskers in large quantities (e.g., fill an order as large as a ton). The properties of these fibers are listed as follows:

Tensile Strength (by bending)

$3-10 \times 10^6$ psi ($0.5-3\mu$ diameter)*
$1.5 - 3 \times 10^6$ psi ($3-15\mu$ diameter)

Density

3.217 g/cc

Elastic Modulus

$70^+ \times 10^6$ psi

Cost/Quantity

Ounce: $100/oz.
Pound: $1000/lb.
2-5 lbs: $750/lb.

Carborundum has also grown plates and flakes of SiC, some with very smooth surfaces of varying sizes and purity (up to few ppm). Most of the flakes are doped for semi conductor applications.

In addition, Carborundum has produced aluminum silicate glass fibers under the registered trade name of "Fiberfrax" for many years. These are available in either bulk fiber (to 1-1/2" long by < 1 to 10μ) or long staple fibers (1/2" - 10" long and various diameters, e.g., $1-8\mu$ and $3-25\mu$) as well as in numerous types of batts and textile forms. Continuous use temperature is limited to 2300°F by devitrification, but short time exposure up to 3200°F is possible.

---

* Extrapolated.

Contact Report No. 19

Central Institute for Industrial Research   May 26, 1964
Oslo-Blindern, Norway

Contacts: Joar Markali
           Hallstein Kjøllesdal
           Normann Bergem

By: L. R. McCreight

Several compositions of oxide whiskers (especially Nb_2O_5 and TiO_2) have been prepared and studied at this Institute. The work was an outgrowth of studies on the oxidation of metals and was primarily aimed at elucidating crystal growth mechanisms. Electron microscopy and x-ray diffraction were the principal tools used in the studies.

αNb_2O_5 whiskers are shown in Figure 44; in Figure 45 there are a number of spherical particles on the whiskers that were caused by electron beam heating during direct observation of the whiskers in the electron microscope.

In general, the whiskers grown in this work were about 400 Å diameter by a few microns long and were quite uniform in diameter as illustrated in the figures.

Reference (c) below contains many more photographs and some discussion of growth mechanisms and apparent strengths (up to 10% elastic strain by bending tests).

Studies on the oxidation of titanium and niobium at the Central Institute for Industrial Research were sponsored in part by Wright Air and Development Center of the Air Research and Development Command, United States Air Force through its European Office under Contract Nos. AF 61(514)-892 and 61(052)-90, and have been reported as follows:

(a) J. Markali, - "On Oxidation Mechanisms of Metals and Alloys," Research (London), Vol. 10, Sept. 1957.

(b) J. Markali - "An Electron Microscopic Contribution to the Oxidation of Titanium at Intermediate Temperatures," Fifth International Congress for Electron Microscopy, Academic Press Inc. , New York, 1962.

(c) J. Markali, - "Observations on Nb_2O_5 Whiskers and Their Strength Properties," pp. 93-102, Mechanical Properties of Engineering Ceramics, edited by W. W. Kriegel and Hayne Palmour III, Interscience Publishers, New York, 1961.

Figure 44. Nb_2O_5 Whiskers (10,900X). Courtesy J. Markali.

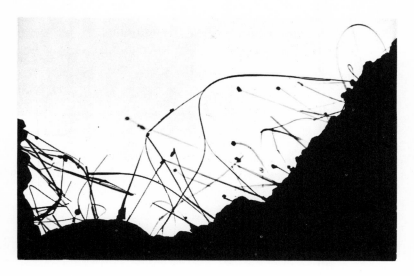

Figure 45. Nb_2O_5 Whiskers Showing Melted Ends Due to Electron Beam Heating in the Electron Microscope (10,900X). Courtesy J. Markali.

Contact Report No. 20

Cincinnati Testing Laboratory   January 5, 1965
Division of Studebaker Corporation
1270 Glendale-Milford Road
Cincinnati 15, Ohio

Contacts: Elmer Warnken
                   President
                   Raymond Silbernagle
                   Manager, Advanced Development

By: L. R. McCreight

    Cincinnati Testing Laboratory has been engaged for several years in the development and production of reinforced plastics, particularly for application to re-entry vehicles and propulsion devices.   Current emphasis is on tailor-making composites, in which several kinds of fibers may be used, to provide a balance among various design factors such as emissivity, specific heat, ablation rate, and back face temperature.

    Among the fibers which have been tested and used are: glass, silica, boron, titania, potassium titanate, carbon, and graphite. In addition, a current program is aimed at evaluating silicon carbide flakes in such composites.

    In the custom formulating of these ablative composites, both computer and experimental techniques are employed to arrive at the desired balance of properties.   In the experimental part of the work, particular emphasis is given to dynamic tests for assessing the chemical compatibility of various mixtures of reinforcements and resins.   Portions of this work are supported by NASA under contract NAS 8-10036.

    The silicon carbide flake program was started during late 1964 under AF33(615)-2112 in cooperation with Carborundum Company as a source of the flakes, and Georgia Institute of Technology where the use of a ceramic matrix with the flakes will be assessed.   Cincinnati Testing Laboratory will work with metallic and organic matrices.

Contact Report No. 21

Compagnie Francaise Thomson-Houston   June 9, 1964
    Division Nucleaire
Dept. Des Techniques Nucleaires
1, Rue Des Mathurins
Bagneux (Seine)
Paris, France

Contact: Dr. J. Schmitt

By: L. R. McCreight

Work on growing alumina whiskers has been initiated at this organization. Samples seen generally ranged in diameter from a few up to about 50μ diameter by about 1/4" long. Tensile values up to 1.5×10^6 psi were reported.

Contact Report No. 22

Cornell University   February 24, 1965
Dept. of Materials Sciences
Ithaca, New York

Contacts: Prof. M. Burton
                Acting Dean
            Prof. E. Scala
            D. M. Schuster

By: W. H. Sutton

D. M. Schuster has been investigating the elastic effects occurring in a photoelastic matrix reinforced by discontinuous fibers (Al_2O_3 whiskers). Significant stress concentrations were observed in the regions of the fiber tips, and the shape of the fiber tips had a profound effect on the magnitude of the stresses in the matrix. The results of this work* have been published recently[425]. Schuster is currently investigating the stress-concentration in adjacent discontinuous boron-fibers, where the position and location of the ends of one fiber are varied with respect to a reference fiber (fibers) in the gage portion of a birefrigent matrix.

---

*For M.S. degree.

Contact Report No. 23

DeBell & Richardson, Inc.   August 25. 1964
Hazzardville, Conn.

Contacts: W. J. Eakins
            R. Humphrey

By: W. H. Sutton

 Exploratory studies to produce glass filaments having a specific cross-sectional geometry are the principal interest of this organization. In one contract, sponsored by NASA (NAS 3-3647), "Glass Microtape Reinforced Research Specimens", they are preparing continuous filaments of glass (soda-lime window, borosilicate, and possibly E) tape 0.005" $\times$ 0.015" of uniform cross-section, and then winding the filaments coated with an epoxy into a densely-packed configuration onto a cylindrical mandrel. In another study sponsored by NASA, (NASw-672), and continuing as NASw-1100, entitled "Studies of Hollow Multi-Partitioned Ceramic Structures", they are producing complex hollow fibers having equivalent diameter of about 0.010" to 0.015". Their goals are to achieve: (1) close tolerance on the cross-sections, (2) complex shape fabrication, and (3) very thin walls in order to obtain maximum stiffness/weight and maximum packing in composite materials. Figures 46 through 48 illustrate some of these shapes.

 These complex fibers are drawn from a preform made by cementing sheets, tubes, etc. together, then drawing them to fibers. The strength of the fibers as shown is in the range of 60-125 $\times 10^3$ psi. By etching the preforms, strength values of 150-325 $\times 10^3$ psi are obtained. Further work is aimed at improving these values and in preparing such fibers from higher strength and modulus glasses.

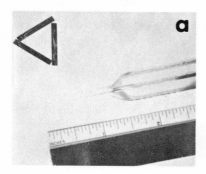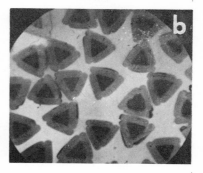

Figure 46. Attenuated Triangular Preform (a) and Photomicrograph at 200X (b) Showing Cross Section Through Imbedded Triangular Fibers.

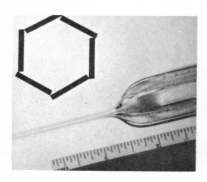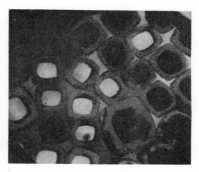

Figure 47. Similar View with Square Preform and Fibers.

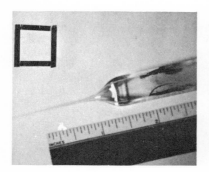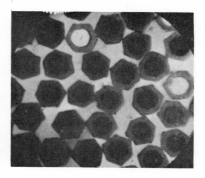

Figure 48. Hexagonal Preform and Fibers.

Note: Wall Thicknesses in All Fibers is No Greater Than the Diameter of Commercial Glass Fibers (.00037'').

Contact Report No. 24

E. I. duPont de Nemours & Company   June 29, 1964
Textile Fibers Department
Wilmington, Delaware

Contact: H. C. Gulledge

By: W. H. Sutton

duPont produces one ceramic fibrous product and has performed considerable research and development on others, as
evidenced by the patent survey. A prime interest in fibers is for
use as pigments in paints where size, chemical compatibility
and resistance, plus optical properties, are important. In addition, the same fibers are of interest for thermal insulation, as
indicated by their response to the questionnaire.

The principal product at this time is potassium titanate which
has the approximate chemical formula $K_2Ti_6O_{13}$ and is sold under
the registered trade name "Tipersul". These are polycrystalline
fibers, 1μ in cross-section by about 100μ long, having a density
of 3.58 g/cc. They are useful at temperatures of 1500° to 2000°F,
depending upon atmosphere, and melt at about 2500°F.

Contact Report No. 25

Explosives Research and Development   May 20, 1964
     Establishment
Ministry of Aviation
Waltham Abbey, Essex
England

Contacts: J. E. Gordon
              N. J. Parratt
              C. C. Evans
              Dr. Bellamy,
                 Director

By: L. R. McCreight

This group has been active for a long time (previously at Tube
Investments, Ltd.) in the ceramic whisker field and is currently
doing a great deal of work on growing and sorting Si_3N_4 whiskers.
This will be described later in this report; however, brief descriptions of several related activities will be given first.

Flaws, and stress concentrations arising from them, have

been studied by Mrs. Margaret Parratt. Her calculations indicate a 90° step on the face of a whisker is essentially as serious as Griffith flaws or notches in decreasing the strength of whiskers.

The strength of Si, ZnO, and other whiskers has been studied to some extent with the usual results showing increasing strength as the diameter or cross-sectional area decreases. This is agreed to be not truly a size dependent strength, but rather a probability that small whiskers will be more nearly free of detrimental defects. In the current work with Si_3N_4 whiskers, the primary measure of the strength of the whiskers is inferred from the strength of composites in which they are used. However, strength values up to 0.5×10^6 psi in tension and $E = 45 - 55 \times 10^6$ psi have been obtained for individual whiskers.

Composites have been fabricated in which Si_3N_4 whiskers have been randomly dispersed in both nickel and a silver - 1% silicon alloy. The whiskers are not stable at high temperatures in nickel, but some composites of 15% whiskers have been made and tested using the silver-silicon alloy as a matrix. These show strengths up to 60,000 psi and chemical stability up to about 1100°C.

Although the major effort has been directed toward Si_3N_4 whiskers, a SiC whisker program has recently been initiated. Equipment for growing the whiskers ranges from small laboratory tube furnaces to a laboratory pilot plant furnace in which 30-60 grams can be grown per week to a 10 times larger furnace which is under construction. The general design features of the larger furnace now in use are briefly as follows:

It is a vertical cylindrical steel shell lined with insulating firebrick and a carbon liner approximately 2' diameter by 3' high. A vertical graphite rod (heated by electrical resistance) is used to heat the furnace to about 1450°C. Nitrogen gas is fed in from near the bottom of one side and out near the top of the opposite side of the furnace. Silicon metal is preloaded into the furnace in a porous graphite container through which argon is passed to carry silicon vapor up into the nitrogen atmosphere. The whiskers which result from the reaction of Si and N_2 are nucleated on either the upper sides of the furnace walls or sometimes onto a graphite rod inserted to serve as a growth surface.

The product of these batches is fine wooly type whiskers about 1μ thick and 10 mm long, but highly entangled so that the other major area of work is on water elutriation to sort and disentangle the whiskers. Then mats or felts are generally prepared which are impregnated with a slurry of metal powders to serve as a matrix. These are hot pressed as 2" diameter × 1/16" thick discs, which are later cut into small rectangular cross-section bars for various test purposes.

Contact Report No. 26

Ferro Corporation   August 1964
4150 East 56th Street
Cleveland, Ohio

Contact: R. W. Pelz
                 Technical Coordinator

By: L. R. McCreight

Ferro produces both E and more refractory glasses such as
S-1014, which is an alkaline earth, alumino-silicate glass having
the following properties:

UTS, Rm. Temp. Median Value for 0.66×10^6 psi
   single filament (1" gage):

UTS, Rm. Temp. Median Value for 0.55×10^6 psi
   strands - (10" gage)

Young's Modulus   (20" gage) 12.8×10^6 psi

Coefficient of Exp.   (0-600°C) 3.89×10^{-6}

Density   2.485 g/cc

Softening Temperature   1735°F

Diameter, as required in the   .0003" to .005"
   range of

Contact Report No. 27

Fritz-Haber Institute   May 28, 1964
Max Plank Gesellschaft
Faradayweg 4-6
Berlin, Dhalem Germany

Contact: Prof. I. N. Stranski

By: L. R. McCreight

Physical chemistry and particularly inorganic chemistry are the main areas of interest at this Institute. Some of these past and current activities are on the vapor deposition vs. structure of inorganic materials and growth mechanisms and only in general are currently related to ceramic and graphite fibers.

Contact Report No. 28

General Electric Co.

The diverse nature of materials research and development throughout the General Electric Company is reflected in the following contact reports from the various General Electric departments and laboratories.

Contact Report No. 28-A

General Electric Company   January 5, 1965
Advanced Engine & Technology Dept.
Cincinnati 15, Ohio

Contacts: L. P. Jahnke
            Dr. Winston Chang
            Dr. Robert Carlson
            Dr. Robert Allen
            George Hoppin

By: L. R. McCreight

This group is concerned with the development of advanced materials for gas turbine engines and accordingly is interested in both low and high temperature fiber reinforced composites. Activities range from preparing some of their own fibers (boron on tungsten) to fabricating various test shapes of fiber reinforced metals and plastics. This laboratory has a strong history of developing super-alloys, brazing materials, and fabrication processes for meeting the long life (usually at elevated temperatures) requirements in gas turbines. Composite materials and metals are therefore tested and evaluated, and compared under similar and realistic conditions.

Some specific recent studies have been conducted on continuous boron fiber reinforced aluminum for which some test results are presented below:

1. Tensile test in air of 16.2% uniaxially aligned boron in aluminum

Temperature	Tensile Strength (psi)
Room	$.045 \times 10^6$
$500^{\circ}F$	$.040 \times 10^6$
$1000^{\circ}F$	$.005 \times 10^6$ (end grip slippage occurred)

2. 100-hour stress rupture tests of 16.2% boron in aluminum gave $.03 \times 10^6$ psi at $500^{\circ}F$ in air.

3. Boron fibers were found to be stable in Al for at least 100 hours at $1050^{\circ}F$ but show reaction in nickel at temperatures in the range of 1200 to $1600^{\circ}F$. This reaction appears to have an activation energy of about 34 kilo calories/mol.

Contact Report No. 28-B

General Electric Company   January 18, 1965
Lamp Glass Department
24400 Highland Road
Cleveland, Ohio

Contacts: K. D. Scott
             R. Sommer
             R. Riddell

By: L. R. McCreight

High purity ($\sim$ 100 ppm impurities) fused silica fibers have
been developed and are produced by this organization. These
fibers are made directly from fused silica rather than by leaching
the alkali components out of lower temperature glasses. The individual solid fibers are made in various diameters in the range
of 0.0004" to 0.001" while hollow silica fibers are available with
0.001" I.D. and .0015" OD, as well as tubing several inches in
diameter. These fibers are available as individual fibers, yarn,
matt, roving, overlay, and wool with various binders or finishes
at prices in the range of $25-$40 per pound.

Some composite fibers have also been made, such as
tungsten wire in a fused silica sheath.

Representative strength data for one of the production fused
silica yarn products (QFY, 150) is as follows:

| Temperature | Tensile Strength psi | | | Remarks |
	Low	Median	High	
Room	78,400	136,500	167,000	Capstan Test Device
Room	72,000	90,400	111,000	Air Grip Tester
500°F	---	79,000	---	Air Grip Tester
1000°F	---	31,900	---	Air Grip Tester
1500°F	---	10,000	---	Air Grip Tester
2000	---	3,500	---	Air Grip Tester

Contact Report No. 28-C

General Electric Company   August 17, 1964
Manufacturing Engineering Services
  Laboratory
Schenectady, New York

Contacts: T. J. Jordan
            D. L. Hollinger
            H. T. Plant

By: H. W. Rauch, Sr.

The program at Manufacturing Engineering Services Laboratory is designed to elucidate the chemical and/or physical processes which influence the measured strength of glass fibers.
Specifically the work is being performed on E-glass filaments to
determine the effects of stress corrosion reactions which involve
water vapor. The principal experimental technique is a systema-
tic investigation of static fatigue, or delayed failure of single virgin fibers in tension under various environmental conditions.

Contact Report No. 28-D

General Electric Company   February 26, 1965
Missile & Space Division
Manufacturing Research Operation
Valley Forge Space Technology Center
P. O. Box 8555
Philadelphia, Pa.

Contacts: W. J. Kinsey
            C. Zvanut

By: W. J. Kinsey

C. Zvanut of the Manufacturing Research Operation has
perfected the making of .0004" diameter pyrolytic carbon-coated
fused-silica filament. These filaments have a coating of 100 to
500 Å of pyrolytic carbon with a resistivity of 4 megohms per
4".

Strengths of 0.5×10^6 psi on single filaments are being
obtained.

Contact Report No. 28- E

General Electric Company   February 26, 1965
Research Laboratory
P. O. Box 1088
Schenectady, New York

Contacts: Dr. A. M. Bueche
            Dr. J. E. Burke
            Dr. W. R. Hibbard
            Dr. L. E. St. Pierre

By: L. R. McCreight

This laboratory, in particular, has had a large and active program of interest to this survey. Numerous references are cited in
other sections of this report. Past work has been on whiskers of
both metallic and oxide compositions, and on bulk glass, silica,
oxides, and intermetallics. Current efforts are primarily on glass
and boron fibers.

Contact Report No. 28- F

General Electric Company   February 28, 1965
Space Sciences Laboratory
Valley Forge Space Technology Center
P. O. Box 8555
Philadelphia, Pa. 19101

Contacts: Dr. Leo Steg
            L. R. McCreight
            Dr. W. H. Sutton
            Dr. F. W. Wendt
            Dr. D. G. Flom

By: L. R. McCreight

This laboratory has one of the largest exploratory research
programs in the United States on advanced whiskers and fibers
for composites. The work is concerned with alumina and boron
carbide whiskers as well as composite filaments such as boron
carbide, and includes preparing the fibers, whiskers and composites using both organic and metallic matrices. In addition, analytical micromechanics studies are being performed in a group
headed by Dr. F. W. Wendt. The programs and sponsors are
listed in Table XII and are referred to and discussed in the text
at the beginning of this book.

TABLE XII. SUMMARY OF PROGRAMS ON
SPACE SCIENCES LABORATORY,

Sponsoring Agency[*] Contract Number	Title
USN(BuWeps) NOw 65-0176-c	Development of Composite Structural Materials for High Temperature Applications
USN(BuWeps) NOw 64-0330-d	Whisker Matrix Bonding Studies
AMRA DA19-066-AMC-184(X)	Investigation of Bonding Oxide-Fiber (Whisker) Reinforced Metals
AFML AF33(615)-1644	Research on High Strength, High Modulus, Low-Density Continuous Filaments of B_4 C.
AFML AF33(615)-1696	Evaluation of Sapphire Wool and Its Incorporation Into Composites of High Strength
AFML AF33(615)-1618	Survey of the Technology of Ceramic and Graphite Fibers and Whiskers
NASA (Hdqtrs.) NASw-937	Study of the Growth Parameters Involved in Synthesizing Boron Carbide Filaments (Whiskers)
NASA (Hdqtrs.) NASw-1144	Properties of Composite Materials

*USN-BuWeps   U.S. Navy, Bureau of Weapons
 AMRA   Army Materials Research Agency

FIBER/COMPOSITE MATERIALS AT THE
GENERAL ELECTRIC COMPANY

Contract Objective
Develop high strength-to-density materials by reinforcing metals such as Ni with whiskers of Al_2O_3.
Develop high strength plastics reinforced with Al_2O_3 whiskers and other discontinuous whiskers.
An investigation of the factors affecting wetting and bonding between Ni and Al_2O_3 with the purpose of developing high strength interfacial bonds in whisker reinforced metals.
Develop a process for producing continuous lengths of B_4C fibers having strengths of 500,000 psi and an elastic modulus of bulk B_4C.
Determine the strength of fine (wool) whiskers of sapphire (α-Al_2O_3) and conduct preliminary studies of metal-wood composites.
The survey shall provide a unified source of authoritative scientific information summarizing the state-of-art of recent progress in the use of whiskers and fibers or fillers, flexible structures and reinforcing media.
Study the growth of B_4C whiskers, document their strength and crystalline structure, and utilize these whiskers in composite materials.
Investigate the relationship of properties of composite materials to their constituents.

AFML   Air Force Materials Laboratory
NASA   National Aeronautics and Space Agency

Contact Report No. 29

General Technologies Corp.   January 25, 1965
Alexandria, Virginia

Contact: J. C. Withers

By: H. W. Rauch, Sr.

General Technologies Corporation is currently working on programs sponsored by the Bureau of Naval Weapons (Contract NOw 64-0176c), the Air Force (Contract AF33(615)-1646), and the National Aeronautics and Space Administration (NASw-1020). These programs cover the study of producing high modulus, high strength filaments by vapor deposition techniques as well as the incorporation of the filaments into composites.

Broadly, they have been studying the deposition techniques for materials lower than atomic number 22 as oxides, carbides, borides, silicides and aluminides. The substrate in most cases has been tungsten wire ranging in diameter from 1 to 5 mils, but recently some studies have been made using carbon-coated fused-silica filaments about 0.4 mil in diameter. Boron fibers have been made continuously. Research is being initiated with the objective of obtaining continuous filaments of boron carbide, boron silicide, silicon carbide, and titanium diboride.

The parameter used to evaluate a given deposition is tensile strength. This property is determined on a tester, designed and built in-house, in which a load is applied to the filament by the flow of water (at a pre-determined rate) into a container. The stress-strain data is then transferred through an electronic sensor to an x-y recorder. The data pick-up is arranged so that no grip slippage is transferred to the read-out.

Contact Report No. 30

Haveg Industries, Inc.   June 1964
900 Green Bank Rd.
Wilmington, Delaware

Contact: John Lux,
                President

By: H. W. Rauch, Sr.

Haveg Industries, Inc. is a producer of large glass fiber reinforced plastic components and structures for industrial and
military applications. In addition, they produce a 98% silica fiber
under the trade name of Sil-Temp for use as thermal and electrical insulation as well as a high temperature filtering media.
No strength or other physical data were made available for
this survey.

Contact Report No. 31

Horizons Inc.   January 18, 1965
2905 E. 79th Street
Cleveland, Ohio

Contact: Dr. Eugene Wainer,
                President

By: L. R. McCreight

Horizons Inc. has been involved since the mid-1950's in process studies on the preparation of single and polycrystalline ceramic fibers and composites. Compositions of fibers studied include alumina, zircon, mullite, zirconia and titania, plus some
electrically conducting compositions. Early and current emphasis
has been on the polycrystalline fibers with the aim of making them
in a continuous form. In between, emphasis was on single crystal
whiskers.
Regarding single crystal whiskers, Dr. Wainer believes that
the alumina whiskers grow by a combination of end extended spiral
screw dislocations and an extrusion process (of alumina) from the
basal plane to form small diameter fibers of uniform cross section. He describes the small whiskers as "fibrils" having a cir-

cular cross-section and not showing crystallographic features if
they are smaller than 2μ. However, they do show high strength
($> 4 \times 10^6$ psi and, in tension, around $2 - 3 \times 10^6$ on the average,
$E = 60 - 125 \times 10^6$ psi). He believes that, to be strong, the whiskers must be of high surface perfection. Internal structural defects can be tolerated if they do not extend to the surface. Indications are that the whiskers grow at rates up to 3"/sec.

These whiskers can be routinely prepared and have been used
to reinforce nichrome in which a strengthening factor of 3 (from
0.7 to 2.5×10^5 psi) was obtained at room temperature with 20%
whiskers by volume. At 1000° C, the strength of the composites
was still twice that of unreinforced nichrome. In some preliminary work with whisker-reinforced ceramics, strengthening by
factors of 2 to 4 have been observed. This was with an alumina
body modified to sinter to full density at 2500° F, in which 20%
whiskers improved the bending strength from $0.28 - 0.30$ to 1.40
$\times 10^5$ psi.

Other whisker work has been on (vapor formed) TiO_2 which
shows strengths of about 2×10^6 psi and appears to be more
easily wetted for bonding purposes by Horizons' techniques than
are the alumina whiskers. In general, Dr. Wainer feels that the
cost of whiskers is not likely to be less than \$100/lb. as produced,
or about \$200/lb. after sorting and classifying. Horizons is
therefore working on polycrystalline continuous fibers with the
idea that these may be more economical, even though they will
have less strength than whiskers.

Their current work is primarily on a γ alumina-spinel composition that is formed in continuous lengths by a modified rayon
fiber process. Sixty fibers are made at a time and one pound is
produced in $4 - 5$ hours. The fibers are then very carefully fired
at 1550° C to yield polycrystalline, transparent to opalescent,
fibers of about $25\,\mu$ diameter. Further work on reducing the
diameter and improving other aspects of the process are expected
to improve the strengths which have been measured up to 5×10^5
psi for an 8" gage length sample.

Dr. Wainer indicated that the initial work with unstabilized γ
alumina was unsatisfactory since γ will convert to α at about
750° C and crystals will grow to become the full diameter of the
fiber. Such fibers are relatively weak ($0.8 - 1.0 \times 10^5$ psi),
easily corroded, and can be rehydrated. By stabilizing the γ
alumina until it can completely convert to alumina spinel, the
strength is retained through the 750° C temperature region, it is
further enhanced by the spinel formed during firing to 1550° C.

Contact Report No. 32

Hughes Aircraft Co.   June 23, 1964
Materials Technology Department
Culver City, California

Contacts: L. E. Gates, Jr.
             W. E. Lent
             W. H. Wheeler

By: W. H. Sutton

    In a program sponsored by NASA* and in a preceding study**, the fiberizing of materials was investigated by using an electric arc process. In addition to being rapid this process has a good potential for fiberizing by a continuous process. It also offers a wide range of glass compositions having high melting points, since the fiberizing process may be 3600° F or higher. The process usually produces short fibers 2 to 4 inches in length, although in one case a 20-foot long fiber was formed. The fiber diameters fall within a range from 14 to 40μ, and the tensile strengths lie between 1.6 and 2.9×10^5 psi. A more detailed listing of the properties of the various fibers is listed in Table XIII.

*L. E. Gates, W. E. Lent, and W. E. Teague, "Development of Ceramic Fibers for Reinforcement in Composite Materials," NASA Contract No. NAS8-50, 1 October 1960 – 2 December 1961.
    Ibid., "Development of Refractory Fabrics," 3 December 1961 – 2 December 1962.
**"Studies on Refractory Fiber Research," Army Ballistic Missile Agency, Contract No. DA-04-495-ORD-1723, U.S. Army Ordnance Corp.

TABLE XIII. PROPERTIES OF HUGHES AIRCRAFT COMPANY REFRACTORY GLASS FIBERS

Fiber No.	Composition (% by wt.)	Average Length (in.)	Average Dia. (mil)	Tensile Strength @ 70°F 10 specimens ($\times 10^6$ psi)			Fusion Temp. (°F)	Thermal Exp. Calculated ($\times 10^{-6}$/°F)
				High	Low	Median		
R 45	SiO$_2$ 36.0 Al$_2$O$_3$ 48.0 MgO 16.0	3–15	1	0.335	0.058	0.163	2685°	2.38
R 74	SiO$_2$ 50.0 Al$_2$O$_3$ 22.5 MgO 7.5 ZrO$_2$ 20.0	2–10	.63	0.362 0.555*	0.177 0.142*	0.272 0.296*	2640°	2.89
R 76	SiO$_2$ 60.0 Al$_2$O$_3$ 7.5 MgO 2.5 ZrO$_2$ 30.0	1–6	1	0.366	0.160	0.257	2920°	2.38
R 86	SiO$_2$ 45.0 ZrO$_2$ 45.0 PbO 10.0	1–2	– –	– –	– –	– –	3055°	– –
R 87	SiO$_2$ 40.0 ZrO$_2$ 40.0 ZnO 20.0	1–2	– –	– –	– –	– –	3055°	– –
R 89	SiO$_2$ 35.0 Al$_2$O$_3$ 32.5 ZnO 32.5	2–4	.98	0.509	0.132	0.261	3290°	2.23
R 91	SiO$_2$ 25.0 Al$_2$O$_3$ 62.5 MgO 12.5	2–4	– –	– –	– –	– –	2950°	– –
R 99	SiO$_2$ 50.0 Al$_2$O$_3$ 27.0 MgO 3.0 ZrO$_2$ 20.0	2–8	.79	0.671	0.02575	0.252	2875°	2.77
R 108	SiO$_2$ 17.9 Al$_2$O$_3$ 35.7 ZnO 35.7 Sb$_2$O$_3$ 10.7	1–2	.53	1.065	0.1035	0.289	>3325°	2.80

Fiber No.	Composition (% by wt.)		Average Length (in.)	Average Dia. (mil)	Tensile Strength @ 70°F 10 specimens ($\times 10^6$ psi)			Fusion Temp. (°F)	Thermal Exp. Calculated ($\times 10^{-6}$/°F)
					High	Low	Median		
R 110	SiO_2	18.5	½–2	--	--	--	--	3325°	--
	Al_2O_3	37.0							
	ZnO	37.0							
	Sa_2O_3	7.5							
R 112	SiO_2	17.9	½–1	--	--	--	--	3180°	--
	Al_2O_3	35.7							
	ZnO	35.7							
	CeO_2	10.7							
R 113	SiO_2	18.5	1–2	--	--	--	--	3180°	--
	Al_2O_3	37.0							
	ZnO	37.0							
	Y_2O_3	7.5							
R 119	SiO_2	60.0	1–3	1	0.404	0.087	0.215	2865°	2.32
	Al_2O_3	10.0							
	ZrO_2	30.0							
R 123	SiO_2	58.2	1–3	.91	0.340	0.071	0.196	2865°	2.49
	Al_2O_3	7.3							
	MgO	2.4							
	ZrO_2	29.2							
	Sa_2O_3	2.9							
R 132	SiO_2	45.0	1–3	--	--	--	--	3255°	--
	Al_2O_3	40.0							
	P_2O_5	15.0							
R 141	Al_2O_3	57.0	1–2	.81	0.496	0.148	0.28	2785°	8.85
	CaF_2	43.0							

*With anhydrous A-1100 finish.

Contact Report No. 33

Illinois Institute of Technology   July 1964
Research Institute
10 W. 35th Street
Chicago, Illinois

Contact: H. Rechter

By: H. W. Rauch, Sr.

    In response to a questionnaire, this organization indicates
that work on extruded and fired polycrystalline alumina filaments
is in the development stage. No data were available at this time,
however.

Contact Report No. 34

Johns-Manville   January 6, 1965
Research & Engineering Center
Manville, New Jersey

Contacts: Dr. Sidney Speil, Chief
                Basic Chemistry Research Section
             Walter K. Hesse, Chief
                Mineral Fibers,
                Aerospace, Specialties Section
             Lee Hedges, Research Associate
                in Aerospace Insulations
             Dr. Fred L. Pundsack, Director
                Basic and General Company Research

By: W. H. Sutton and H. W. Rauch, Sr.

    At Johns-Manville, primary interest lies in the area of fi-
brous thermal insulation. Thus, relatively little effort has been
spent in measuring strengths of individual fibers. The compositions of some of their major products are:

    Micro-Quartz - high silica (98$^+$%) glass obtained by leaching
                        away alkali

Micro-Fibers - various glasses including the well-known
                 E-glass

Thermoflex   - one of the many alumina silicate composi-
                 tions being used today

Dynaflex   - a modified alumina silicate fiber in felted
                 form for steady-state service up to 2700° F.
                 It is anticipated that public release of this
                 material will be made in May 1965.

An interesting new development is the "TX" hydrous magnesium oxide fiber recently announced. It is primarily intended as
a reinforcement for thermosetting resin systems, and has a high
specific heat as well as temperature resistance.

While Johns-Manville is not actively engaged in whisker
growth, they have been closely following this activity and are
aware of the potential this kind of reinforcement offers.

Contact Report No. 35

Lexington Laboratories   November 1964
84 Sherman St.
Cambridge, Mass.

Contact: W. Campbell

By: W. H. Sutton

Lexington Laboratories is working under the sponsorship of
the Army Materials Research Agency (Contract DA-19-020-AMC-
0068), on the "Feasibility of Forming Refractory Fibers By a Continuous Process," with the purpose of investigating and scaling-up
the growth of high-strength-α - Al_2O_3 whiskers. They have
achieved some success in developing a process involving the decomposition of $AlCl_3$ and the nucleating of whiskers on very finely
divided powders, which are suspended in the growth chamber.
They are optimistic that this method can lead to a commercial
process for producing α-Al_2O_3 whiskers, costing about 0.3% of
current experimental samples.

Contact Report No. 36

P. R. Mallory and Company, Inc.   February 26, 1965
Laboratory for Physical Science
Northwest Industrial Park
Burlington, Mass.

Contacts: Dr. S. P. Wolsky
                  Director of Research
               Dr. R. H. Krock
                  Group Manager
               R. H. Kelsey
                  Project Leader
               M. Ginsberg
               F. Glock
               C. Jones

By: W. H. Sutton

   P. R. Mallory and Company, Inc. has recently established a
group for investigating whiskers and composite systems.   While
the company's primary interests are the reinforcement of electrical contact materials such as copper and silver with whiskers,
basic studies are underway of the effects of (tungsten) fiber length-
to-diameter (aspect) ratio and fiber volume fraction on composite
strength.   Whiskers of Si_3N_4 and Al_2O_3 are being studied for the
reinforcement of metal matrices.   A tensile testing apparatus was
designed and constructed which can be adapted to record automat-
ically the load-elongation behavior of the whiskers.   The elongation is determined by a capacitive displacement sensing device,
capable of measuring strains below 0.1%, based on a 1-mm gage
length.

Contact Report No. 37

Marquardt Corporation   April 12, 1965
16555 Saticoy Street
Van Nuys, California

Contact: Arnold Brema

By: L. R. McCreight

The Marquardt Corporation, under a company-sponsored program, is currently engaged in the development of high strength filament materials. The program consists of two phases; filament research, and filament-matrix studies. The first phase has received the major emphasis to date. In this first phase of the program, a filament material designated as RM-015 has been developed.

The RM-015 filament material is based on silicon carbide. Additions to the basic SiC composition have been made to improve filament properties. The filaments are produced by chemical vapor deposition onto heated substrates (0.5 - 0.1.0 mil tungsten wire) at a reduced pressure. Substrate temperature, reactant flow rates, reactant partial pressures, total pressures, and time are the important controlling variables. By properly controlling these variables, filaments with a specific composition and thickness can be produced. So far, the full composition range of the RM-015 filament material has not been completely investigated, but the results of mechanical property tests to date have been quite encouraging.

Representative room temperature fracture strengths as determined in tensile-type pull tests are reported in Table XIV. Strengths as high as 385,000 psi for 1 in. gage lengths, and 417,000 psi for 1/4 in. gage lengths have been observed. Even higher strengths are expected for filaments with the optimum composition, grain size, and surface finish. A comparison of these data with that reported for boron shows that the RM-015 filament strengths are comparable with those of boron filaments. Initial elevated temperature pull-type tensile tests have also been conducted. After a 1/2-hour soak period at 1800° F in air, the filaments retained 70% - 80% of their room temperature strength. The observed modulus of elasticity is about 70×10^{6} psi. The density of the RM-015 coating can vary from 2.4gm/cc to 3.2gm/cc, depending on the exact composition. Obviously the density of the filament will also be greatly influenced by the density and relative volume of the core material.

Only a limited amount of work has been done on the fabrication of other filament materials; however, graphite, ZrC, HfC, ZrB_2, TiB_2, and TiC have been deposited onto various substrates.

TABLE XIV. TENSILE STRENGTH OF RM-015 FILAMENTS

Specimen Number	Gage Length (in.)	Optical Diameter (in.)	Breaking Load (lbs.)	Tensile Strength ($\times 10^6$psi)
BX-10-4-1	0.125	0.00410	3.54	0.268
BX-11-1-1	1.10	0.00494	6.68	0.349
BX-11-1-2	0.125	0.00494	6.90	0.36
BX-11-1-3	0.250	0.00494	7.98	0.417
BX-12-1-1	1.08	0.00600	5.29	0.184
BX-13-2-1	0.250	0.00183	0.497	0.188
BX-14-4-1	0.250	0.00197	0.675	0.222
BX-15-4-1	0.125	0.00248	1.95	0.403
BX-15-4-2	0.250	0.00248	1.16	0.24
BX-16-4	0.250	0.00313	1.66	0.215
BX-4-3	1.00	0.00486	5.36	0.289
BX-5A-1	1.00	0.00535	5.60	0.249
BX-5A-2	1.00	0.00500	5.30	0.27
BX-5A-3	1.00	0.00535	6.80	0.302
BX-8-2-2	0.125	0.00446	4.25	0.273
BX-9-2-1	1.07	0.00446	6.00	0.385
BX-9-2-2	0.250	0.00446	5.60	0.359

NOTES: 1. Results were determined by a standard pull-type tensile test.
2. Tungsten wire substrates were 1 mil in diameter.
3. All data were obtained at room temperature.
4. Strain rate during tests was 0.005 in./in./min.

Contact Report No. 38

Narmco Research and Development Div. July 1964
Whittaker Corp.
3540 Aero Court
San Diego, California

Contacts: R. A. Long
           W. H. Otto
           R. A. Jones

By: W. H. Sutton

    The fiber work at Narmco is concerned with a study of the ef-
fective incorporation of fibers (i.e., filament winding technique)
into a plastic matrix*, the development of large diameter (5-mil)
fibers for composition-loaded composites, and the development of
new and improved fibers.** The studies on the fibers include investigations of the design and building of apparatus for drawing
fibers from high temperature bushings (up to 2500° C), and of the
feasibility of drawing a duplex filament consisting of a glassy
sheath (silica, vycor, or pyrex) and an inner ceramic core. This
approach appears both novel and promising.
    In another program sponsored by the U.S. Army (Frankford
Arsenal) an investigation of the polycrystalline W wire, single
crystal linked tungsten wire, WC wire, and single crystal WC
wire of 0.5-mil diameter is underway. The purpose is to develop
high modulus, high strength fibers for reinforcing aluminum
alloy.***

---

*D. W. Stevens, W. H. Otto, and C. Y. Chin "Potential of
   Filament Wound Composites," USN, Contract NOw 61-0623
   1 March 1962 - 28 February 1963.
** W. H. Otto and R. B. Vidanoff "Silica Fiber Forming and
   Core-Sheath Composite Fiber Development," 1 January 1963 -
   31 December 1963.
*** S. Rodney and Roger A. Long, "Fiber-Reinforced Metal Composites," U.S. Army - Frankford Arsenal, Contract DA-04-
   495-AMC-431(A), December 1964.

Contact Report No. 39

December 30, 1964

National Aeronautics
  & Space Administration
Office of Advanced
  Research & Technology
600 Independence Avenue, S.W.
Washington, D.C.

Contacts: G. Deutch
            J. Gangler
            M. Rosche
            N. Mayer

By: L. R. McCreight

NASA Headquarters sponsors several contracts on various aspects of ceramic fibers and composites. Many of these contracts are referred to in this and other sections of this report based on reports and contacts with the performing organization. In addition, this office maintains overall liaison and cognizance of in-house work at the various NASA centers. The largest effort directly pertinent to this survey is at NASA-Lewis and is reported separately. Other work in NASA is briefly summarized here:

NASA-Goddard has done some work and has an interest in the whisker field. In particular, a review of the "Physico-Chemical Studies on Crystal Whisker Growth Parameters" has been done.

NASA-Langley has a strong interest in, and is actively working on the mechanical properties of fibers such as boron, and has plans to study the fabrication and utilization of composites in aerospace vehicles.

JPL has participated in the whisker and fiber field for several years. Current work is on the Taylor process of drawing metal fibers encased in glass.

Contact Report No. 40

National Aeronautics   April 16, 1965
  & Space Administration
Lewis Research Center
Cleveland, Ohio 44135

Contact: John W. Weeton

By: L. R. McCreight

      Metal wire reinforced metal composites have been the pri-
mary subject of the investigations in this laboratory. Their work
in the field is considered outstanding and basic to many other activities in the fiber reinforced composites field. In particular, the
successful demonstrations of reinforcing metals with discontinuous metal wires has provided support to the efforts to reinforce
metals with ceramic whiskers.
      The NASA-Lewis work is not confined to metal wire rein-
forcements. Some effort is being devoted to preparing in situ
ceramic fibers in refractory metals and to preparing aluminum
oxide whiskers for reinforcing metals. The in situ formation of
ceramic fibers has been done by mixing tungsten powder with about
8 v/o of one of the following ceramic powders: HfO_2, ThO_2, ZrO_2,
HfC, Y_2O_3, HfB_2, HfN_2, TaC, then hot extruding it (4200° F) to
both consolidate and fiberize the ceramic. Some of the ceramic
additives react with the tungsten and others form fibers rather
well, but both results have yielded some very high-strength-at-
high-temperature composites. This work will be reported in a
NASA report and will also probably be submitted for publication in
a metallurgical journal. The presently proposed NASA report is
titled, "Studies of Tungsten Composites Containing Fibered or
Reacted Additives", by Max Quatinetz, John W. Weeton and
Thomas P. Herbell.
      In the area of principal activity (metal wire reinforced met-
als) the work has been largely with tungsten wires in copper.*

----

*Jech, R. W., McDanels, D. W., and Weeton, J. W.: Fiber
  Reinforced Metallic Composites. Composite Materials and Com-
  posite Structures, Proc. Sixth Sagamore Ordnance Materials
  Conf., Aug. 18-21, 1959, pp. 116-143.
  McDanels, D. L., Jech, R. W., and Weeton, J. W.: Stress-
  Strain Behavior of Tungsten-Fiber-Reinforced Copper
  Composites, NASA TN D-1881, Oct. 1963.
  Petrasek, D. W. and Weeton, J. W.: Alloying Effects on
  Tungsten-Fiber-Reinforced Copper-Alloy or High-Temperature-
  Alloy Matrix Composites. NASA TN D-1568, Oct. 1963.

This work has been concerned with strengthening mechanisms and has shown that the law of mixtures is valid, over a considerable range of compositions, for both continuous and discontinuous fiber reinforced composites. Recent work, soon to be published, will also show that the relationship applies to high temperature tensile and stress-rupture conditions. The most imminent paper is tentatively titled, "Elevated Temperature Tensile Properties of Alloyed Tungsten Fiber Composites", by D. W. Petrasek. The stress-rupture results will be published at a later date. Other work of a fundamental nature being done includes studies of size effects and length-to-diameter ratio effects. Many of the composites produced for the investigations described were made by liquid infiltration fabrication techniques. Powder metallurgical methods are being utilized also for composites with high temperature alloy or refractory metal matrices.

Some of the work that has been published by NASA and some recent accomplishments were orally reported at the ASM Symposium on Composites held October 17-18, 1964 in Philadelphia. A publication which will include a survey of the state-of-the-art will be published soon as a NASA Technical Note titled, "Fiber Reinforced Metallic Composites" by John W. Weeton.

Contact Report No. 41

National Beryllia Corp.   June 1964
Haskell, New Jersey

Contact: P. Hessinger

By: W. H. Sutton

Beryllium oxide whiskers made by vapor deposition have been studied under Air Force, Navy and NASA contracts.* The work was summarized and reported in response to a questionnaire.

Whiskers can be grown either by using vapor from molten beryllium or by condensation from $Be(OH)_2$ in $10-30\mu$ diameters and 1/8" to 1/2" long. They show room temperature tensile strengths of 300,000 to 2.8×10^6 psi with a median value of 1×10^6 psi. The Young's modulus is reported as ranging from 20 to 117×10^6 psi with 60×10^6 as a median value.

The whiskers show a density of 3 g/cc and are primarily "c" axis types. They are stable to 2000°C (where a transformation occurs) but only to 1200°C in the presence of water vapor.

---

* NOw 63-0662-f and 64-0512-f, AF 33(616)-8066, NASw-685.

Contact Report No. 42

National Bureau of Standards   January 26, 1965
Washington 25, D.C.

Contact: Dr. H. C. Allen, Jr.

By: H. W. Rauch, Sr.

    The National Bureau of Standards, with partial support from
ARPA, is continuing a wide program of studies involving crystalline materials. These include investigation of methods and theory
of growth, study of detection and effects of defects, determination
of physical properties, refinement of chemical analysis, and determination of stability relations and atomic structure. The types
of materials range from organic compounds, through metals and
inorganic salts to refractory oxides. There are no current programs devoted to studies of "whisker" crystals, only macro-
crystals and, according to Dr. Allen, "immediate" plans do not
include single crystal whisker studies.

Contact Report No. 43

Naval Ordnance Laboratory   January 26, 1965
U.S. Navy
Silver Springs, Maryland

Contacts: F. R. Barnet, Chief
               Non-Metallic Materials Division
            S. Prosen
            Dr. P. W. Erickson

By: H. W. Rauch, Sr.

    Although there is no current effort on whisker growth at NOL,
past work in this area was done by Edwards and Happel* and re-

---

* Edwards, P. L. and Happel, R. J., Jr., "Beryllium Oxide
  Whiskers and Platelets," J. Appl. Phys. $\underline{33}$ (3), pp. 943-948,
  1962.
  Edwards, P. L. and Happel, R. J., Jr., "Alumina Whisker
  Growth on a Single Crystal Alumina Substrate," J. Appl. Phys.
  $\underline{33}$ (3), pp. 826-827, 1962.

ported about three years ago. Current work is on filament wound
structures and recent emphasis is on the incorporation of "whiskers" in filament wound NOL rings to reduce interlaminar shear.

Prosen will issue a report in several weeks which describes
the use of sapphire whiskers in filament wound NOL rings. A reduction in interlaminar shear of about 21% was observed, but additional work is required to substantiate this.

NOL is particularly concerned with materials for hydrospace
applications.

Contact Report No. 44

Naval Research Laboratory   January 25, 1965
U. S. Navy
Washington, D.C.

Contact: J. A. Kies

By: H. W. Rauch, Sr.

Mr. Kies has been engaged for several years in studying the
strength of glass fibers. Some of the testing and data analysis was
done at NRL but most was done on contracts for which Mr. Kies
was the scientific officer for NRL. Contract research has included the Solar Division of International Harvester (G. Schmitz
and A. Metcalfe) and the General Electric Co. (T. Jordan and D.
Hollinger). Some of the fibers were drawn by the contractors but
others were supplied by the principal fiber manufacturers. Experimental work on GRP at NRL was supplemented by contract work,
especially at Goodyear Aerospace Company, Narmco, and the
University of Vermont.

A recent program on S-994 glass and E glass has shown that
with good design, the mechanical damage inflicted on fibers when
they are made into roving, is not reflected in a correspondingly
large size effect in the strength of pressure vessels.

NRL is devoting a large effort to the investigation of materials
suitable for hydrospace applications.

Contact Report No. 45

Owens-Corning Fiberglas Corporation   January 7, 1965
Technical Center
Granville, Ohio

Contacts: Dr. A. C. Siefert
          G. R. Machlan
          E. M. Lindsay
          R. E. Lowrie
          R. J. McEvoy
          F. M. Veazie

By: L. R. McCreight

In general, the laboratory has concentrated on fiber glass; however, some effort has been devoted to producing both TiO_2 and ZrO_2 fibers by nucleation and growth from a melt. Of these two, ZrO_2 is further developed. Methods of forming acicular single crystals of ZrO_2 have been defined. TiO_2 forms twins and parallel growths. Fibers of ZrO_2, up to 3/4" long with an 1/d ratio up to 500, have been produced; these show tensile strengths of about 1.5×10^5 psi and modulus of elasticity of about 25×10^6 psi. Figures 49 and 50 illustrate the appearance of single crystals of ZrO_2. Potentially the process used for making these fibers should permit economical production, although more effort is needed to achieve greater crystal perfection for higher strengths and moduli.

Owens-Corning has been primarily involved in glass fiber-research, development, and production. Properties of some of the glass fibers which are now offered commercially are shown in Table XV. In addition, past and current efforts have been devoted to preparing higher modulus glass fibers. One such fiber which has been developed is designated as "M", or as YM-31A. This fiber has a modulus of 15.9×10^6 psi; however, the fact that it contains BeO has retarded its application. Current efforts are being devoted to other compositions which have the same or higher values of modulus and up to 1.0×10^6 psi tensile strengths.

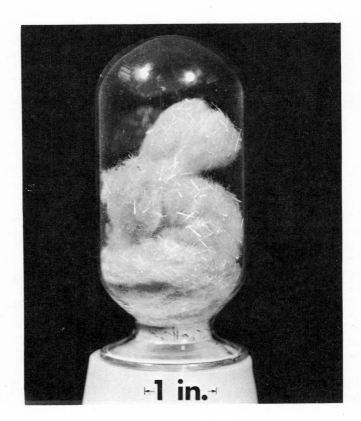

Figure 49.   ZrO$_2$ Crystals. Length ½''-¾''. Diameter 30-200 microns.

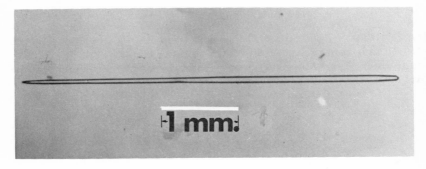

Figure 50.   Monoclinic Zr O$_2$ Crystal Elongated Parallel to C Crystallographic Direction.

TABLE XV. PRODUCT COMPARISON CHART

Properties	S-Glass	E-Glass	C-Glass
Tensile Strength (virgin) at temperature			
75°F	700,000 psi	500,000 psi	380,000 psi
600°F	600,000	425,000	N.D.
1000°F	350,000	250,000	N.D.
1200°F	240,000	150,000	N.D.
1400°F	180,000	Yields	N.D.
1500°F	120,000		N.D.
1600°F	Yields		N.D.
Modulus of Elasticity × 10^6	12.4 psi	10.5 psi	10.0 psi
Top Temperature Limitation	1400°F	1000°F	800°F
Dielectric Constant @ 10^{10} cycles /72°F	5.6	6.1 - 6.4	N.D.
Specific Gravity (Fibers)	2.49	2.54	2.49
Chemical Durability (Bulk)			
Acid(Na_2O Equivalent)	0.035	0.10	0.030
Water (Na_2O Equivalent)	0.001	0.003	0.006

Contact Report No. 46

Rolls-Royce Ltd.   May 21, 1964
Advanced Research Laboratory
Old Hall, Littleover
Derby,
England

Contacts: Dr. A. M. Smith
          Dr. John Morley

By: L. R. McCreight

　　　Several activities on fibers and composites are under study at
Rolls-Royce. A group of two dozen people representing several
disciplines are effectively performing both research and development on ceramic fibers and composites. While the principal
current work is on the fused silica reinforced aluminum, past
work has involved precipitation of filamentary intermetallics in
metal matrices, and stainless steel wire in aluminum. Current
work also includes flame polished alumina rods (as models for
whiskers) and alumina whiskers. These later items will be discussed first, then the greater effort on fused silica-aluminum
will be described.
　　　Stainless steel wires (non-continuous) in aluminum were
studied at Rolls-Royce and reported by D. Cratchley (Powder
Metallurgy, 1963, No. 11) in a paper entitled, "Factors Affecting
the UTS of a Metal/Metal-Fiber Reinforced System". A linear
relation between strength and volume percent of wires up to 20%
was obtained. Actual strengths obtained in the composites however were in the area of 40×10^3 psi.
　　　Cratchley has also worked in the field of precipitation or in
situ formation of fibers in a metal matrix, particularly studying
Ni_3Al as the precipitate. Two general sizes of precipitated fibers
tend to form; the smaller size is easily redissolved under the
effects of temperature and even more rapidly under combined
stress and temperature.
　　　This result, plus other limitations imposed by the in situ
precipitation of fibers and whiskers have caused Rolls-Royce to
drop this approach to composites.
　　　In the ceramic fiber area, work has been underway since 1963
on growing alumina whiskers, and for some time on preparing
flame polished alumina rods for use in model studies of composites
in which alumina would be used. Little information was available
on the alumina whiskers; however, the flame polishing work was
shown and discussed. An automatic, well-controlled laboratory
machine has been built to do the flame polishing. Sapphire rods
about 1 mm diameter × 4" long are mounted vertically and rotated

about their axis while a gas flame traverses the surface to provide flame polishing. Some of these rods have been tested in bending with a 1/2" gage length and values of about 1×10^6 psi obtained.

Finally, the most extensive work is on 15-25μ diameter fused silica fibers, both plain and coated with a proprietary diffusion barrier coating and aluminum, as well as on composites made with the coated fibers. In general, very large numbers (hundreds) of data points have been obtained on each kind of test being performed on the fibers. Samples for much of this work are therefore prepared and measured in automated laboratory equipment to help insure uniformity and reproducibility. Among the types of studies discussed are several relating to the strength of the fibers as a function of temperature and exposure to various atmospheres. These results are well covered in Rolls-Royce publications such as "Strength of Fused Silica" by J. G. Morley, P. A. Andrews and I. Whitney, Physics and Chemistry of Glass, Volume 5, No. 1, February 1964, pp. 1-10. Highlights of the data are that they find 0.85×10^6 psi as the average room temperature strength and that this increases to about 2.1 to 2.2×10^6 psi at -196° C. The strength also increases significantly when the fibers are tested in vacuum and in dry air, as compared to moist air. Extensive data have also been obtained for temperatures to 700° C and times to one hour.

The possible mechanisms of degradation of these fibers as shown by decreased strength are also discussed in the above paper. In general, they agree with the stress corrosion mechanism proposed by Charles and Hillig of the General Electric Research Laboratory. At temperatures about 300° C the Rolls-Royce authors feel that a second type of weakening process occurs; i.e., time, temperature, and atmosphere dependent even in the absence of applied stress.

Composites of fused silica reinforced aluminum have also been prepared and extensively studied to provide basic information on fiber reinforced metals. The preparation of these composites is in two steps. First, the fused silica fiber is drawn and immediately coated with both the barrier and the aluminum to give about 50% silica - 50% aluminum. The second step is to consolidate large numbers of oriented fibers (fibers are full length of the specimen) in a mold by hot pressing. The preparation, behavior (fracture mechanisms, crack propagation etc.), and properties of these composites, as well as the individual coated fibers, are described in a recent paper, "The Tensile Strength of a Silica Fiber Reinforced Aluminum Alloy" by D. Cratchley and A. A. Baker, Metallurgia, pp. 153-158, April 1964. Some significant results have been obtained with these composites including: (1) strengths of about 70 tons/sq. in. as reported in the paper referenced above and verbally reported as sometimes approaching or exceeding 100 tons/sq. in., (2) finding that creep, fatigue, and crack propagation behavior is favorable in this composite.

A limiting aspect of this combination (SiO_2 - Al) in terms of technological applications, however, is the detrimental chemical reaction which occurs very rapidly at 565° C, in spite of the diffusion barrier coating. This reaction and the general drop in strength at elevated temperature limits the possible use of this composite to about 300° - 400° C.

A list of the Rolls-Royce publications on Materials Research in this field follows:

Date	Name of Author/s	Title of Paper and References
1959	Dr. J. G. Morley	Strength of Glass Fibres. (Nature 184, p. 1560)
1960	Mr. B. A. Proctor	Strengths of Acid-etched Glass Rods (Nature 187, p. 492)
1962	Mr. B. A. Proctor	Strengths of Acid-etched Soda Glass Rods (C. R. du symposium sur la resistance mecanique du verre et les moyens de l'ameliorer. Union Scientifique Continentale du Verre).
1962	Mr. B. A. Proctor	The Effects of Hydrofluoric acid etching on the Strength of Glass. (Phys. Chem. Glasses 3, p. 7)
1962	Mr. B. A. Proctor Mrs. B. Wilkinson	The Development of Defects on Etched Glass Surfaces (Phys. Chem. Glasses 3, p. 203)

1962	Dr. J. G. Morley Mr. P. Andrews Mr. I. Whitney	Strength of Fused Silica C. R. du symposium sur la resistance mecanique du verre et les moyens de l'ameliorer. Union Scientifique Continentale du Verre, p. 417)
1963	Mr. B. A. Proctor Dr. J. G. Morley	Strength of Sapphire Crystals (Nature 196, p. 1082)
1963	Dr. D. Cratchley	Factors affecting the UTS of a Metal/Metal Fibre Reinforced System (Powder Metallurgy No. 11, p. 59)
1963	Dr. J. G. Morley	Reinforcing Metals with Fibres (New Scientist 17, p. 122-125)
1963	Mr. R. G. C. Arridge	Orientation effects in fibre re- inforced composites (Proceed- ings 18th Annual Conference S. P. I. , Section 4A)
1964	Dr. J. G. Morley	Fibre Reinforced Metals (Royal Society Discussion Meeting, June) Proc. Roy. Soc. 282A Oct. 1964
1964	Mr. R. G. C. Arridge Miss K. Prior	The Cooling Time of Silica Fibres (Nature) 203, 386, July 1964
1964	Mr. R. G. C. Arridge	The Coating of Glass Fibres with Molten Metal Union Scien- tifique Continentale du Verre.
1964	Mr. B. A. Proctor	Fracture of Glass (Appl. Mat. Research, January, p. 28)
1964	Mr. R. G. C. Arridge Dr. D. Cratchley Mr. A. A. Baker	Metal Coated Fibres and Fibre Reinforced Metals (Journal of Scientific Instruments 41, p. 259-261, May.)
1964	Mr. B. A. Proctor Mr. F. P. Mallinder	Elastic Constants of Fused Silica as a Function of Large Tensile Strain (Phys. Chem. 5, August).

1964	Dr. D. Cratchley Mr. A. A. Baker	The Tensile Strength of a Silica Fibre Reinforced Aluminum Alloy (Metallurgia, p. 153, April, Vol. 69)
1964	Dr. D. Cratchley Mr. A. A. Baker	The Effect of Soaking at Elevated Temperature on the Mechanical Properties of Aluminum-Silica Composition (Union Scientifique Continentale du Verre Conference) (In press)
1964	Dr. J. G. Morley Mr. P. Andrews Mr. I. Whitney	Strength of Fused Silica (Phys. Chem. Glasses $\underline{5}$, p. 1-10)
1964	Dr. D. Cratchley Mr. A. A. Baker	Metallographic Observations on the Behavior of Silica Reinforced Aluminum under Fatigue Loading (Applied Materials Research) October 1964

Contact Report No. 47

Royal Aircraft Establishment   May 19, 1964
Farnborough Hants,
England

Contacts: W. Watt
          P. Forsythe
          R. Ryder

By: L. R. McCreight

 In addition to some work on carbon fibers, silicon carbide whiskers have been grown on a limited basis. No data were available for either program.
 Other work at Farnborough, while not specifically on ceramic fibers, is nevertheless of interest in the composites field. This is the work of Forsythe and Ryder on aluminum sheets reinforced in one direction with stainless steel wires. The motivation for the work was in improving creep, fatigue, and crack propagation rather than strengthening the aluminum per se. This was very adequately demonstrated. For example, creep resistance was improved by a factor of 10, fatigue life by a factor of 15, and the tensile strength was also improved ($\sim 20\%$). Some data on this work, including brief descriptions of the samples, are shown in Tables XVI and XVII, as supplied by Forsythe and Ryder.

TABLE XVI. FATIGUE CRACK GROWTH FOR Al ALLOY PANEL WITH WIRE REINFORCEMENT

Material	0.1% Proof Stress (tsi)	Tensile Strength (tsi)	Elongation (%)	Young's Modulus ($\times 10^{-6}$ psi)	Fatigue Life* Cycles to Failure	Remarks
1. Commercial DTD 687 Al Zn Mg Cu Alloy	29.2	33.7	12.5	10.5	120,000	Fatigue life an average value for many tests
2. DTD 687 sandwich with one central layer of conventionally woven 80 ton stainless steel mesh at center.	30.3	33.1	1.5	10.6	312,000	Wires broke up during manufacture of composite
3. As 2. but wire mesh aluminum sprayed prior to incorporation in sandwich	25.6	30.1	2.0	10.6	289,000	" "
4. As 3. but mesh at 45° to tensile axis of sheet	32.2	36.5	8.0	10.5	544,000	" "
5. DTD 687 incorporating 2 layers of "Elgiloy"-wire-aluminum wire mesh. Strong wires parallel to sheet axis.	19.7	28.6	5.0	10.7	214,000	" "
6. As 5. but strong wires to 450° to sheet axis and at 90° to each other	20.3	28.1	7.3	10.2	296,900	" "
7. Commercial L73 Al. Cu. Mg Alloy	26.2	29.7	10.0	10.0	455,000	" "

Material	0.1% Proof Stress (tsi)	Tensile Strength (tsi)	Elongation (%)	Young's Modulus ($\times 10^{-6}$ psi)	Fatigue Life* Cycles to Failure	Remarks
8. L73 sandwich. 1 layer of Al sprayed stainless steel mesh at 45° to sheet axis. Conventional woven mesh 0.015" dia. wires approx. 80 tsi	25.0	28.0	7.0	10.8	2,000,000	Wires broke up during manufacture of composite
9. L73 sandwich. 1 layer of STA1 1.0% Carbon Spring wire-aluminum wire mesh at centre, 25 strong wires 0.007" dia. per inch	23.4	29.4	7.7	10.7	1,120,000	Wires broken. STA1 wire initially about 160–170 tsi tensile strength reduced to about 80 tsi during heat treatment of the composite
10. As above but 5 layers of STA1 mesh interleaved with L.73	24.0	27.9	7.5	10.7	1,460,000	Wires broken. STA1 wire initially about 160–170 tsi tensile strength reduced to about 80 tsi during heat treatment of the composite
11. As above with 4 layers of STA1 mesh but wires unbroken (area occupied by wires about 1.5% of total cross sectional area of panel.)	28.0	31.0	8.5	11.0	1,600,000	Wire unbroken during manufacture of composite
12. As 11 but 12 layers of STA1 mesh	24.0	30.5	7.0	11.0	2,600,000	Strong wires unbroken
13. L73 incorporating 8 layers of "Elgiloy" mesh. 25 0.015" dia. wires per inch parallel to sheet axis	25.3	27.0	2.0	11.1	7,340,000	Strong wires unbroken

* Fatigue life of slotted panel at 1800 ±2000 psi.

Courtesy Forsythe and Ryder, RAE, Farnsborough, England.

XVII. TENSILE STRENGTHS OF L73 AND L73 WITH WIRE REINFORCEMENT
AT ROOM AND ELEVATED TEMPERATURE

No.	Specimen	Test Temp. (°C)	0.1% Proof Stress (tsi)	Tensile Strength (tsi)	Elongation (%)	Young's Modulus (×10⁶ psi)	Specific Strength (TS/e)	Specific Modulus (E/e)	Remarks
1	L73	RT	26.0	28	10	10.0	10.0	3.6	
2	L73	200°	17.1	19.5	12	8.5	6.5	3.0	
3	L73	250°	6.4	8.0	20	7.0	3.0	2.5	
4	L73	300°	3.6	4.5	23	5.7	1.7	2.0	
5	L73 + 8% 0.007 diameter tungsten wire	RT	38.0	44.0	–	14.2	10.6	3.5	No elongation figure could be measured because fractures could not be fitted together. Probably 2%.
6		200°	22.5	29.0	–	13.3	7.3	3.2	
7		250°	14.4	19.2	3.5	12.4	4.7	3.0	
8		300°	10.6	16.2	5	11.1	4.0	2.7	
9	L73 + 6% 0.007" diameter FV. 520 stainless steel wire	RT	33.0	36.1	8	11.8	11.7	3.8	
10		200	22.3	26.3	8	11.4	8.5	3.7	
11		250	13.9	16.2	4.5	10.5	5.2	3.4	
12		300	7.5	13.1	4.5	9.6	4.2	3.2	

Contact Report No. 48

Rutgers University   January 6, 1965
School of Ceramics
New Brunswick, N. J.

Contact: Prof. G. J. Phillips

By: W. H. Sutton and H. W. Rauch, Sr.

Prof. Phillips had worked on high strength glasses in the
early 1930's at the Corning Glass Works, but much of the data
was proprietary. He exhibited a photograph showing a large
glass rod in bending where the stresses were over 4×10^5 psi.
Much of his current work has been devoted to high strength
glasses and is being published; see

(1) Sigma-Xi Lecture, published in the March 1965 issue
of American Scientist

(2) December issue of Glass Technology (1964), where the
modulus of elasticity of the bulk glass can be calculated on the basis of its chemical composition

Some of his recent work has been based on the elastic
moduli of TeO_2 glasses compared to silicate glasses. His studies show that Te glasses have low elastic moduli and that the
TeO_2 acts as a network former much like SiO_2, B_2O_3 and P_2O_5.

Contact Report No. 49

Solar   August 1964
Division of International Harvester Co.
2200 Pacific Highway
San Diego, California

Contact: G. K. Schmitz

By: H. W. Rauch, Sr.

Solar is studying the effect of environments, primarily
moisture, on the strength of virgin glass fibers by means of the
length effect. Fibers of different gage lengths are subjected to
various strain rates and to static fatigue conditions. Strength
characteristics of new, high-modulus glasses also are being
determined by these methods. A paper* presented by Solar
personnel to the 67th ASTM meeting described the effect of
length on the strength of E- and S- glass fibers. Among other
things, this study determined the influence of two kinds of flaws,
one governing the strength of long filaments, and the other influencing the strength of short fibers.

In a second paper** presented to the 20th Anniversary
Technical Conference of the SPI, the above work has been extended to describe additional characteristics of the two kinds of
(surface) flaws. A third (structural) type of flaw and its role in
stress corrosion has been recognized. The stress corrosion of
glass fibers will be treated in more detail in a paper to be presented to the American Chemical Society (Ind. Eng. Chem).

---

*Metcalfe, A. G. and Schmitz, G. K., "Effect of Length on the
    Strength of Glass Fibers, "ASTM Preprint No. 87, presented
    at the 67th Annual Meeting, June 1964.
**Schmitz, G. K., and Metcalfe, A. G. "Characterization of
    Flaws on Glass Fibers," Proceedings, SPI Reinforced Plastic
    Division, Society of Plastics Industries, Inc., February 1965.

Contact Report No. 50

Swedish Institute for Metals Research   May 27, 1964
Drottning Kristinas Vag 48
Stockholm, Sweden

Contact: Prof. Roland Kiessling
           Dr. A. Ronnquist

By: L. R. McCreight

    While the primary emphasis at this Institute is placed on
metals research, an important part of this field receiving considerable attention is the study of oxidation of metals. Under
certain conditions, the oxidation of metals produces oxide
whiskers. The nucleation of copper oxide whiskers has been
studied in the recent past by Ronnquist[*]. Whereas others have
proposed a nucleation theory in which the whiskers are postulated
as growing from a dislocation in the base metal, his work corrob-
orates the theory that they nucleate at stress-induced dislocations
in the oxide layer. In particular, this theory is advanced for
explaining low temperature (300° C) growth of copper oxide
whiskers since, in this temperature range, the rapid increase in
the ratio of CuO/CuO_2 gives rise to an equally rapid increase of
stresses in the oxide layer due to formation of a less dense oxide.

[*]A. Ronnquist and H. Fischmeister, "The Oxidation of Copper,
A Review of Published Data", J. Inst. of Metals 89, 1960-61,
pps. 65-76.
A. Ronnquist, Discussion of the paper, "An Electron Optical
Study of the Effect of Temperature and Environment on the
Growth of Oxide Whiskers on Cold-Rolled and Annealed Copper,"
by W. R. Lasko and W. K. Tice (J. Inst. of Metals Vol. 109,
No. 3, pp. 211-215).
A. Ronnquist, "The Oxidation of Copper: A Kinetic and Morpho-
logical Study of the Initial stage," J. Inst. of Metals 91, 1962-
63, pps. 89-94.

Contact Report No. 51

Texaco Experiment, Inc.   January 25, 1965
Richmond, Virginia

Contact: Claude P. Talley

By: Claude P. Talley

Texaco Experiment, Inc. has been studying the preparation and properties of boron for various applications for over ten years with emphasis in the last three to four years on boron filaments.

Although boron filament is not yet a standard production material and its properties as well as manufacture are under continuing development, some general information on these properties can be given.

At present the filament is continuously produced by a vapor deposition process, in circular filament form, with diameters between 3 and 6 mils. The boron is of the microcrystalline or amorphous type, with a purity of about 99%, excluding the substrate. Density has averaged 0.09 lb/in^3 for 5-mil diameter filament.

Average tensile strength of the filament for 1-inch gauge length has ranged from about 3 to 5 $\times$ 10^5 psi with a coefficient of variation ranging from about 10% to 30%. The modulus of elasticity has been approximately 55 to 60 $\times$ 10^6 psi and the stress-strain curve is linear up to the breaking point.

Contact Report No. 52

Thermokinetic Fibers, Inc.   January 5, 1965
135 Washington Street
Nutley, New Jersey

Contact: J. V. Milewski,
            Vice President

By: W. H. Sutton and H. W. Rauch, Sr.
    Visited Thermokinetic Fibers January 1965

    Thermokinetic Fibers, Inc. is currently producing whiskers
of Al_2O_3 and SiC on a commercial basis. The whiskers are
offered in a variety of forms, clusters, chopped needles, wool,
mats, etc. The diameters vary from 1-3 to 10-30 microns. In
conjunction with the University of Florida, they have developed a
"paper" of Al_2O_3 about 0.050" thick by 8" × 8", which has
sufficient strength to tolerate mild handling. This paper can be
impregnated with a plastic, metal, or ceramic, resulting in a
high strength material.
    Although the price of whiskers ranges between $1,000 and
$20,000/lb., Thermokinetic Fibers believes that with moderate
scale production the price will be reduced to $100/lb. and
eventually, with large scale production, to below $10/lb. The
properties of the whiskers are summarized in Table XVIII.

Contact Report No. 53

Thiokol Chemical Corporation   October 1964
Alpha Division
Huntsville, Alabama

Contact: A. O. Kays

By: H. W. Rauch, Sr.

    Fe - Fe_2O_3 and Cu in polycrystalline fiber form have
been investigated at Thiokol, as their reply indicates. These are
made by an undisclosed "chemical conversion" technique in the
range of 1-10μ diameter by 1/4 to 1 1/2" long. The Fe - Fe_2O_3
fibers show ultimate tensile strengths at room temperature of
0.03 to 1.8 × 10^6 psi with a median value of 0.6 to 0.8 × 10^6 psi.
The Young's modulus is estimated at 40 - 50 × 10^6 psi. However,
these fibers completely oxidized in air at 77°F in three weeks.

TABLE XVIII. PROPERTIES OF THERMOKINETIC FIBERS, INC., WHISKERS

Material	Whisker Diameter (μ)	Aspect Ratio (L/D)	Tensile Strength ($\times 10^6$ psi)	Elastic Modulus* ($\times 10^6$ psi)	Density (gm/cc)	Melting Point ($^{\circ}$F)
1. Sapphire (Al_2O_3) Whiskers						
(A) Type 2B, loose needles	1-10	10-200	0.6-2.0	80-150	3.97	3780°
(B) Type 3B, loose needles	1-30	10-200	0.2-0.6	60-100	3.97	3780°
(C) Type 1B, needles in powder form	1-3	10-200	2.0-3.5	100-300	3.97	3780°
(D) Type 4A, Cluster Ball (a 3D cluster of fine whiskers) also contains minor amount of AlN whiskers.	5-15	10-100	--	---	3.97	~3780°
(E) Type 1A, Mat Form 3/16" × 4" × 15"	1-3	500-5000	2.0-3.5	100-300	--	~3780°
(F) Type 2A, Mat Form 3/16" × 4" × 15"	1-10	150-200	0.6-3.5	80-150	--	~3780°
(G) Type 3A, Mat Form	1-30	100-2000	0.2-3.5	60-100	--	~3780°
(H) Type 1C, Paper 8" × 8" × 0.010"						

Material	Whisker Diameter (μ)	Aspect Ratio (L/D)	Tensile Strength ($\times 10^6$ psi)	Elastic Modulus* ($\times 10^6$ psi)	Density (gm/cc)	Melting Point ($^{\circ}$F)
2. β-SiC Whiskers						
Type 5A, Wool	1-10	1000-10,000	0.2-3.0	>100	3.2	~4400° **
3. Mixed Whiskers						
(A) 50% SiC + 50% SiC/Al$_2$O$_3$ bulk crystals Type B	1-5	--	--	--	3.2	
(B) AlO$_3$ Maj + AlN Minor (Submicron rods) Type 4B	~0.2-1.0	--	--	--	--	~3780°
(C) Al$_2$O$_3$ Maj, AlN Minor, Type 4A Cluster Ball-Polycrystalline	5-15	--	--	--	--	~3780°
(D) AlN Major, Al$_2$O$_3$ Minor, Type 6A	3-30	100-200	0.5	50	3.6	~4000°

*Estimated

**Sublimes

Contact Report No. 54

H. I. Thompson Fiber Glass Company (HITCO)   June 1964
Research and Development Division
Gardena, California

Contacts: J. P. Sterry,
                Vice President

By: W. H. Sutton
      Visited H. I. Thompson June 1964

    The H. I. Thompson Fiber Glass Company produces several
fibrous materials and has been investigating the development of
polycrystalline ceramic fibers by an extrusion technique. These
studies include the development of the process* and the optimi-
zation of fiber properties with investigation of their incorporation
into various matrices.** Emphasis has been placed on the development of ZrO_2-based fibers, including ZrO_2 stabilized with additions of CaO, Yb_2O_3, MgO, B_2O_3, Cr_2O_3, La_2O_3, ThO_2, Nd_2O_3
and SiO_2. The chief problems with these fibers are shrinkage
during drying and firing, and the control of the grain growth during firing operations. Single fiber tensile strengths in the 2 to
3.4×10^5 psi range have been observed, with the maximum values
in the 5×10^5 psi range. Elastic moduli of the order of 50×10^6
psi have been observed.
    A summary of data on several products is listed in Table
XIX; some of the thermal conductivity values at various temperatures are shown in Figures 51, 52, and 53.

---

*"Non-Vitreous Non-Metallic Fiber Synthesis", J. P. Sterry,
  G. P. Wetterburg, W. Lachman; AF 33(616)-8080, (1 April
  1961-30 Sept., 1962).
**"Polycrystalline Ceramic Fiber Reinforcement for High-Temp-
  erature Structural Composites", J. P. Sterry and D. Newman,
  Contract AF33(616)-8080, (1 October 1962 - 31 March, 1963).

TABLE XIX. PROPERTIES OF HITCO PRODUCTS

Material	Composition	Form	Density (gm/cc)	Limiting Temp. (°F)	Fiber Diameter (μ)	Tensile Strength ($\times 10^6$ psi)	Young's Modulus ($\times 10^6$ psi)	Process
Refrasil	Glass 99.0–99.5% SiO_2	Continuous filament, chopped, batt	2.1–2.2	3100°	~1	--	--	Formed by leaching impurities from fiber.
Poly-crystalline Oxides	A. 94% ZrO_2/CaO B. 89% ZrO_2/SiO_2 C. 67% ZrO_2/SiO_2	Continuous, batt	~4.5	<2000° (grain growth)	1.2 to 2.4	0.25 to 0.30 (0.587 max)	45 to 60	Extrusion method cost: ~ $10/lb.
Amorphous Carbon #CCA-1	96–99% Carbon	Continuous filament	1.8–1.9	<6000° (non-oxidizing atmosphere)	~10	--	1.6	Controlled decomposition of rayon fiber.
Graphite #G1550	98–100% Carbon	Continuous filament	~2.0	<6000° (non-oxidizing atmosphere)	~10	--	--	Controlled decomposition and graphitization of rayon fibers. Cost: $30/lb.

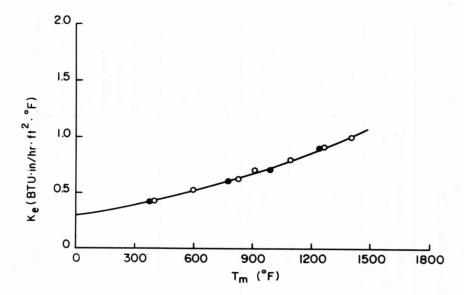

Figure 51. Effective Thermal Conductivity vs Mean Temperature Refrasil A-100 Batting; Density, ρ_b= 6.4 pcf (HITCO, Formerly H. I. Thompson Fiberglass Co.).

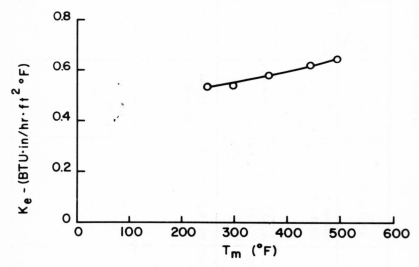

Figure 52. Effective Thermal Conductivity vs Mean Temperature, HITCO Carbon Cloth CCA-1; Density ρ_b = 23.6 pcf (HITCO).

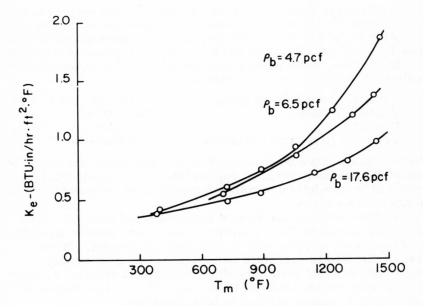

Figure 53. Effective Thermal Conductivity vs Mean Temperature, Zirconia-A Batting at Three Densities, ρ_b (HITCO).

Contact Report No. 55

Union Carbide Corporation   February 9, 1965
Parma Research Center
12900 Snow Road
Parma, Ohio

Contacts: John C. Bowman,
              Director
          Dr. E. Epremian
          Dr. Roger Bacon

By: L. R. McCreight

A relatively large and successful effort is being devoted to the development of carbon fibers. Improved properties have been obtained for the type of fibers made in quantity for ablative applications (now 0.2×10^6 psi in tension and 6-9 $\times$ 10^6 psi modulus of elasticity).

In addition, some preliminary properties of new laboratory scale carbon fibers were reported as:

Tensile Strength (2 cm gage)	0.35×10^6 psi
Young's Modulus	30×10^6 psi
Density	1.4 g/cc
Diameter (irregular cross-sec)	$\sim$ 5 microns

Higher values, have been obtained in the laboratory and are to be reported in various technical papers this year. These fibers are prepared by carefully controlled degradation of synthetic organic fibers such as rayon.

Further work on mechanical properties and behavior, such as creep, stress rupture, elevated temperature tensile strength, is planned and, to some extent, underway at present. In the case of elevated temperature tensile tests, the preliminary indications are that the fibers show a relatively constant strength versus temperature behavior to about 1500° C; it then begins to increase in a manner similar to bulk graphites.

Some preliminary results on epoxy matrix composites indicate that about 50% of the strength of the (0.2×10^6 psi initially) fibers is demonstrated in the composite. However, the wet strength and dry strength values are about equal. The fibers have both open and closed pores as well as a very high surface area, partially due to a very irregular cross-section that leads to an affinity for moisture absorption in the range of 5%-6%.

Other fiber work at this laboratory includes: (1) the use of the carbon fibers as a substrate for coatings of such materials as pyrolytic graphite and boron, and (2) some work on the impregnating of cellulose with various substances which when fiberized and heated yield ceramic (e.g. Al_2O_3) fibers.

Contact Report No. 56

United Aircraft Research Laboratory   August 25, 1964
East Hartford, Conn.

Contact: Dr. J. A. Ford
           W. Lemkey

By: W. H. Sutton

    United Aircraft has conducted much of the pioneering studies
on unidirectionally solidified melts, in which rod-like (single
crystal whiskers) can be grown parallel to the axis of the specimen. Much of their work has been on the Cu-Cr system, where
Cr whiskers extracted from the solidified specimen have exhibited
tensile strengths over 10^6 psi. Similarly, in the system Al/Ni,
Al_3Ni whiskers have shown tensile strengths as high as 0.4×10^6
psi and bending strengths of 0.8×10^6 psi. The volume fractions
of the Cr whiskers in the matrix was about 2 v/o, whereas that
of the Al_3Ni whiskers was about 10 v/o. Recent work has been
concerned with thermal stability, elevated temperature tensile
behavior, and creep behavior of Al-Al_3Ni whiskers and Al-$CuAl_2$
platelets – reinforced composites. These composites have exhibited remarkable thermal stability at elevated temperatures.
The elevated temperature and creep behavior have been shown to
be somewhat better than SAP. Work on elevated temperature
systems has yielded several refractory alloy systems containing
more that 25% whiskers by volume and exhibiting true whisker
reinforcement. This work has been sponsored by the Navy under
contract NOw-64-0433-d and by the Army under contract
DA-19-020-AMC-00434(x).

Contact Report No. 57

University of Grenoble   June 5, 1964
Laboratory for Electrostatique and
    Physics of Metals
Grenoble,
France

Contacts: Prof. Weil
           Dr. R. Conte

By: L. R. McCreight

    Graphite, iron, and copper whiskers have been studied by
R. Conte and others at this Institute during the past several years.
Conte, in particular, has used torsional tests as a measure of
effects of irradiation and cold working on the metal whiskers.
This work is described in his thesis, "Torsion plastique de poils
monocristallins de fer et de cuivre", which was published in Acta
Metallurgica during the past two years.

    Similar studies are being made on graphite as whiskers be-
come available. However, the supply is apparently very much a
limiting condition to the work. These whiskers are grown by
cracking methane at 900° C, just above atmospheric pressure, in
a tube furnace. So far only a few tests have been conducted, but
quite a few electron microscope pictures have been made of cross
sections (Figure 54). These pictures show both single and twin
whiskers of the order of 20μ in diameter. Strength values have
generally been obtained only up to 100 kg/mm^2; however, elastic
strains of as much as 5% were reported.

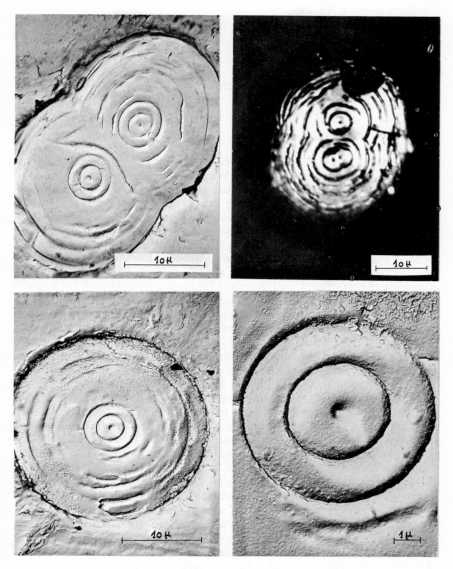

Figure 54. Typical Electron Microscope View of the Cross-Sections of Graphite Whiskers. Courtesy R. Conte, University of Grenoble.

Contact Report No. 58

Watervliet Arsenal   February 25, 1965
Watervliet, New York

Contacts: F. Schmiedeshoff,
                Director of Research
            I. Ahmad,
                Head, Phys. Chem. Section

By: W. H. Sutton

    Much of the U. S. Army in-house research on fibers and whiskers is being performed by Watervliet Arsenal personnel. Mr. Fred Schmiedeshoff, Director of Research, is also serving as chairman of an Army working group on composite materials.

    In the Benet Laboratory, one group under the leadership of M. A. Sadowsky, has been investigating the effect of Poisson's ratio, and of couple-stresses on the force transfer between embedded microfibers and the thermal stress discontinuities in microfibers.* Their results are showing some interesting and significant structural and thermal factors which should be considered in the design of fibrous composite materials.

    Dr. Ahmad engaged in the study of the growth techniques and the mechanical properties of various whisker materials including Fe, Cu, B, Mg, Al_2O_3, NiO, $W_{18}O_{49}$, $W_{20}O_{58}$, WO_2, SnO_2, and B_4C. One goal of these studies is to elucidate the mechanisms of their growth. For example, the role of hydrogen in the growth of Al_2O_3 whiskers in the Webb and Forgeng's ** method; and transport mechanism of the vapour species (Al_2O, AlO, etc.) prior to nucleation and growth are being investigated. Alpha-Al_2O_3 whiskers are grown in the absence of H_2 in argon atmosphere containing varying partial pressures of oxygen. Two forms of NiO whiskers have been grown; the normal green variety and a white variety which is oxygen deficient. The crystallographic orientation of the green variety has been found to depend on the preferred orientation in the substrate nickel sheet. Out of ten whiskers grown on a cold rolled nickel sheet, nine were found to have <110> growth direction. Values of elastic modulus and the tensile strength were found to be $15-34 \times 10^6$ and $190-410 \times 10^3$ psi respectively. (Modulus and tensile strength values for bulk NiO are, respectively, 16×10^6 psi and 60×10^3 psi.

    The mechanical properties of other whisker materials are also being investigated. For the measurement of moduli, resonace frequency (in which the specimen is mounted as a cantilever) and ultrasonic pulse echo techniques, are also being explored.

Prior to this visit, reports on whiskers and candidate matrix materials to be reinforced by whiskers were written by M. J. Salkind,*** who is now at United Aircraft Corporation.

---

*Sadowsky, M. A., and Weitsman, Y., "Effect of Poisson's Ratio of an elastic filler on force transfer between embedded microfibers," Watervliet Arsenal, Watervliet, N. Y. Rpt. WV7-RR-6108-R (Sept. 1961).

Sadowsky, M. A. et al, "Effect of Couple Strains on Force Transfer Between Embedded Microfibers, "Watervliet Arsenal, Watervliet, N. Y., Rpt. WVT-RR-6407 (June 1964).

Sadowsky, M. A., and Hussain, M. A., "Thermal Stress Discontinuities in Microfibers, "Watervliet Arsenal, Watervliet, N. Y., Rpt. WVT-RR-6401, AD 601-057 (April 1964).

**Webb, W. W., and Forgeng, W. D., "Growth and Defect Structure of Sapphire Microcrystals, " J. Appl. Phys. 28 (12) 1449-54 (1957).

***Sears, G. W., and DeVries, R. C., "Growth of Alumina Crystals by Vapor Deposition," General Electric Co., Research Laboratory, Schenectady, N. Y., Rpt. 60-RL-2377M (March 1960).

Salkind, M. J., "Whiskers and Whisker Strengthened Composite Materials," Watervliet Arsenal, Watervliet, N. Y. Rpt. WVT-RR-6315 (Oct. 1963).

Salkind, M. J., "Candidate Materials for Whisker Composites," Watervliet Arsenal, Watervliet, N. Y., Rpt. WVT-RR-6411 (May 1964).

TABLE XX. ΓABULAR SUMMATION –

Composition	Contact Report	Length	Cross-Sectional Shape and Size
Alumino-Silicate Glass	15	to 10 ft.	Round $\sim$ 5 - 50μ
E-Glass	44	Contin.	" $\sim$ 10 μ
E-Glass	26	"	" $\sim$ 10 μ
S-Glass	45	"	" $\sim$ 10 μ
4-H1-Glass	1	"	" $\sim$ 10 - 12μ
R-108 Glass	32	"	" $\sim$ 12μ
S-1014 Glass	26	"	" $\sim$ 10 - 100μ
Fused Silica	17	"	" $\sim$ 35 μ
Fused Silica	5	"	" $\sim$ 10 - 100μ
Fused Silica	13	"	"
Fused Silica, Carbon Coated	28-D	"	" $\sim$ 25 μ diameter
Sil-Temp. (98% SiO_2)	30	"	" $\sim$ 10μ
Vitreous Monofilament	9	"	" $\sim$ 10μ
Dyna Quartz (98% SiO_2)	34	Discont	" $\sim$ 1μ
Fiberfrax (Alumino-Silicate)	18	Shorts to 1½"	" $\sim$ 2μ (Mean)
Kaowool (Alumino-Silicate)	9	Staple to 10"	" $\sim$ 3μ (Avg.)
Micro Quartz (98% SiO_2)	34	Discont.	" $\sim$ 1μ
Refrasil (99.5% SiO_2)	54	Cont./ Discont.	" $\sim$ 10 - 12μ

GLASS AND SILICA FIBERS

Tensile Strength ($\times 10^{-3}$ psi)			Young's Modulus ($\times 10^{-6}$ psi)	Density (lb/in^3)	Elongation (%)
High	Low	Median			
360	50	150	--	--	--
650	450	500	10.5	0.092	--
450	390	420	10.5	0.092	4.5
700	--	--	12.6	0.092	--
790	650	730	14.0	0.092	4.5
1065	103	289	--	--	--
--	--	660	12.6	0.090	--
600	400	450	10.1	0.079	4.5
600	50	30	--	0.079	--
600	--	400	--	--	--
--	--	500	--	0.079	--
--	--	--	--	0.080	--
1032	100	500–700	10.8	0.084	--
--	--	--	--	0.078	--
--	--	--	--	--	--
--	--	--	--	0.096	--
--	--	--	--	0.078	--
--	--	--	--	0.080	--

TABLE XXI. TABULAR SUMMATION —

Composition	Contact Report	Length	Cross-Sectional Shape and Size
Al_2O_3	14	4-8"	$8-20 \times 10^{-7}$ sq. in.
Al_2O_3	11		Circ., 7 mil dia.
Al_2O_3	13	Contin.	Round, 40 μ
Al_2O_3	9	Staple	Round, 3 μ
Boron Carbide on Tungsten Core	28-F	Contin.	Round, > 5 mil
Boron on Tungsten Core	51	Contin.	Round, 3-6 mil
Carbon	47	Contin.	Round, 15 μ
Carbon (Laboratory)	55	1"	Irreg. round, 5 μ
Carbon (Production)	55	Contin.	Irreg. round, 5 μ
$Fe-Fe_2O_3$	53	$1/4 - 1\frac{1}{2}$"	Ellipse or Rect.
Potassium Titanate (Tipersul) $(K_2 Ti_6 O_{13}$	24	100 μ	Ribbons, 1 μ
Zircon	54	2-12"	Round, $\sim 25-50 \mu$
ZrO_2	9	Staple	Round, 6 μ
ZrO_2	45	$3/4$"	Various
ZrO_2 - A	54	2-12"	Round, $\sim 25-50 \mu$
ZrO_2 - C	54	2-12"	Round, $\sim 25-50 \mu$
Stabilized ZrO_2	14	to 3"	$40 \times 80 \mu$

[*]No data available on Elongation (%) of polycrystalline ceramic and carbon-graphite fibers.

POLYCRYSTALLINE CERAMIC AND CARBON-GRAPHITE FIBERS

Tensile Strength (× 10⁻³ psi)			Young's Modulus (× 10⁻⁶psi)	Density (lb/in³)	Elongation* (%)
High	Low	Median			
140	2	100	20 (median)		
80 (in bending)					
70	20	35	70 (median)		
300	20	125		0.143	
300	--	200	70	0.09	
		300-500	55-60	0.09	
135	110	--	12-20	0.061	
350	--	--	30	0.54	
--	--	200	6-9	0.54	
1800	3	700	40-50 (est.)		
--	--	--		1.29	
587	--	300	45-60	~1.62	
500	20	300		1.8	
150	--	--	25		
587		300	45-60	~1.62	
587		300	45-60	~1.62	
190	150	100			

TABLE XXII. TABULAR SUMMATION – SINGLE

Composition	Contact Report	Length	Cross-Sectional Shape and Size
Al_3Ni	56		Whiskers, ellip. to circular, $0.2-2\,\mu$
Al_2O_3	31		Round, hex., few μ
Al_2O_3	6		Round
Al_2O_3	21	$1/4$"	Round and rhomb. $1-50\,\mu$
Al_2O_3	35		Hex. or rhomb., sub $\mu-\mu$
Al_2O_3	16		Hex. or rhomb., sub $\mu-\mu$
Al_2O_3	28-F	$1/16-1$" 6" max	Hex. or rhomb., $0.3-50\,\mu$ (diag.)
Al_2O_3	52	$1/250-1$"	Rhomb., $1-30\,\mu$
BeO	41	~ 13mm	$10-30\,\mu$
BeO	7		$7\,\mu$
B_4C	28-F	$1/16-3/4$"	Triang. and rhomb. $1-25\,\mu$ (diag.)
Cr	56		Circular, $0.3-1\mu$
Graphite	55	3 cm.	Cylinder., $\sim 5\,\mu$
MgO	8	$1-8$ mm	Sq., $2-10\,\mu$ (diag.)
SiC	52	$1/8-1$"	$-- 2\,\mu-8\,\mu$
Si_3N_4	25	10 mm	Rhomb., $1\,\mu$ across
Si_3N_4	6		Hex.
α SiC	18	$10-300\,\mu$	$0.5-15\,\mu$
$W_{18}O_{49}$ with ≤ 5 W/O WO_2 and WO_3	8	$3-10$ mm	$0.1-1.0\,\mu$

CRYSTAL CERAMIC AND GRAPHITE WHISKERS

Tensile Strength ($\times 10^{-3}$ psi)			Young's Modulus ($\times 10^{-6}$ psi)	Density (lb/in^3)	Elongation (%)
High	Low	Median			
400	240	350	20 (Median)	--	1.5
4000	--	2000 3000	40 - 75	0.143	--
2000	400	--	70	0.143	2.0-5.0
1500	--	--	--	0.143	--
2500	--	--	~300	0.143	--
1180	--	--	67 (c-axis)	0.143	2.3
3230	--	--	180 (a_1- axis)	0.143	1.6
2220	--	--	330 (a_2- axis)	0.143	0.7
6200	--	1000	60 avg.	0.143	1.0-8.0
2000	600	1500	80- 150	0.143	1.3
2800	800	1000	20- 117 60 (Median)	0.103	--
2700	(in bending)		54.5- 59.5	--	--
930	--	300	70	0.091	~1.2
1340	580	1200	36 (Median)	--	2.4
2834	--	--	102	0.051	2.0
81.3	11.8	--	--	0.129	2.78
3000	1000	2000	100	0.115	1- 3
500	--	--	45- 55	--	--
2000	400	--	55	--	2.0-5.0
3000	(in bending)		70$^+$	0.115	--
--	--	--	--	0.277	--

VI. EVALUATION AND DISCUSSION

There is an increasing abundance of literature and activity related to ceramic and graphite fibers and whiskers. The motivation appears to be both the potential strength of these materials and the recent developments in preparing high-strength, high-modulus fiber-reinforced composites.

Some research efforts on these materials are directed toward gaining a better understanding of the role of dislocations and of gross structural imperfections on their strength. Other programs are devoted to the application of fiber and whiskers in specialized requirements for electronic, insulation, and other uses. However, the prime application appears to be as reinforcements in composite materials. For this reason, emphasis is being placed on characterizing the following: optimum growth or manufacturing techniques, accurate determination of tensile strength and modulus of elasticity, choice of a suitable matrix (or matrices), and the requirements for obtaining strong bonds between the fibers and the matrix.

An evaluation of the data obtained thus far reveals some trends in compositions and processes as well as some areas urgently requiring further study. These items are discussed in the remainder of this section.

1. Test methods: While it is recognized that one test method for determining a specific property will probably not satisfy all the requirements, some standardization of test methods and of reporting data is needed. For example, Subcommittee D-5 of ASTM Committee D-30 recommends that all tensile data for fibers and whiskers be accompanied by gage length, and if possible various gage lengths should be tested. Other information such as strain rate and ambient conditions would also be very useful information.

The importance of standardized test methods and complete data reporting cannot be over-emphasized. Large variations in strength data appear in the literature because, for example, bending may have been the test method in one case and tensile tests in another.

2.   Composition and Processes: This survey indicates a trend toward concentration on a relatively few classes of compositions and fabrication methods. This is based on a finding that only a few compositions have the balance of properties and behavior needed for various applications, and that relatively few fabrication methods yield high performance fibers. The general compositions and processes of interest are:

Glass: Although glasses offer modulus values in the 10-18 × 10^6 psi range, and usable tensile strengths of about 500,000 psi, they are easily made by drawing from a melt and, therefore, are economical and widely useful. Based on current research activities, however, glass fibers having tensile strengths of 10^6 psi or higher and modulus values of 20 × 10^6 psi or higher appear to be possible with future efforts.

Carbon Fibers: Current carbon fibers show tensile strengths of about 250,000 psi and modulus of elasticity values of about 6-9 × 10^6 psi. However, laboratory feasibility for increasing these values to 350,000 psi and 30 × 10^6 psi, or more, respectively, have been demonstrated. These fibers generally are made by the controlled degradation of organic fibers (as rayon) which, like glass, yield a relatively economical product. Due to the ease of oxidation of carbon, the applications of the fibers must be confined to suitable environments.

Ceramic and Intermetallic Fibers and Whiskers: In this class of materials, oxides (alumina), carbides (boron and silicon), and boron appear to be the most promising. The availability, high strength, and good thermal properties of these materials are important technical and economic considerations.

Although several different processes can be used to make filaments of these materials, vapor deposition is the most widely used. Not only does it produce strong filaments, but it can also be applied to both continuous and discontinuous filaments. Examples of the former are the polycrystalline filaments (sometimes referred to as amorphous because of the very small crystal size), produced by depositing the desired substance onto a heated filamentary substrate. Single crystal whiskers are an example of the short, discontinuous fibers produced by vapor deposition. While the modulus and several other properties are about equal in both forms of the material, the tensile strength of the whiskers can be greater by about an order of magnitude than the tensile strength usually obtained from continuous fibers (about 400,000 to 500,000 psi for the continuous fibers and in the range of 3-6 $\times$ 10^6 psi for whiskers). However, the average strength, rather than the maximum strength of individual fibers or whiskers, is more important. Both types of fibers are likely to be of value particularly for use in composites where the short fibers or whiskers are applicable to the reinforcement of metals but not as useful in reinforcement of plastics. Since the two types of fibers or whiskers are still expensive, as a result of the vapor deposition process, consideration is given either to using them in conjunction with lower cost fibers or only where the properties yield a significantly higher performance in the product. Meanwhile, research to improve the production methods is being pursued in an attempt to significantly reduce costs.

3.   Applying Composites: In the course of this survey, as well as through other contacts, it was found that there is an obvious need to give greater consideration to engineering and manufacturing problems of fiber-reinforced composites. While there are indeed some capabilities in these two areas, there appear to be a great number of engineering and manufacturing personnel who expect these materials to be designed and processed into useful shapes by the same methods used for more homogeneous materials. The nature of fiber-reinforced systems does not appear amenable to this approach, but favors the fabricating processes currently being used to produce glass fiber reinforced plastics.

VII. CONCLUSIONS AND RECOMMENDATIONS

There is a widespread cognizance of the potential for fiber-reinforced composites. The rapid growth in both the patent and technical literature, together with the willingness of many organizations to respond to this survey, firmly support this observation. The rapidly increasing interest in filamentary materials has made it virtually impossible to include all of the most recent developments in this report. In fact, many programs now in progress may yield results that change some of the views expressed in preceding sections.

Relatively few materials possess the necessary balance of properties to satisfactorily perform in aerospace environments. Among those which appear to be capable of meeting the requirements are: the newer, high-strength, high-temperature fiber glasses, carbon filaments, the polycrystalline filaments of oxides, carbides, and boron, and the single crystal whiskers.

The trends to develop stronger, stiffer, more refractory filaments has arisen largely from an urgent need for materials having very high strengths and low weights. Filaments, of the materials mentioned above, in addition to having desirable strength-to-weight ratios, can be combined with other materials to form systems with "tailor-made" properties for a specific application.

On a strength-to-weight basis, fiber reinforced composites appear to offer great promise for providing high performance structural materials for aerospace applications. However, the widespread utilization of the more advanced filaments depends primarily on the development of suitable means for economical production and handling, a full evaluation of their properties, and identification of characteristics. In addition, technology must advance sufficiently to permit the use of whiskers in a variety of matrices so that new high-performance composites become available.

In view of these conclusions, the following recommendations are presented:

1. That the interested reader continue to seek the latest information in the patent and technical literature.

2. That effective techniques for testing large populations of fibers be developed in order to establish data confidence limits.

3. That strength data be accompanied by such other information as gage length, rate of loading, test method, test environment, and data scatter.

4. That the various factors likely to affect the strength of fibers be identified and that the strength retention of fibers exposed to these factors be determined.

5. That standardized test methods be adopted, so that comparative evaluation of the same material is possible. (This is particularly necessary in whisker data where one investigator will use a bend test and another tensile tests.)

VIII. PATENTS ON CERAMIC AND GRAPHITE FIBERS

A. Introduction

Patents sometimes serve as the only published result of technical work. Even when technical articles are published on patented items, the patents usually are valuable supplements to the technical papers. Accordingly, a patent survey on ceramic and graphite fibers was conducted and the results are presented in this section.

It must be emphasized that this patent search and the resulting abstracts and indexing of patents pertinent to the survey are presented for purely technical information relative to ceramic and graphite fibers. The inclusion or exclusion of any patent is not to be construed as an endorsement or lack thereof of the validity of the patent. Neither is any indication of the value of a patent intended by the length of the abstract.

The search spanned thirty years of patents, even though a patent will normally only provide a 17-year period of protection for the inventor. Over 500 patents were initially considered for inclusion in this section, of which over 200 were abstracted and are included in numerical order. Of these, about 35% were considered to be the most pertinent to this survey and were particularly diligently sought; however, the authors chose to include others of related interest, since many of them may conceivably provide solutions or ideas for solutions to technical problems which arise with ceramic and graphite fibers. Among the patents of related interest are some having to do with metallic and organic fibers, some on sizes and coatings for fibers, some on crystal growth methods, and many on composites and products in which fibers are used as at least one constituent.

It is interesting to note that the rate of granting of patents on ceramic and graphite fibers is increasing exponentially.

Other aspects of this survey strongly indicate that a very large
number of patents will be applied for and issued in the next few
years. The reader interested in this source of technical information would be well advised to follow the patent abstracts published
by the U.S. Patent Office or the selected abstracts, such as are
published monthly in the Journal and Abstracts of the American
Ceramic Society. Ceramic and graphite fiber patents are not as
well classified as are glass fiber patents, so the patent searcher
must be careful to search many categories.

The following patent abstracts were prepared by Dr.
Louis Navias and Mr. L. R. McCreight.

B. Index to patent abstracts

Over 200 patents are abstracted in this section. They
are listed in numerical order, which also corresponds to the date
of issue as a patent. The categories and sub-categories used in
this patent index are as follows:

Compositions
    Glass and Mineral
    Ceramic Fibers
    Ceramic Whiskers
    Carbon and Graphite
    Surface Treatments and Coatings
        (a) Organic
        (b) Metal
        (c) Others
Fiber Forming Processes, Apparatus, and Products
    From Melts
    Crystal Growing
    Colloidal
    Paper, Mat, and Textile Techniques
Composite Products
    With Resins
    With Ceramics
    With Metals
    Insulation
    Fillers
    Building Materials

For convenience, the index sometimes lists a given patent under
more than one heading.

Compositions

Glass and Mineral

2,461,841	2,685,527	2,946,694
2,494,259	2,693,668	2,978,341
2,557,834	2,710,261	3,007,806
2,640,784	2,733,158	3,053,672
2,664,359	2,772,987	3,060,041
2,674,539	2,823,117	3,084,054
2,681,289	2,870,030	3,127,277
2,685,526	2,908,545	3,132,033

Ceramic Fibers

2,744,074	3,030,183	3,080,241
2,816,844	3,031,417	3,096,144
2,833,620	3,031,418	3,104,943
2,915,475	3,039,849	3,108,888
2,968,622	3,056,747	3,110,545
2,980,510	3,065,091	3,129,105
3,012,856	3,069,277	
3,012,857	3,077,380	

Ceramic Whiskers

2,813,811	3,023,115	3,094,385
3,011,870	3,063,866	

Carbon and Graphite

2,765,354	2,822,321	3,107,152
2,796,331	2,957,756	3,107,180

Surface Treatments and Coatings

(a) Organic

2,238,694	2,728,972	2,932,587
2,390,190	2,776,910	2,940,875
2,676,898	2,793,130	2,951,772
2,694,655	2,834,693	2,953,478
2,710,290	2,845,364	2,958,614
2,720,470	2,860,450	3,025,588
2,723,208	2,874,135	3,039,981
2,723,215	2,895,789	3,143,405

(b) Metal

2,699,415	2,797,469	2,938,821
2,731,359	2,818,351	2,956,039
2,749,255	2,860,450	3,023,490
2,772,987	2,880,552	3,046,170
2,782,563	2,895,789	3,078,564
2,791,515	2,907,626	

(c) Others

2,446,119	2,767,519	3,019,122
2,723,211	2,776,910	3,025,588
2,723,215	2,779,136	3,039,981
2,727,876	2,793,130	3,045,317
2,728,740	2,834,693	3,046,084
2,728,972	2,838,418	3,125,428
2,731,359	2,872,350	3,152,006
2,739,077	2,874,135	
2,739,078	2,932,587	

Fiber Forming Processes, Apparatus, and Products

From Melts

2,020,403	2,453,864	2,714,622
2,048,651	2,461,841	2,823,117
2,234,986	2,477,555	2,908,545
2,245,783	2,489,508	2,939,761
2,267,019	2,494,259	3,041,663
2,300,736	2,495,956	3,135,585
2,313,296	2,527,502	3,142,551
2,331,944	2,569,700	3,142,869
2,331,946	2,582,919	3,145,981
2,335,135	2,629,969	3,152,878
2,339,928	2,686,821	3,155,475
2,360,373	2,691,852	3,155,476
2,398,808	2,693,668	3,159,475

Crystal Growing

2,852,890	3,139,653	3,152,006
2,854,364	3,147,085	3,152,992
3,055,736	3,147,159	3,157,541
3,073,679	3,148,027	3,160,476
3,075,831	3,149,910	3,161,473
3,135,585	3,150,925	3,168,423

Colloidal

3,024,088	3,082,051	3,108,888
3,024,089	3,082,099	3,110,545
3,031,418	3,096,144	

Paper, Mat, and Textile Techniques

2,401,389	3,012,289	3,059,311
2,552,124	3,012,856	3,125,404
2,700,866	3,016,599	3,144,687
3,007,840	3,031,322	

Composite Products

With Resins

2,245,203	2,875,474	3,043,796
2,306,781	2,920,992	3,050,427
2,311,613	2,940,875	3,053,713
2,552,124	2,951,780	3,062,682
2,658,849	2,975,503	3,063,883
2,683,697	2,980,982	3,067,482
2,694,660	2,996,411	3,082,143
2,762,739	3,008,913	3,102,835
2,794,238	3,013,915	3,118,807
2,827,099	3,024,145	3,141,809
2,866,769	3,041,131	3,142,598

With Ceramics

2,731,359	3,019,117	3,131,073
2,793,130	3,109,511	3,141,786
2,816,844	3,118,807	3,157,722
2,902,379	3,121,659	

With Metals

2,699,415	2,907,626	3,095,642
2,772,987	2,938,821	3,098,723
2,782,563	2,956,039	3,110,571
2,791,515	3,026,200	3,110,939
2,797,469	3,046,170	3,114,197
2,818,351	3,047,383	3,127,668
2,848,390	3,078,564	3,157,722
2,849,338	3,084,421	
2,880,552	3,085,876	

Insulation

    2,674,539   2,941,904   3,129,105
    2,699,397   3,125,404   3,154,463

Fillers

    2,751,366
    3,080,256

Building Materials

    2,221,945   2,849,338   3,019,117
    2,793,130   2,902,379   3,121,659
    2,848,390   3,970,127

C. Patent abstracts

2,020,403   11/12/35

Engle

Process for Producing Mineral Fiber

9 Claims (9 Process, 0 Product), 4 Figures

Shapes are made of inexpensive materials, such as waste limestone, waste clay and waste coal fines, and these shapes are charged into a retort or cupola and melted. A stream of the melt is caused to form fibers by means of an (air) fluid blast under pressure.

An example gives a range of compositions for such mineral fibers.

2,048,651   7/21/36

Norton
Massachusetts Institute of Technology

Method of, and Apparatus for Producing Fibrous or Filamentary Material

11 Claims (9 Method, 2 Apparatus), 2 Figures

An electrostatic field and an air blast are maintained to propel a viscous liquid or a molten viscous liquid toward a collecting target, and fibers are formed during this travel. There are a number of variations of this method described.

No substances are mentioned in the claims, but the text mentions substances such as resin, glass wool, slag wool.

2,221,945   11/19/40

Hanson
Union Carbide and Carbon Corporation

Moldable Sheet Composition and Process of Preparing Same

12 Claims (7 Process, 5 Product)

Process of making sheets of long wood fibers, mixed with blended short fibers and (phenol-formaldehyde) resin, by hot-rolling. Short fibers may be wood flour, ground pulp, asbestos fibers.

Impact strength of product given as 7.6-10 ft.lbs. per sq. in.

2, 234, 986   3/18/41

Slayter and Thomas
Owens-Corning Fiberglas Corp.

Mechanically Drawing Fibers

22 Claims (15 Method, 7 Apparatus)

Detailed description of apparatus for drawing fine glass fibers from Pt-Rh
alloy orifices, 0.02-0.08" dia., electrically heated for control; gas cooled
means for chilling a stream of glass and drawing fibers from it, by mechanical means.

2, 238, 694   4/15/41

Graves
.duPont

Polymeric Materials

13 Claims (8 Process, 5 Product)

Synthetic (organic) linear filaments are coated with elastic organic
coatings, etc.

2, 245, 203   6/10/41

Kuzmick
Raybestos-Manhattan Inc.

Manufacture of Molded Compositions for Brake Linings or Similar Articles

6 Claims (3 Process, 3 Product)

Process of making friction bodies (brake linings) by mixing asbestos fibers
with heat convertible resin and a sugar compound. No Strength Data.

2, 245, 783   6/17/41

Hyde
Owens-Corning Fiberglas Corp.

Method of Coloring Glass Fibers

5 Claims (5 Method, 0 Product)

Glass fibers are colored in several ways (using glasses containing alkalies): (1) dipping the fibers in aqueous solutions of metals of higher groups
(II, III) to replace the ions of Group I of the glass, Na, K. In the case of
Mn and K permanganate the fibers become light tan; (2) taking the product
of (1) and dipping in solutions of anions to produce color. Ex. First solu-

tion – lead acetate, second solution – K ferrocyanide, color of fiber – dark blue; (3) taking the product of (1) and dipping in solutions of organic dyes; (4) dipping fibers in organic dyes as the only step. 15 examples.

Replacing most of the alkali in glass fibers by Cu or Pb, increases electrical resistance from 5 megohms to 1 million megohms.

2, 267, 019   12/23/41

Esser (Germany)
Oscar Gossler Glasgespinst-Fabrik (Germany)

Apparatus for the Production of Glass Threads

3 Claims (3 Apparatus, 0 Product), 3 Figures

A circular glass furnace has a central opening in which is located a (gas) flame pointing downwards towards a plate with bushings or orifices through which glass forms filaments. Various modifications of apparatus are described including rotating parts, etc.

2, 300, 736   11/3/42

Slayter and Thomas
Owens-Corning Fiberglas Corp.

Method of Making Filamentous Glass

4 Claims (4 Method, 0 Process), 2 Figures

Glass fibers are made from a very fluid glass, 70-700 poises (preferably 100-300 poises), using an orifice opening 0.02-0.08" (preferably 0.03-0.04"), and mechanically drawing the fibers at more than 5000 ft. per min.

The text says fibers of 0.0001" can be so made, with strengths greater than 350,000 psi.

2, 306, 781   12/29/42

Francis, Jr.
Sylvania Industrial Corp.

Product Containing Siliceous Fibers and Method of Making the Same

17 Claims (6 Process, 11 Product)

A process of making a fibrous product, a textile, a yarn, a felt, by mixing siliceous fibers with resin fibers, and treating the mixture to form the product. The siliceous fibers may be long or short glass fibers, asbestos fibers, mineral fibers such as rock wool, mineral wool, slag wool, etc. The resin fibers may be thermoplastic or thermosetting. Examples show a content of 2-20% resin fibers by volume in the mixture.

2,311,613   2/16/43

Slayter
Owens-Corning Fiberglas Corp.

Transparent Composite Material

8 Claims (0 Process, 8 Product)

Transparent articles made of glass fibers and plastic organic materials,
the glass fibers and the plastic having the same refractive index, the glass
fibers to be less than 10 microns in diameter. Examples give glass compositions in weight %, and trade names of resins, for combinations of refractive index, 1.466, 1.47, 1.48, 1.51, 1.544 and 1.60.

2,313,296   3/9/43

Lamesch (Germany)
Alien Property Custodian

Fiber or Filament of Glass

6 Claims (6 Product, 0 Process), 3 Figures

Glass fibers are made of two different glasses, one forming the core and
the other the shell. Methods are described to obtain such fibers; as pulling
out a rod having a casing of a different glass; or by having two reservoirs
of molten glass, one inside the other, so that the stream of one glass forms
an outside layer onto the inner glass core, as both are pulled out together
to form the finished double fiber. The text mentions that the fiber will be
strengthened if the outer glass is in compression and the core in tension by
having the core with a higher coefficient of expansion than that of the shell.

2,331,944   10/19/43

vonPazsiczky and Steingraeber (Germany)
Alien Property Custodian

Production of Fibers from Minerals and Like Materials

9 Claims (9 Method, 0 Product)

Glass fibers with a rough surface are made by subjecting the streams of
glass, as they are attenuated, to vapors of hydrofluoric acid or phosphoric
acid, by mixing these vapors with the compressed air or steam used to
attenuate the fibers, etc.

2,331,946   10/19/43

vonPazsiczky (Germany)
Alien Property Custodian

Manufacture of Glass Fibers

2 Claims (2 Method, 0 Product)

A specially designed glass furnace, having an attached chamber with a head
of molten glass leading to the bushings, etc.

2,335,135   11/23/43

Staelin
Owens-Corning Fiberglas Corp.

Manufacture of Fibrous Glass

6 Claims (6 Apparatus, 0 Product)

The bushings or nipples leading from the molten glass to form fibers are
arranged in rows, in circles, at various levels, etc. to form groups of
fibers, for collection on a drum, etc.

2,339,928   1/25/44

Hood
Owens-Corning Fiberglas Corp.

Method of Treating Glass Fibers and Article Made Thereby

6 Claims (6 Product, 0 Process)

Colored glass fibers are obtained by adding PbO and/or CuO to the glass
batch, melting the batch, drawing fibers, and then heating the fibers in a
hydrogen atmosphere at about 400°C to obtain colored fibers due to colloidal
distribution of (metallic ?) particles. PbO additions give grey to black
colors. CuO additions give red to black colors. 10 compositions given in
text.

2,360,373   10/17/44

Tiede
Owens-Corning Fiberglas Corp.

Apparatus for Feeding Glass in the Manufacture of Fibers

11 Claims (11 Apparatus, 0 Product), 3 Figures

Baffles are placed in the chamber containing the molten glass leading to
the bushings - also electrical heating of them.

2,372,433   3/27/45

Koon
Columbian Rope Co.

Moldable Plastics Composition and Method of Preparing Same

4 Claims (2 Product, 2 Process), 3 Figures

Long vegetable fibers and resin binders are treated and molded, etc.

2,390,190   12/4/45

Soday
The United Gas Improvement Co.

Chemical Process and Product

8 Claims (6 Product, 2 Method)

Detailed description of preparation of hydrocarbon resin polymers for the
coating of glass fibers; also fibers of mineral wool, slag wool, quartz,
(to eliminate abrasion etc.) for batts, etc.

2,398,808   4/23/46

Slayter and Fletcher
Owens-Corning Fiberglas Corp.

Apparatus for Forming Fibrous Strands

9 Claims (9 Process, 0 Product), 4 Figures

Means for mechanically collecting fibers, using a hollow rotatable support
with spiral grooves, etc.

2,401,389   6/4/46

Truitt
American Viscose Corporation

Asbestos Yarn

5 Claims (5 Product, 0 Process), 1 Figure

Yarn is made with asbestos fiber and 10-25% of cellulosic fibers.

2,446,119

7/27/48

White, Steinman and Biefeld
Owens-Corning Fiberglas Corporation

Glass Fiber Reinforced Plastics

4 Claims (1 Article, 3 Process)

Coating glass fiber with dextrinized starch to obtain a better bond with synthetic resins, etc.

2,453,864

11/16/48

Schlehr
Glass Fibers, Inc.

Method for Drawing Fibers

9 Claims (9 Method, 0 Product), 2 Figures

Apparatus is described wherein molten glass in a closed container is caused to move through a series of bushings to form filaments, by means of gas pressure developed in the glass and from outside gas pressure.

2,461,841

2/15/49

Nordberg
Corning Glass Works

Method of Making Fibrous Glass Articles

10 Claims (10 Method, 0 Product)

Method of making an article composed of refractory glass fibers: (1) by melting a glass containing no more than 70-75% SiO_2, or by melting a borosilicate glass containing no more than 56% SiO_2 or by melting a glass containing no more than 15% ZrO_2, etc.; (2) drawing the glass into fibers, and making an article out of them; (3) leaching the glass fiber article with acid (pH no greater than 7) and washing – leaving a porous fibrous structure; (4) reheating the fibers at 600-800°C for various lengths of time, to revitrify the fibers. The fibers so treated should have a diam. less than 0.001". A few examples given of initial and final compositions.

2,477,555

7/26/49

Roberts and Metzler
Owens-Corning Fiberglas Corp.

Mineral Fiber Mat and Process of Manufacture

15 Claims (11 Product, 4 Process), 5 Figures

Glass fibers are arranged in certain "haphazardly arranged" fashion, etc. to form a mat--.

2,489,508

Stalego
Owens-Corning Fiberglas Corp.

Apparatus for Producing Fibers

12 Claims (12 Apparatus, 0 Product), 8 Figures

Improvements in bushing structure of glass fiber making devices.

2,494,259

Nordberg
Corning Glass Works

Fibrous Glass Articles

8 Claims (8 Product [article], 0 Process)

Glasses containing soluble constituents (in mineral acids) are drawn into fibers, and woven into an article. Lubricants are removed by heating at 300°C or by solvents, and the fibers are treated in hot mineral acid to remove the soluble constituents, as alkali, boric oxide. After washing to remove the soluble salts, the fibers are dried -- they are porous in this state. However the leached porous fibers may be heated to some temperature as 600-800°C for various lengths of time to cause them to vitrify into non-porous glass fibers. The treatments of leaching in acids, and of heating, may be carried out on the finished fibrous glass article.

Four examples give the original glass compositions and the compositions of the leached products -- these latter compositions being given in the claims for porous and non-porous vitreous fibrous articles made with them. In all claims, the diameter of the fibers is stated to be less than 0.001".

The leached glasses contain SiO_2 88-96%, with minor amounts of alkaline earths, boric oxide, alkalies, etc., and up to 10% alk. earths or 12% ZrO_2.

2,495,956

Cook
Glass Fibers Inc.

Method of Producing Glass Fibers

12 Claims (12 Method, 0 Product), 5 Figures

Detailed descriptions of equipment for drawing of glass fibers, by heating a secondary reservoir of molten glass by high frequency, and causing the molten glass to flow through orifices, by means of differential mechanical pressure, differential gas pressure, vacuum, etc.

2, 527, 502   10/24/50

Simison, Billman and McKelvy
Owens-Corning Fiberglas Corp.

Method and Apparatus for Producing Glass Fibers

17 Claims (13 Apparatus, 4 Method), 8 Figures

Winding glass filaments in strand form on a rotable spool, etc. Detailed
description of apparatus.

2, 552, 124   5/8/51

Tallman
Owens-Corning Fiberglas Corp.

Fibrous Glass Fabric

8 Claims (8 Product, 0 Process), 3 Figures

Glass fibers of two different diameters are interwoven in preferred layers
and patterns in a continuous mass of organic plastic, to give stretch.

2, 557, 834   6/19/51

McMullen
The Carborundum Company

Refractory Glass Wool

15 Claims (10 Product, 5 Batch compositions)

Glass fibers are made from a batch containing alumina and silica, and containing borax glass, or the equivalent constituents.

Range	Al_2O_3	45 - 55 parts
	SiO_2	55 - 45
	Borax glass $(Na_2B_4O_7)$	1 1/2 - 6
	also ZrO_2 to replace up to 7 1/2% of SiO_2	

Maximum diameter of fibers 10μ, and resistant to breakdown at 1400°C
over a period of 24 hours.

2,569,700   10/2/51

Stalego
Owens-Corning Fiberglas Corp.

Method of Making Colored Glass Fibers

8 Claims (8 Process, 0 Product), 4 Figures

A slurry of coloring material is used to put a coloring layer onto a glass
rod or glass fibers, which are then reheated to cause adherence. The
rods are heated sufficiently to be drawn into fibers. The coloring material may be an oxide of a metal, as Co, Cr, Mn, etc., and it may be
mixed with glass powder in the suspension. Organic binders may also be
used.

2,582,919   1/15/52

Biefeld
Owens-Corning Fiberglas Corp.

Coloring Glass Fibers

7 Claims (7 Method, 0 Product)

The surface of glass fibers is attacked with mineral acids to make the surface porous, and after leaching to remove extraneous material, is impregnated with (organic) coloring matter.

2,629,969   3/3/53

Peyches (France)
Soc. Anon. Man. Glaces et Produits
Chimiques de Saint-Gobain (France)

Manufacture of Fibers Such as Glass Fibers

15 Claims (10 Apparatus, 5 Method), 5 Figures

Description of apparatus having centrifugal means to disperse molten glass
into fibers, mechanical obstructions to break up the fibers, means of
collecting them, etc.

2,640,784

6/2/53

Tiede and Tooley
Owens-Corning Fiberglas Corp.

Composition of Glass, Especially for Fibers

5 Claims (5 Composition, 0 Product)

Range –			
	SiO_2	50 – 62%	
	TiO_2	2.5 – 16	TiO_2 and ZrO_2 together
	ZrO_2	2.5 – 16	not to exceed 25%
	B_2O_3	up to 12	
	Na_2O	10 – 20	
	Al_2O_3	up to 10	
	F	0.5 – 8	

2,658,849

11/10/53

Lew
Atlas Powder Company

Fiber Bonding Resin and Bonded Product

12 Claims (6 Resin Composition, 3 Product, 3 Process)

Organic resins for use in bonding glass fibers into a mat.

2,664,359

12/29/53

Dingledy
Owens-Corning Fiberglas Corp.

Glass Composition

13 Claims (11 Composition, 2 Articles [Fibers])

Glasses are selected for their high fluidity and ability to be drawn into fibers:

Range		
	SiO_2	40 – 60%
	Al_2O_3	4 – 18
	CaO	15 – 22
	MgO	4 – 12
	Na_2O	10 – 15
	B_2O_3	Up to 6
	TiO_2	0.5 – 6

2,674,539

4/6/54

Harter and Norton, Jr.
The Babcock and Wilcox Co.

High Temperature Refractory Products

7 Claims (7 Product, 0 Process)

Mineral wool is made from aluminum silicate minerals, especially kaolin, and mixtures of kaolin and bauxite, by melting in the temperature range 2900-3350°F and blown into fibers.

Range of composition	Al_2O_3	45-69%	covers kaolin and
	SiO_2	26-51	mullite compositions
	Alkali	<0.75	made into fibers

Kaolin fibers, 2-4μ	Young's Modulus	12.9×10^6 psi
	Breaking Stress	308,000 psi
	Permissible use	long time 2,000°F
		short time 2,300°F
	Melting Temp.	3,180°F

(compared to properties of glass wool)

2,676,898

4/27/54

Folger and Roberts
Owens-Corning Fiberglas Corp.

Method of Treating Glass Fiber Bats with Resin and Product

2 Claims (1 Method, 1 Product)

Especially relates to removing the calalyst from a resin used to bond fiber glass.

2,681,289

6/15/54

Moore
Gustin-Bacon Manufacturing Company

Glass Composition

4 Claims (2 Composition, 2 Product)

It is stated that these glasses have good chemical corrosion resistance.

Range	SiO$_2$	46 - 52%
	Al$_2$O$_3$	14 - 18
	CaO	18 - 24
	B$_2$O$_3$	9 - 14

Viscosity data given.

2,683,697   7/13/54

Newell and Atticks
United Aircraft Corporation

Nylon-Reinforced Copolymer Resins

8 Claims (2 Compositions, 6 Process), 3 Figures

Mechanical properties of some copolymer resins are improved when the
resins are reinforced with nylon fibers.

2,685,526   8/3/54

Labino
Glass Fibers Inc.

Glass Composition

7 Claims (7 Compositions for fibers)

It is stated that fibers of diameter 10-20 millionths of an inch (instead of
normal 30) can be made from glasses containing CuO and/or PbO. Claims
give range of compositions and specific compositions.

Range	SiO$_2$	54.3 - 54.75%
	CaO	6.35 - 7.43
	MgO	3.98 - 4.12
	B$_2$O$_3$	10.2 - 10.5
	PbO } CuO }	9 - 10.5
	Al$_2$O$_3$ } Fe$_2$O$_3$ } TiO$_2$ }	balance

2,685,527   8/3/54

Labino
Glass Fibers, Inc.

Glass Composition for High Tensile Fiber

7 Claims (3 Compositions, 4 Articles of glass fiber)

Glass fibers are made for high electric resistivity and for high tensile
strength for filter paper purposes.   Paper strength 250 psi, and porosity
only 1 part of smoke per million by test.   General range of composition:

SiO_2	52 - 56%
CaO	6 - 16
B_2O_3	8 - 13
Al_2O_3	12 - 16
ZnO	5 - 15
Alkalies	minor

2,686,821   8/17/54

McMullen
The Carborundum Company

Apparatus for Melting and Fiberizing Refractory Materials

8 Claims (7 Furnace Construction, 1 Apparatus), 3 Figures

A furnace is described having a main melting chamber, and a forehearth
attached with its own set of electrodes to maintain the molten condition
needed for pouring the melt through a spout or tilting it to pour over a lip -
to obtain a stream which is fiberized by means of a hot gas stream.   No
compositions or materials mentioned.   No technical data.

2,691,852   10/19/54

Slayter and Drummond
Owens-Corning Fiberglas Corp.

Method and Apparatus for Producing Fibers

14 Claims (9 Method, 5 Apparatus), 10 Figures

Detailed descriptions of apparatus designed to collect glass fibers as they
issue from the furnace, on supports from which the fibers are severed
into short lengths; apparatus to coat the issuing fibers continuously with a
binder, etc.

2,693,668   11/9/54

Slayter
Owens-Corning Fiberglas Corp.

Polyphase Systems of Glassy Materials

10 Claims (10 Product [Composite Structure], 0 Process)

A glassy composite is made by surrounding glass fibers of a given thermal expansion coefficient with a glassy matrix of a lower coefficient, so as to put the fibers under tension and the matrix under compression.

The claims require differences in thermal expansion for the two glass components, in some instances, and in other instances require differences in surface tension and modulus of elasticity as well.

Example: Higher expansion glass fibers   10 - 60% by weight
            Lower expansion matrix   90 - 40%

12 examples, with glass compositions of 3 pairs.

Example: leaching of surface of fibers to get a different thermal expansion on the surface. Applying bentonite, etc. to the surface of fibers and firing it into place to get different expansions.

2,694,655   11/16/54

Pullman and Reynolds, Jr.
American Cyanamid Co.

Process for Sizing Glass Filaments and Product Produced Thereby

13 Claims (11 Process, 2 Product)

Glass filaments while still hot from formation are covered continuously with an aqueous dispersion of one or more of (1) a halohydrin, and aliphatic primary amine (and other combinations), (2) formaldehyde resin, (3) polyvinyl alcohol, and heating to drive off volatiles.

2,694,660   11/16/54

Schwartz
Vibradamp Corporation

Fiber Glass Mat

2 Claims (2 Product, 0 Process), 4 Figures

Glass fibers are treated with a thermosetting resin binder, to have binder not only at the fiber contacts but also within the mass, which is heat treated under some pressure, to form a springy mat.

<u>2,699,397</u>   1/11/55

Hahn
Johns-Manville Corp.

Refractory Mineral Fiber

5 Claims (5 Product, 0 Process)

Refractory mineral wool is made by melting a mixture of kyanite 100 parts by weight and silica 6-12 parts, and fiberizing the molten stream.

Several analyses are given in the claims, limiting content of Fe_2O_3, TiO_2, alkalies, alkaline earths.

It is claimed that the fibers will have good chemical and strength stability at least as high as $2000^\circ F$.

Example:			
Al_2O_3	51 - 55	52.3%
SiO_2	43 - 47	45.5
Fe_2O_3	Bal.	0.9
TiO_2	Bal.	0.9
CaO & MgO		0.3
Alkali		<0.1

<u>2,699,415</u>   1/11/55

Nachtman
Owens-Corning Fiberglas Corp.

Method of Producing Refractory Fiber Laminate

26 Claims (26 Method, 0 Product)

Refractory fibers issuing from bushings attached to a molten bath are coated continuously with a single coating, or with a series of coatings. The first coating may be inorganic or metallic, and subsequent coatings may include organic coatings.

The coated fibers may be bonded together to form a laminate, by application of heat and pressure.

The fibers, in the text, are made of glass.

Inorganic coatings may be metal, metallic oxide, metallic salt, a salt, a phosphate, otherwise the claims are general as regards compositions.

<u>2,700,866</u>

2/1/55

Strang

Method of Concatenating Fibrous Elements

4 Claims (4 Method, 0 Product), 2 Figures

In textile manufacture, fibers are suspended in a liquid and caused to form a yarn, a ribbon, roving, etc.

<u>2,710,261</u>

6/7/55

McMullen
The Carborundum Company

Mineral Fiber Compositions

8 Claims (7 Product, 1 Batch), 1 Figure

Mineral fibers are made from melts containing fair amounts of TiO_2, with small or no amounts of alkali, and limiting the B_2O_3, CaO and MgO contents.

A triaxial diagram of the system SiO_2 - Al_2O_3 - TiO_2 gives a general area of compositions, and a smaller area of preferred compositions.

A slag containing Al_2O_3 and TiO_2 is claimed as a raw material.

Example:				
SiO_2	47%	30 – 70	40 – 60
Al_2O_3	43	5 – 50	10 – 45
TiO_2	7.5	2.5 – 50	10 – 40
B_2O_3	1.25	< 5	–
CaO	–	< 10	<10
MgO	–	< 5	< 5
Alkali	Rem.	< 5	–

These fibers are described as being resistant to leaching by acids.

<u>2,710,290</u>

6/7/55

Safford and Bueche
General Electric Company

Organopolysiloxane-Polytetrafluoroethylene Mixtures

18 Claims (11 Composition, 7 Method), 2 Figures

A composition made by mixing and reacting (1) an organopolysiloxane (2) polytetrafluoroethylene dispersed in the form of randomly distributed fibers, (3) a filler, (4) a curing agent, etc.

8/2/55

2,714,622

McMullen
The Carborundum Company

Method and Apparatus for Fiberizing Refractory Materials

12 Claims (7 Apparatus, 5 Method), 9 Figures

A water cooled orifice (of special shape) is a part of a forehearth in which refractory materials can be melted by arc-heating electrodes. One electrode is located directly over (or near) the orifice, 1/2" to 3" away, to help maintain the temperature needed for keeping the material fluid as it flows through the orifice. The orifice can be mild steel, water cooled, and capable of delivering a molten stream of at least 1/8" diameter, and at temperatures above 1500°C.

The molten stream meets a high velocity gas stream which fiberizes the molten stream.

Numerous modifications are mentioned in the text.

No materials are mentioned in the claims, but the text describes the preparation of fibers of 50% Al_2O_3 - 50% SiO_2, with power consumption for electrodes, etc.

10/11/55

2,720,470

Erickson and Silver
U.S.A.

Allylaroxydichlorosilane and Method of Its Preparation and Application to Glass

23 Claims (16 Method, 7 Product)

New compounds are described and claimed for use in treating fiber glass reinforcements for plastics; the compounds being of the class allylaroxydichlorosilane. Specific compounds are named.

11/8/55

2,723,208

Morrison
Owens-Corning Fiberglas Corp.

Method of Sizing Mineral Fibers with a Suspension of a Polyamide Resin and Product Produced Thereby

9 Claims (7 Method, 2 Product)

Mineral and glass fibers are sized with cationic suspensions of a polyamide resin and butadiene-acrylonitrile copolymer, etc.

2,723,211   11/8/55

MacMullen and Marzocchi
Cowles Chemical Co.

Stable Silane Triol Composition and Method of Treating Glass Therewith

11 Claims (9 Method, 2 Compositions)

Glass fiber and glass surfaces are made water repellant by treating them
with an aqueous alkaline solution of the hydrolysis products of hydrocarbon
substituted trihalosilane, etc.

2,723,215   11/8/55

Biefeld and Lydic
Owens-Corning Fiberglas Corp.

Glass Fiber Product and Method of Making Same

8 Claims (4 Articles, 4 Method), 4 Figures

A waxy surface layer is applied to glass fibers, continuously, from a
molten mixture of a mineral wax, with either or both silanes and polyethylene, to minimize rubbing and scratching. 14 examples.

2,727,876   12/20/55

Iler
duPont

Composition and Process

23 Claims (22 Product and Composition, 1 Process)

Natural rubber, synthetic rubber, organic elastomers, are mixed with an
(inorganic) pulverulent solid, such as silica gel (inorganic siliceous material), which has been treated on the surface with organic radicals and
materials to form chemically bound groups (--O R groups) by means of
Si-O-R bonds, which R is a hydrocarbon radical containing 2 to 18 carbon
atoms, etc.

2,728,740   12/27/55

Iler
duPont

Process for Bonding Polymer to Siliceous Solid of High Specific
Surface Area and Product

6 Claims (2 Process, 4 Composition), 3 Figures

An (organic) polymer is made with an inorganic solid of large surface area
(colloidal silica, silica gel), by treating the silica powder with unsaturated
alcohols (ethylenically-unsaturated compounds), and forming polymers,
which are attached to the silica by Si-O-C bonds, with coatings of - O R
groups, where R contains 2-18 carbons, etc.

2,728,972   1/3/56

Drummond and Smock
Owens-Corning Fiberglas Corp.

Method and Apparatus for Coating Fibers

11 Claims (9 Apparatus, 2 Method), 4 Figures

Detailed description of apparatus required to coat continuously glass fiber
with a liquid or flowable material (as lubricant, size, etc.)

2,731,359   1/17/56

Nicholson
The Carborundum Company

Refractory Fiber Body and Method of Making Same

15 Claims (10 Product, 5 Method)

Refractory fibers in mat or sheet form are bonded together, by means of
a silicide of carbon (SiC), or of nitrogen (SiN).

Example: Refractory fibers in a bat are filled with Si metal powder, and
subjected to a nitrogen atmosphere at 1350° - 1400^oC, whereby Si com-
bines with N_2 to form SiN, which bonds the fibers.

Example: Fibers with Si powder, heated at 1250^oC in an atmosphere of
CO gas, are bonded by SiC.

The text gives an example of a refractory fiber product made of

$$Al_2O_3 \quad 52.67, \; SiO_2 \; 45.77, \; B_2O_3 \; 1.06, \; Na_2O \quad 0.50\%.$$

<u>2,733,158</u>

1/31/56

Tiede

Owens-Corning Fiberglas Corp.

Glass Composition

4 Claims (2 Composition, 2 Fibrous Glass)

Glass compositions and fibers made from these glasses are claimed.

Specific Comp.		
SiO_2	55.7%	
Al_2O_3	5.6	
MgO	2.6	
CaO	6.0	
B_2O_3	4.4	
Na_2O	2.5	
K_2O	0.2	
SrO	11.4	
CuO	9.7	
Li_2O	1.9	

It is stated that the presence of CuO causes the fibers to become non-abrasive toward each other.

<u>2,739,077</u>

3/20/56

Goebel

duPont

Product and Process

6 Claims (3 Product, 3 Process)

Colloidal silica is treated with fluorocarbon alcohols, to form compounds of esters on the surface, having at least 50 $-OCH_2 (CF_2) nX$ groups per sq. millimicron of surface area of the siliceous material. X is H or F. n is 1-17.

<u>2,739,078</u>

3/20/56

Broge

duPont

Product and Process

7 Claims (4 Material, 3 Process)

Colloidal silica is treated with organic compounds, as a glycol, to form surface esters -- to form ORO groups on the surface.

2,744,074   5/1/56

Theobald
duPont

Polymeric Organic Aluminum Oxides and Method for Preparing Same

11 Claims (7 Method, 4 Product)

Compounds (and methods of preparing them) of the type sketched below are
described:

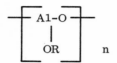

where R is an alkyl group
where n is 16 or more

2,749,255   6/5/56

Nack and Homer
The Commonwealth Engineering Co. of Ohio

Method of Producing Metalized Glass Fiber Rovings

11 Claims (9 Process, 2 Product), 6 Figures

Roving of glass fiber is exposed to the gas of a metal bearing compound
(as Ni) decomposable at the proper temperature to deposit the metal as a
continuous coating on the glass fiber, and vibrating the roving at 520-1750
cycles per minute by a mechanical device, to separate the filaments for
more complete coverage.

2,751,366   6/19/56

Braendle
duPont

Rubbers with Methoxy Containing Silica Fillers

10 Claims (5 Composition, 5 Product)

Improved dispersions of fillers in rubber are obtained with methoxy groups
bound to surface of filler.

2,762,739   9/11/56

Weiss
U.S.A.

Fiber-Reinforced Structural Panel and Method of Making Same

9 Claims (6 Method, 3 Article), 7 Figures

(Glass) fibers in parallel arrays are strung between two sheets of plastic
material, and anchored, to form lightweight reinforced structural panels.

2,765,354

10/2/56

Carpenter, Blanchard and Buselli
Air Reduction Co.

Polymeric Carbon and Method of Production

9 Claims (3 Product, 6 Method)

Acetylene, in the polymeric form (cuprene), is caused to dehydrogenate in
an inert atmosphere, such as nitrogen or argon, at temperatures of 320°-
900°C, to form polymeric carbon.

The electric volume resistivity can be varied between 10^4 and 10^{13} ohm.
cm. Shaped articles may be so treated. Examples are given.

2,767,519

10/23/56

Bjorksten

Methods for Treating Glass Fibers

6 Claims (6 Process, 0 Product), 1 Figure

Vinyl silane vapor is applied to issuing hot glass filaments, and then a
lubricant, to give better adherence to synthetic resins.

2,772,987

12/4/56

Whitehurst and Otto
Owens-Corning Fiberglas Corp.

Glass Composition and Metal Coated Glass Fiber

5 Claims (5 Product, 0 Process), 3 Figures

Composition:
$\begin{cases} SiO_2 \ 56.9\%, \ Al_2O_3 \ 5.4, \ MgO \ 2.5, \ CaO \ 6.1, \ B_2O_3, \ 6.4, \\ Na_2O \ 0.5, \ K_2O \ 0.2, \ SrO \ 11.3, \ CuO \ 9.0, \ Li_2O \ 1.7. \end{cases}$

It is claimed that the presence of CuO in the glass helps the adherence of a
superimposed metal obtained by passing the hot filaments through baths
(in slots) of molten metal, as Zn plus 1% Ti, Zn, Pb. 7 examples. Other
compositions of glasses and metals given in text.

2,776,910

1/8/57

Erickson and Silver
U.S.A.

Method of Coating Glass Fibers with Reaction Product of Organo Halsilanes
and Glycidol Derivatives and then with Epoxy Resins and the Product per se

18 Claims (12 Composition, 6 Method)

2,779,136   1/29/57

Hood and Stookey
Corning Glass Works

Method of Making a Glass Article of High Mechanical Strength and
Articles Made Thereby

15 Claims (5 Method, 10 Product)

The process of introducing Li ions (from a salt bath) into the surface layer
of a glass object to replace Na and/or K ions of the glass composition at a
temperature between the strain point and softening point of the glass, to
form a glass surface layer of lower linear expansion than that of the glass
itself. 11 examples.

2,782,563   2/26/57

Russell
Owens-Corning Fiberglas Corp.

Method and Means for Producing Metal-Coated Glass Fibers

11 Claims (9 Method, 2 Apparatus, 0 Product), 3 Figures

Glass fibers issuing hot from a series of bushings come in contact with
thin sheets of metal, picking up a thin layer of metal. The temperature
conditions of the various components are described.

The text mentions Al, Pb, and Sn as the metal coatings.

2,791,515   5/7/57

Nack
The Commonwealth Engineering Co. of Ohio

Metal Coated Glass Fiber and Method of Its Formation

11 Claims (6 Process, 5 Product), 1 Figure

Glass fiber is treated with a chrome complex, as methacrylate chromium
chloride at a temperature of about 600°F, to form an adherent coating of
Cr on the glass fibers, and (at the same time) treated with a metal carbonyl
(like Fe, Ni, Cr) to decompose at about the same temperature and deposit a
second coat of metal. 1 example.

2,793,130   5/21/57

Shannon and Mitchell
Owens-Corning Fiberglas Corp.

Pressure Molded Cement Products and Methods for Producing Same

9 Claims (6 Product, 3 Method)

Glass fiber with specified protective coatings is used as filler in products made with inorganic cements, such as gypsum, Portland cement. etc., using pressure molding. 5 examples.

2,794,238 6/4/57

Dildilian and Wood
Fiber Glass Industries, Inc.

Fiber Glass Mat

4 Claims (4 Product, 0 Process), 4 Figures

A low density sheet of glass fibrous material is described. It is made with continuously operating apparatus, some parts of the mat being bonded by adhesive and others not so bonded.

2,796,331 6/18/57

Kauffman, Griffiths and Mackay
Pittsburgh Coke and Chemical Company

Process for Making Fibrous Carbon

7 Claims (7 Process, 0 Product)

Methane with hydrogen, or coke oven gas, is heated in the range of 1150-1450°C to break down by pyrolysis, with a duration time of 0.4-15 sec., to form carbon fibers. The addition of H_2S gas, 0.3-4%, increases the yield of fibers. 6 examples.

Example: CH_4 plus H_2 gases at 1350°C (obtained from coke oven gas), in 2 sec. residence time, yielded 7% (of total carbon fed into the reaction vessel) as "long whiskers of carbon fibers," around 4 microns in diameter.

The addition of H_2S increases the yield several fold.

2,797,469 7/2/57

Kahn
The B.F. Goodrich Company

Metallized Glass Fibers and Products Thereof

19 Claims (19 Product, 0 Process), 5 Figures

To lead is added small quantities of another metal, such as Zn, Ti, Al, etc. to give better wetting of the second metal and to provide a protective coating to the glass fibers.

2,813,811   11/19/57

Sears
General Electric Company

High Strength Crystals

15 Claims (14 Method, 1 Product), 2 Figures

A material is heated in an evacuated vessel to form vapor of that material,
and the vapor rises to another part of the vessel held at a lower temperature, where the vapor deposits as rod-like crystals with a single, axially
disposed screw dislocation. The temperatures required to make such
crystals are given for Cd, Zn, Ag, CdS.

2,816,844   12/17/57

Bellamy

Ceramic Compositions

10 Claims (10 Product, 0 Process)

Fibrous talc is used in mixtures to make refractory objects, with SiC,
zircon, etc., and with clays, fluxes (as sodium silicate), etc. 4 examples.

2,818,351   12/31/57

Nack and Whitacre
The Commonwealth Engineering Company

Process of Plating Glass Fiber Rovings with Iron Metal

2 Claims (2 Process, 0 Product), 1 Figure

Glass fiber roving is passed through several chambers, reaching one containing gases including iron pentacarbonyl, which at temperature breaks
down to deposit iron metal on the fibers.

In this last chamber the roving is vibrated at 500-1750 cycles per second,
to expose all fiber surfaces to the decomposing iron containing gas.

2,822,321   2/4/58

Pickard (England)

New Carbon Product and Method for Manufacturing the Same

5 Claims (4 Method, 1 Product)

Peat or peat moss with a water content of about 40% is milled into shreds,
passed through a 60 mesh screen, and then carbonized at about 450°-$950^\circ C$
($800^\circ C$ preferred) to form fibrous, fernlike branching, carbon product.

2,823,117   2/11/58

Labino
L. O. F. Glass Fibers, Inc.

Glass Paper - Calcium Silicate

15 Claims (11 Product, 4 Process)

Calcium silicate fibers are made in the folowing steps –

(1) an alkali containing glass is melted and drawn into filaments –
     0.002-4".

(2) the filaments are reheated and blown into fibers – 0.01-1μ.

(3) the fibers are soaked in a solution of Ca, or Ba, or Zn chloride
     (or Ag nitrate) with heat to remove Na, and replace with Ca in the
     fiber glass structure.

(4) the fibers can be heated to several stages, to remove water, until
     above 1000°C, where anhydrous calcium silicate fibers result.

(5) the fibers can be made into a batt or other article.

1 Example:

	Glass Melt	Leached fibers (100°C)	Heated fibers (100°C)
SiO_2	74.5%	74.5%	85.4%
CaO	–	5.86	6.71
R_2O_3 (Al_2O_3, Fe_2O_3)	0.5	1.7	1.95
H_2O in combination	–	Balance	Balance
Ha_2O	25.0	–	–

2,827,099   3/18/58

Youngs
Dow Corning Corporation

Siloxane Rubber Glass Cord Tire

2 Claims (1 Product, 1 Method), 4 Figures

Glass fiber (cords) reinforced siloxane rubber tires are described.

2,833, 620   5/6/58

Gier, Salzberg and Young
duPont

An Inorganic Flexible Fibrous Material Consisting of the Asbestos-Like
Form of an Alkali Metal Titanate and its Preparation

27 Claims (11 Method, 16 Product)

Inorganic flexible fibers, not soluble in water, are made by subjecting alkali compounds and titania containing compounds in water to high pressures
(200-4000 atmospheres) and at high temperatures (400-800°C) at the same
time. The alkali compounds may be hydroxides or carbonates of Na, K,
Rb, Cs. The titania compounds may be TiO_2, or an oxygen-containing
titanium compound.

The fibers have the general formula - $M_2 O (TiO_2)_n$ where n is 6 to 7, and
also the formula (general) $M_2 Ti_6 O_{13}$, where M is an alkali metal.

The fibers have the ratio 1/w of at least 10/1.

Example: TiO_2, NaOH, $H_2 O$ at 625°C and 2800 atmospheres gives fibers
25μ long, and ratio 1/w 100/1.

10 examples. Chemical analysis, X-ray diffraction, etc.

2,834,693   5/13/58

Jellinek
Union Carbide Corporation

Vinyl Silane Composition and Process for Treating Fibrous Glass Material
Therewith

17 Claims

2,838,418   6/10/58

Starkweather, Jr.
duPont

Adhesion of Resins to Glass

10 Claims (10 Method, 0 Product)

Glass fibers are treated with a solution of tetra-alkyl titanate, and an organic polymeric material in an inert organic solvent, and heated to remove
solvent and convert titanate to TiO_2 and add stratum of organic material.

2,845,364   7/29/58

Waggoner
Owens-Corning Fiberglas Corp.

Process for Burning Size from Glass Fabric and Coating the Resulting
Fabric

10 Claims (10 Process, 0 Product), 6 Figures

A continuous process of burning off organic size and then coating with
silicone product.

2,848,390   8/19/58

Whitehurst and Otto
Owens-Corning Fiberglas Corp.

Method and Apparatus for Applying Metal to Glass

10 Claims (10 Method, 0 Product), 11 Figures

The application of metal(s) to glass fiber is done in 3 distinct steps:

(1) the glass fibers as they issue from the molten glass through bushings go past a molten bath of low melting point metal, and pick up
a coating.

(2) the coated fibers are then collected in bundles and in other con-
figurations after having a layer of water-soluble lubricant applied.

(3) the bundles are then submerged in a metallizing solution, or subjected to an electroplating bath, to pick up a coating of a second
metal.

No specific metals are mentioned in the claims. Text mentions Sn, Bi, Cd,
Pb, Zn, etc. for the undercoat and Cu, Fe, Brass, Cr, Ag, etc. for the
second coat. 14 examples.

2,849,338   8/26/58

Whitehurst and Otto
Owens-Corning Fiberglas Corp.

Metal Coated Fibrous Glass

4 Claims (4 Product, 0 Process)

Glass fibers as they issue from the molten bath of glass through bushings
are coated with metal through slots in the bath. The claims cover metal
coated fibers of a definite glass composition, and specifically for the
metals zinc and lead.

The glass contains CuO to enhance wetting of glass fibers by metal. The
text mentions other metal oxides for wetting purposes. 7 examples.

2,852,890   9/23/58

Drost and Kebler
Union Carbide Corporation

Synthetic Unicrystalline Bodies and Methods for Making Same

10 Claims (10 Process, 0 Product), 7 Figures

In the classical Verneuil process of making single crystal boules (of sapphire) alumina powder is dropped through an oxy-hydrogen flame onto a stalk of refractory or sapphire, held in a vertical position, and which may be rotated in that position.

In this patent the stalk or rod of sapphire is held horizontally and rotated to pick up the molten droplets of alumina falling down through the burner, onto its periphery to form a disk of ever increasing size. In addition, the horizontally rotating rod may be caused to move back and forth in reciprocal motion to make thick shapes of various contours.

The claims mention a synthetic unicrystalline body, and sapphire in particular.

A disk 2 1/4" in diameter of sapphire is mentioned having been grown on a rod 0.200" diam.

2,854,364   9/30/58

Lely (Netherlands)
North American Phillips Co., Inc.

Sublimation Process for Manufacturing Silicon Carbide Crystals

6 Claims (6 Process, 0 Product), 1 Figure

A graphite vessel is partially filled or lined with SiC, and heated to about 2500°C, whereby the SiC sublimates and forms crystals of SiC in the space, a protective gas being present.

Desirable impurities may be added to the SiC lining, which impurities will be transferred to the sublimated SiC crystals, to form n type crystals with N, P, As, or p type with B, Al, or p-n junction, by controlling the vapor pressure of the impurity gas, which in this case contains the impurity desired. Thus, the conductivity and the conductivity type can be predetermined.

5 examples, with data on composition of impurities, resistivity, Hall calculation for electrons/cm^3, with process details.

2,860,450   11/18/58

Case
(U.S. Government)

Method for Coating Glass Fibers

3 Claims (3 Method, 0 Product), 4 Figures

Two distinct steps are involved in the coating process: (1) Adding an impurity to the glass melt from which fibers are to be drawn - the impurity
accumulating at the surface of the glass fiber, as it is drawn from the
bushing. The impurity can be reduced to metal in a reducing atmosphere
and examples are Cu, Fe, Ni. (2) The issuing nascent fibers immediately
pass through a container, totally enclosed, having a reducing atmosphere,
and in the presence of a coating material which coats the glass fibers from
gas, particles, etc.

No particular materials are mentioned in the claims, and the text of the
patent mentions metals (as above), and organic coatings.

2,866,769   12/30/58

Happoldt
duPont

Process for Making a Mat Composed of Aluminum Silicate Fibers and Poly-
tetrafluoroethylene

2 Claims (2 Process, 0 Product)

Stock for gaskets is made from a mixture of fibrous aluminum silicate (as
"Fiberfrax"), colliodal poly-tetrafluoroethylene in water, aluminum sulfate
as a coagulating agent, etc.

2,870,030   1/20/59

Stradley and Beck
Minnesota Mining and Manufacturing Co.

High-Index Glass Elements

9 Claims (9 Products, 0 Method)

Glass beads and elements of high refractive index (2.2, 2.4, 2.5) are
made from glasses whose composition contains:

Bi_2O_3 ·	70-95% by wt.
TiO_2	5-30
Bi_2O_3 plus TiO_2	90-100
ZnO	0-10
Alk. earth oxides	0-10
Strong glass formers	0-5

29 examples.

2,872,350   2/3/59

Homer and Whitacre
The Commonwealth Engineering Co.

Gaseous Deposition of Tungsten Carbides

2 Claims (1 Process, 1 Product), 5 Figures

Tungsten carbide is applied directly to a metal tool tip by decomposing
tungsten carbonyl at 3 mm. pressure in an evacuated atmosphere, while
the tool is at 650-770°F.

The carbide layer consists principally of WC and W_2 C with less than 10%
W metal.

2,874,135   2/17/59

deMonterey
General Electric Company

Glass Coating Compositions

5 Claims (5 Product [Compositions])

For coating glass fibers, mixtures of several methylpolysiloxane liquid
resins with finely divided silica, linoeic acid, etc.

2,875,474   3/3/59

Lauterbach
duPont

Process of Shaping Resin-Impregnated Filamentary Material

2 Claims (2 Process, 0 Product)

Impregnating synthetic organic filamentary material with organic thermosetting resin.

2,880,552   4/7/59

Whitehurst
Owens-Corning Fiberglas Corp.

Heat Treatment of Metal-Coated Glass Fibers

4 Claims (4 Method), 7 Figures

Glass fibers issuing in parallel array from a group of orifices (from a molten mass of glass) are metallized continuously, cooled somewhat, and
then reheated by means of electric current passed through the metal layer
(such as by means of an induction coil), followed by rapid cooling, to give
greater strength and adherence.

2,895,789   7/21/59

Russell
Owens-Corning Fiberglas Corp.

Method for Producing Coated Glass Fibers

7 Claims (7 Method, 0 Product), 6 Figures

A group of glass fibers issuing hot from a series of orifices set in a line, pass by one or two sheets of "heat softenable coating material", in close contact, so as to pick up this material and form a coating on the line of fibers.

No coating materials are mentioned in the claims. The body of the patent mentions Al, Pb, Sn, and resins such as polyethylene, etc.

2,902,379   9/1/59

McCollum and Gindoff
International Minerals and Chemical Corporation

Fibrous Agglomerate

11 Claims (4 Product, 7 Method)

Expanded perlite (volcanic glass) and asbestos fibers are used in the making of concrete and plaster. 4 examples. Methods, properties.

2,907,626   10/6/59

Eisen, Nachtman, Bjorksten and Roe
Bjorksten Research Laboratories

Metal Coating of Glass Fibers at High Speeds

6 Claims (6 Process, 0 Product), 3 Figures

The claims refer to the process of metallizing "ceramic fiber" as the fiber issues from a mass of molten ceramic.

The hot fiber passes through a flame formed by a gaseous carbonyl of Fe or Ni, and air.

2,908,545   10/13/59

Teja
Montecantini (Italy)

Spinning Nonfused Glass Fibers from an Aqueous Dispersion

4 Claims (3 Method, 1 Product), 2 Figures

A "non-fused glass fiber" is produced by the following sequence of operations - mixing (1) glass forming oxides (5-140μ) or a glass frit in water; (2) an organic mixture and reaction products of siliconate, a hydrophilic material - like polyvinyl alcohol, a hydrophobic material - like rubber latex. This mixture acts as a stabilizing agent; (3) colloidal silica in solution.

After several treatments, the mixture is turned into a syrup from which a filament is drawn vertically upwards from the end of a pointed body.

Claim 4 refers to a "glass fiber" with Young's Modulus of 21 $\times$ 10^6 psi and a softening point of about 1950°F.

3 examples.

2,915,475   12/1/59

Bugosh
duPont

Fibrous Alumina Monohydrate and its Production

16 Claims (13 Product, 3 Process), 14 Figures

The process consists of two distinct phases: (1) making a water solution of alumina with the proper pH and acid radical present, and degree of polymerization and (2) subjecting this solution, in the proper dilution, to temperature and time to precipitate or grow fibrous alumina monohydrate, which has the boehmite crystal lattice.

The solution of alumina may be obtained by using a number of salts and materials, such as Al chloride, Al nitrate, Al metal with acids, gelatinous Al hydroxide, sodium aluminate, alum, Al sulfate, and certain combinations of these, to obtain an aqueous amorphous solution of alumina, using chemical techniques to remove the undesirable ions.

The (opalescent) aqueous solution of alumina is then subjected to a temperature of 160-170°C (range 80-375, 120-250, 150-220°C mentioned in claims) for a number of hours, and fibrous monohydrate alumina is formed.

The crystal structure of the fibers is determined by X-ray diffraction to be that of boehmite, and this structure is claimed in all cases.

Fibrous alumina hydrate fibrils have dimensions:

eg. 1/d 50:1 to 150:1 1-100-700 mμ, d- 3-10 mμ. Surface area 250-350 m^2/gm.

eg. 1/d 20:1 to 300:1 1-100-1500 mμ, d- 3-15 mμ. Surface area 200-400 m^2/gm.

eg. 1/d 20:1   1-200-2000 mμ.

Claims 1-5 cover the product fibrous alumina and aquasols of them.

Claims 6-13 cover the product fibrous alumina hydrate and aqueous and organic dispersions and sols.

Claims 14-16 cover processes of making fibrous alumina monohydrate having the boehmite crystal lattice.

In these fibers the boehmite crystal structure exists at 200°C, and changes to kappa alumina in the range 400-900°C.

34 examples of detailed processing, properties and some uses.

This is an exceptionally long, detailed patent, running to 56 columns and 14 figures.

2,920,992   1/12/60

Hubbard
duPont

Article of Commerce

9 Claims (4 Product, 5 Process), 2 Figures

A synthetic paper sheet made of polymers of acrylonitrile is described.

2,932,587   4/12/60

Labino
L.O.F. Glass Fibers Co.

Colored Glass Fibers

4 Claims (3 Method, 1 Product), 1 Figure

Water soluble resin, stabilizers, and Prussian Blue are sprayed on glass fiber, and cured to give colored glass fiber.

2,938,821   5/31/60

Nack
Union Carbide Corporation

Manufacture of Flexible Metal-Coated Glass Filaments

1 Claim (1 Method, 0 Product), 4 Figures

Glass fibers issuing from furnace bushings pass through a chamber in which a decomposable metal bearing compound in gaseous form deposits metal on the fibers, in a continuous process. The metal coated fibers pass through rolls to form a solidified mass, having cold welded the metal coatings together.

The product may be a cable. The gaseous metal carrying vehicle may be carbonyl of Ni, Mo, Cr, Fe; the chloride of Mo, the iodide of Zr, the bromide of Ti.

2,939,761   6/7/60

Stein
A.O. Smith Corporation

Method of Producing Glass Fibers

11 Claims (11 Method, 0 Product), 1 Figure

Glass fibers issuing from the furnace bushings are immediately and continuously coated with a thin layer of a fluoride, to give a better bond with a resin to form reinforced articles.

The fluoride may be BF, AlF_3, HF, ethyl fluoride, preferably in the gaseous state, and free of water and moisture.

It is stated that the intermediate bond so effected reduces the change in modulus of elasticity, existing between fiber and resin.

2,940,875   6/14/60

Smith and Welch
Union Carbide Corporation

Silicone-Coated Fibrous Products and Process Therefor

7 Claims (3 Product, 4 Process)

A product such as tape is made with a fibrous material, both organic and inorganic, with a silicone elastomer, such as dimethyl siloxane, the product being non-adhesive or "abhesive" to another layer of material in intimate contact.

The fibrous material may be glass cloth, asbestos cloth, cotton, rayon, nylon, etc. 5 examples.

Scotch Tape is used as an example, referring to the backing material, from which an adhesive layer is stripped.

2,941,904   6/21/60

Stalego
Owens-Corning Fiberglas Corp.

High Temperature Glass Fiber Insulation Product and Method for Manufacturing Same

12 Claims (8 Product, 4 Method)

A bonded glass fiber insulation product is made with glass fiber, organic binders of alkyd resin and urea-borate condensation reaction product, silicic acid filler, etc. 20 examples, and indications of applications such as molded insulation, pipe wrap, etc.

2,946,694   7/26/60

Labino
L.O.F. Glass Fibers Co.

Glass Composition

5 Claims (5 Product [compositions], 0 Process)

Glass compositions in the range:

Li_2O	0.0 - 4.5%
Al_2O_3	16.0 - 28.0%
Na_2O	31.0 - 42.0%
B_2O_3	10.5 - 14.0%
P_2O_5	18.0 - 25.0%
F_2	8.0 - 10.5%

Batch compositions and glass compositions of definite proportions and
materials also given. Body of patent mentions the use of these glasses as
low melting glasses and for fibers. One example gives melting behavior,
temperature range of fluidity, devitrification, rigidity.

2,951,772   9/6/60

Marzocchi and Janetos
Owens-Corning Fiberglas Corp.

Treatment for Fibrous Glass used to Reinforce Resins

11 Claims (4 Resin Claims, 5 Process, 2 Product), 2 Figures

Improved bonding between resin and glass fibers is obtained by using aqueous
suspensions of mixed resins, as polyester, styrene, etc., wetting agents, etc.

Coating constitutes 3-10% of total weight of product.

2,951,780   9/6/60

Bushman
General American Transportation Corp.

Methods of Making Decorative Glass Fiber Reinforced Resin Bodies

7 Claims (7 Method, 0 Product), 3 Figures

Methods of preparing decorative glass fiber-resin products for use as
luggage, counter tops, etc. are given.

2,953,478   9/20/60

Harvey and Rosamilia
Harvel Research Corp.

Glass Fibers Coated with Acetone-Forming Aldehyde Reaction Products

7 Claims (3 Process, 4 Product)

Glass fibers coated with an organic composition comprised of a mixture
of water soluble thermo-setting resins, urea, melamine, starch, dextrin;
and cured.

2,956,039   10/11/60

Novak and McCallum
Union Carbide Corp.

Method of Making Quick Curing Metal Containing Epoxy Resin Composition

2 Claims (2 Method, 0 Product), 6 Figures

A heat-curable epoxy resin composition is made by including a metal carbonyl of Ni, Fe, Cu, Cr, Mo in an inert atmosphere to deposit metal in the
resin in order to reduce curing time.

Glass fibers, rock wool, with metal deposited on them may be included.
7 examples.

2,957,756   10/25/60

Bacon
Union Carbide Corporation

Filamentary Graphite and Method for Producing the Same

2 Claims (1 Product, 1 Process), 2 Figures

Graphite filaments are grown in a pressure vessel by striking an arc between a graphite rod and a carbon block. The pressures are critical, and
must be maintained below the triple point of graphite - 1470 psi, and
preferably in the range 1150-1400 psi.

An inert gas is used in the chamber (such as Ar, He, Ne, Xe, Kr). As
the rod is moved away from the block under voltage and gas pressure, a
boule containing the graphite fibers is grown on the block.

The fibers are substantially monocrystalline, giving sharp diffraction
patterns by X-rays and electrons, $0.5-5\mu$ in diam. and 0.5-3 cm. long.
The c-axis of the crystals is exactly perpendicular to the axis of the
filament.

Tensile strength, 3×10^6 psi; strain, 0.4%; electrical res., 6×10^{-5}
ohm cm.

Example: 1/2" diam. graphite rod, Ar, 1360 psi, 80 volts, 60 amps. per
sq. in. area, 30 minute run; result, a boule 6" long containing filaments
as noted above.

2,958,614

11/1/60

Perry

General Aniline and Film Corp.

Resin Coatings for Glass Materials

6 Claims (6 Process, 0 Product)

The process of sizing a glass surface (fibers, strands, sheet, bottles, etc.) giving "excellent adhesion" by mixing two copolymers, such as vinyl halide and N-vinyl pyrrolidone.

2,968,622

1/17/61

Whitehurst

Owens-Corning Fiberglas Corp.

Magnetic Ceramic Fibers and Method of Making Same

14 Claims (8 Product, 6 Process), 5 Figures

A continuous flexible fiber of ferromagnetic ceramic materials is made by melting a batch of the material and drawing the molten mass through an orifice into a cooler region where it pulls out into a rod or fiber, where at the same time it passes through a magnetic field to orient the crystallites as they form.

Where crystallites are not formed in the first step, the uncrystallized fiber is reheated and passed through a hot zone and a magnetic field, to cause devitrification and orientation of the crystallites.

The composition of the fibers is of the general formula $MeO-F_2O_3$, where Me is divalent, as Mn, Fe, Co, Ni, Cu, Zn, Cd, and Mg, being of the class of ferrites and of the more general class of spinels. The fibers are to be less than 0.1" in diameter.

2,970,127

1/31/61

Slayter and Shannon

Owens-Corning Fiberglas Corp.

Glass Reinforced Gypsum Composition and Process of Preparation

6 Claims (5 Process, 1 Product), 9 Figures

"A plaster mass comprising gypsum, a synthetic resin compatible with the gypsum, and from about 50 to 70% by weight of flakes of glass---".

Processes are given to make plaster board, etc. Resins, coupling agents, buffers, required in the processes are discussed.

Glass flakes are about 8μ thick, and their method of making is given in USP 2,457,785, Slayter et al. December 28, 1948

2,975,503   3/21/61

Bacon and Nemeyer
Owens-Corning Fiberglas Corp.

Glass Fabric Reinforcement for Plastics

7 Claims (7 Product, 0 Process), 8 Figures

Glass fibers are made into a mat by being strung in parallel strands with
loops at the edges of the mat, and pressed into the final articles with moldable plastics. Numerous variations of this construction are given.

2,978,341   4/4/61

Bastian and Ottoson
Imperial Glass Corporation

Glass Composition

11 Claims (8 Product [Fibers] , 3 Compositions)

Unleached glass fibers, with diam. less than 0.001" and a Young's modulus
of at least 15×10^6 psi are made from given compositions and ranges of
composition. It is stated that BeO, CaO, MgO and ZrO_2 increase the
modulus, and Na_2O, K_2O and B_2O_3 lower the modulus of the fibrous glass.
Li_2O is advantageous. The melting range of the glasses - 2500-2600°F.

2,980,510   4/18/61

Berry
duPont

Preparation of Fibrous Titanium Dioxide

5 Claims (5 Process, 0 Product)

$TiCl_4$ gas (carried by N_2) passes over molten B_2O_3, to form rutile TiO_2
crystals on the surface, at a temperature of at least 580°C. The crystals
are separated by dissolving the bath salts in hot water.

$TiCl_4$ may be replaced by titanium bromide, iodide, or fluoride, and
mixtures. B_2O_3 may be supplemented with an alkali metal borate, such as
Na, K, borates and mixtures. The temperature range may be 580-1000,
600-900, 650-900, 650-850°C. 4 examples are given.

Example: $TiCl_4$ with N_2 gas, oven molten B_2O_3 at 600°C in Pt. rutile
fibers were $1-2\mu$ in diameter and 100μ long.

Example: $TiCl_4$ with N_2 gas, oven molten B_2O_3 and sodium tetraborate at
800°C. Rutile fibers -- less than 5μ in diameter and 0.5-0.7 mm. long.

In some cases a minor amount of anatase (TiO_2) is found with the rutile
fibers.

Claims cover compositions of materials in a general manner only.

2,980,982   4/25/61

Costa, Le Boeuf and Lefevre
Dow Chemical Co.

Fibrous Article

5 Claims (5 Product, 0 Process), 2 Figures

Fiber and film are made from an aqueous latex of synthetic (organic) polymers (vinylidene chloride, acrylonitrile, etc.) to provide useful and more economical fibers for textile use. The film is 0.001-0.01" thick.

2,996, 411   8/15/61

Lauterbach
duPont

Resin-Impregnated Filamentary Material

4 Claims (0 Process, 4 Product)

Articles (trays, truck bodies, etc.) made of batts of synthetic organic filaments, bonded with thermo-setting resins, for deep drawing operations.

Ratio of breaking strength/elongation for several filamentary materials included - needled glass mat, 3.6; bonded glass mat, 8.5.

3,007,806   11/7/61

Hartwig
The Babcock and Wilcox Co.

High Silica Refractory Wools

5 Claims (3 Product, 2 Method)

Continuous monofilament glassy fibers are made by melting SiO_2 70-95%, with a modifier such as MgO, CaO, Al_2O_3, TiO_2, Cr_2O_3 and ZrO_2 (5-30%), at a temperature to have a viscosity of 100-1000 poises.

It is claimed that these fibers are useful above $1500°F$, and that their compositions in the crystalline form are above $2800°F$. in melting point.

Examples of fibers with composition SiO_2 90% and Al_2O_3 10%, the batch having been melted, drawn into fibers (as a long filament) were 6μ in one case and 20μ in another.

Young's modulus of $10-30\mu$ fibers average 8×10^6 psi. Fibers can be woven into cloth, and the coarser fibers used for mineral wool.

3,007,840   11/7/61

Wilcox
duPont

Process of Dispersing Fibrous Material in a Foam and Resulting Product

15 Claims (15 Process, 0 Product), 6 Figures

Fibers (organic like nylon) which are hydrophobic, are made into a foam
condition, without liquid remaining, and spread on a screen and formed
by vacuum (into paper) into a non-woven web. Fibers are at least 1/4"
long. 12 examples.

3,008,913   11/14/61

Pangonis
duPont

Reaction Product of Fibrous Alumina Monohydrate, Tetraethyl Ortho-
silicate, and Diphenylsilanediol, Process for Making, and Solvent
Solution Thereof

9 Claims (5 Process, 4 Product)

Film is made from a mixture of fibrous (boehmite) alumina hydrate 10-
90%; tetraethylorthosilicate 5-45; diphenylcilanediol 5-45; at 60-80°C.

The fibers have diameters less than 150 mμ. 2 examples.

3,011,870   12/5/61

Webb, Wissler and Forgeng
Union Carbide Corp.

Process for Preparing Virtually Perfect Alumina Crystals

10 Claims (6 Method, 4 Product)

Alpha-aluminum oxide (α-Al_2O_3), (corundum), is formed on a substrate of
alumina, by allowing vaporous aluminum oxide to come in contact with the
substrate and deposit thereon single crystals of alpha-alumina, in an
atmosphere containing hydrogen and water vapor, at temperatures 1300-
2050°C, preferably 1300-1450°.

The vaporous aluminum oxide is obtained by reacting Al, or aluminides of
Zr, Hf, V, Nb, Ta, Cr, Mo, W, and particularly of Ti, at the above
temperatures and in the atmospheres given there. The body of the patent
states that the gaseous AlO so produced disproportionates to solid Al_2O_3
and liquid Al, the solid Al_2O_3 assuming the crystal structure of alpha-
alumina.

The crystal products obtained are both needles and platelets, some of each
having tiny holes, to form a tube in one case and a cavity in the other.

The needle-like crystals with hexagonal cross-section of width $3\text{-}50\mu$ and length 1-30 mm, and cylindrical centerline cavity $0.5\text{-}30\mu$ in diameter.

The platelets-thickness $0.5\text{-}10\mu$, and up to 10 mm long., and cavity $0.5\text{-}3\mu$ diam.

Strength of fibers found by measuring distortion and using Young's modulus and Hook's law, are in two cases – 738,000 and 1,700,000 psi.

Elastic strain on the average – 1% for needles and platelets, and the highest obtained 2.3%.

3,012,289   12/12/61

Weber
Carborundum Company

Method and Apparatus for Blending Ceramic Fibers with Carrier Fibers

8 Claims (8 Process, 0 Product), 5 Figures

Detailed description of apparatus which removes shot from ceramic fibers, and mixes these fibers with carrier fibers (5-25% of total) by means of carding equipment, to make yarn and fabrics.

Body of patent shows:

(1) ceramic fibers as aluminum silicates, sodium-calcium silicates, calcium aluminates, sodium silicates, and glass.

(2) carrier fibers as acrylic fibers, rayon, cotton, wool, asbestos, glass, fluorethylene fibers, polyamide fibers, and polyester fibers.

(3) low carrier fiber content permits fire curtains of aluminum silicate yarn to withstand temperatures about 2000°F compared to about 1000°F for conventional asbestos cloth.

3,012,856   12/12/61

Berry
duPont

Flexible Feltable Fibers Fibers of Titanium Nitride and Their Preparation

7 Claims (5 Product, 2 Process)

Flexible feltable fibers of TiN are produced by passing a mixture of gases, namely $TiCl_4$ and N_2, at specified rates, over surfaces of a siliceous material, or an aluminum silicate, at temperatures of 1225, 1300-1500°C, and in the presence of a reducing agent, such as carbon.

The gas mixtures may contain Ar, in addition to N_2, and NH_3 which may be used as a source of N_2. The aluminum silicate used in many examples is mullite (boat or furnace tube) or a mullite bonded aluminum tube.

The reducing agent may be carbon (in many cases), Al, Mg, etc. , lower valent oxides -- TiO, SiO, etc. , metal nitrides, carbides --. The statement is made that SiO from siliceous sources is an effective and important intermediate for the production of TiN fibers. 12 examples of procedures and products are given.

TiN fibers are golden in color, with diameters 0.05-5μ, up to 10μ , and 5 mm to 1 cm in length. Tensile strength of one fiber, hexagonal in cross-section (11.6μ across corners) was 1 million psi.

X-ray diffraction methods are used to determine pattern of structure, and to distinguish TiN from closely related substances. Several claims state that the fibers are flexible to the extent that they will not break when bent around a mandrel of 1 mm. diameter.

3,012,857   12/12/61

Pease
duPont

Preparation of Fibrous Titanium Dioxide

12 Claims (12 Process, 0 Product)

Titanium subhalide ($TiCl_2$ to $TiCl_3$, and mixtures of them) are heated in a salt bath, with air or oxygen, to form TiO_2 fibers. The salt bath may contain single metal halides or mixtures of these halides, as the chlorides, bromides, iodides of Na, K, Li, Ca, Ba, Mg, and the like -- including fluorides of the alkali metals. Temp. of melting 550-1000°C (600°-800°C preferred).

The fibers are found to be rutile in type, with diam. 1-5μ, and 0.205 mm in length.

11 examples of compositions of melts and procedures.

3,013,915   12/19/61

Morgan
Owens-Corning Fiberglas Corp.

Reinforced Polyolefins and Process for Making Same

25 Claims (15 Product, 10 Method)

Glass fiber of glass flake are bonded together with organic resins, particularly the polyolefins, by means of two substances in either one or two steps. The glass surfaces are cleaned, then treated with a coupling agent like vinyl silane and then with a source of "dehydrogenating radicals", or at one time. The radical source may be an organic peroxide as dicumyl peroxide.

There are 5 examples, with directions for preparation, and strength of bond measured on multiple tapes made from polyethylene strips attached to glass fiber woven strips.

Organic formulae and detailed discussion are presented.

3,016,599   1/16/62

Perry, Jr.
duPont

Microfiber and Staple Fiber Batt

1 Claim (1 Product, 0 Method), 1 Figure

Two different types of organic fiber are interwoven, and later made into a
batt, all in a continuous process.

One type of fiber is a continuous filament, man-made, known as a staple
filament, such as cellulose, its derivatives, etc. Its thickness is expressed in denier, at least 1 denier. (1 denier equals 0.05 gm. and the
length of fiber which weighs 1 denier must be 450 meters.) The other
type of fiber is called microfiber, also synthetic organic, made by blowing
a liquid polymer into a stream of short fibers, less than 1μ.

In actual practice the staple fiber is cut into lengths continuously and forced
into the stream of microfibers, onto a drum to form a continuous sheet.

The mixture contains 25-70% microfibers.

3,019,117   1/30/62

Labino
Johns-Manville Fiber Glass Inc.

Refractory Blocks

5 Claims (5 Method, 0 Product)

Refractory Blocks are made by compressing and heating glass fiber saturated with nitrates of Co, Ni, Fe, Cu, Cr.

Example: Glass fiber is steeped in a saturated aqueous solution of the
nitrate, dried and heated to 1600°F; then compressed into a block and
fired in the range 2900-3500°F (preferably 3300°-3500°).

The blocks are claimed to be rigid, solid refractory bodies. The body of
the patent says they are light-weight, gas impervious.

3,019,122   1/30/62

Eilerman
Pittsburgh Plate Glass Co.

Glass Fiber Surface Treatment

4 Claims (4 Process, 0 Product)

Treating glass fibers with water soluble silanes, such as vinyl triacetoxy
silane, and affixing the silicon compound by heating at 350°F.

Flexural strength of treated glass fiber cloth, bonded with styrene polyester resin:

Dry	24,451 psi
Wet	18,054 psi

3,023,115   2/27/62

Wainer and Cunningham
Horizons Inc.

Refractory Material

6 Claims (5 Method, 1 Product), 2 Figures

Fibers of Al_2O_3 (95-99%) SiO_2 (1-5%) are prepared by heating

(1) Al plus SiO_2 (or silicates of Al, Zr, Al-Ti) in a hydrogen atmosphere at 1200-1500°C (1380°)

(2) a mixture of gaseous AlO or Al_2O with gaseous SiO (prepared separately) at 1200-1500°C, with or without the presence of molten Al, in a hydrogen atm.

The Al may contain Ti, Zr, etc.

The claims stipulate that the proportion of Al/Si or Al compound/Si compound in the reaction mixture shall be at least 2 moles/3 moles. 5 examples.

Fibers obtained are $0.5-1.5\mu$ in diameter and 1/2-3" long, and have tensile strengths of 2×10^6 psi.

It is stated that the presence of SiO_2 causes an increase in rate of growth of fibers, and it is given in one place as 300-400 mg. of fiber per sq. inch of Al starting surface.

3,023,490   3/6/62

Dawson
Dawson Armoring Co.

Armored Metal Articles with a Thin Hard Film Made in situ and Conforming to the Exact Contour of the Underlying Surface

14 Claims (14 Product, 0 Process), 8 Figures

The working surfaces of armored steel saw teeth, ferrous tools, etc. are given hard thin protective overlay surfaces, by applying first a thin layer of adhesive, (e.g. shellac-alcohol) and immediately sprinkling onto it some powdered metal, alloy, carbide, etc. The combination layer is then heated to form a permanent bond. Fluxes, brazing alloys, etc., are also included.

3,024,088   3/6/62

Palmqvist (Sweden) and Brudin (Egypt)
Tholand Inc. & Brudin

Making Micro-Porous, Discrete, Few Micron Size Oxide Compounds of
Metals, and the Products

12 Claims (3 Product, 9 Process)

Finely divided micro-porous powder-like products, such as Al_2O_3, $Al(OH)_3$
(in particular), and carbonates, hydroxides and oxides of Al, Fe, Mg are
formed solutions of salts and compounds in water, in which reaction CO_2
is formed as well as a precipitate of the sought after material. Hence a
carbonate must be present and an acid forming chemical to liberate the CO_2.

Inorganic metal salts, such as chlorides, nitrates, sulfites, sulfates --
also organic acids as formate, acetate, citrate of Mg.

Example: Sodium carbonate, aluminum sulfate when dissolved in water and
mixed give off CO_2 and precipitate $Al(OH)_3$. On drying at $50^{\circ}C$, this hy-
droxyde is powdery, $1-3\mu$, and with a surface area of 1-2 sq. m/gm.
5 examples.

Claims cover several chemical combinations for preparation of products
of Al, Fe, Mg.

3,024,089   3/6/62

Spencer, Smith and Cosman
Cabot Corp.

Process for Producing High Surface Area Metal Oxides

9 Claims (9 Process, 0 Product)

Finely divided carbon black is used as the base material on which to deposit
finely divided reaction products produced by hydrolyzing certain metal
compounds. These compounds must be readily hydrolyzed, and they may
be metal chlorides as of Si, Ti, Al, Fe -- also ethylorthosilicate, Al
ethoxide and chlorsilane.

Example: Carbon black of surface area 131 sq. m/gm treated with water
vapor will cause dimethyldichlorosilane vapor carried in nitrogen gas to
form a siloxane coating on the carbon, of 5.8% by weight of carbon black.
The dried coating is heated to $500^{\circ}C$ in air to remove all the carbon black
by oxidation. The resulting powder is SiO_2 in aggregates of about 1 mm.
in diam. and of ultimate size 25-50A, with a surface area of 1094 sq.
m/gm.

Variations of process are given, but ultimately the metal salt must come
in contact with water vapor on the surface of the carbon black. 5 examples
are given.

The claims do not mention particle size nor surface area, although these
properties are stressed in the body of the patent, namely about 1000 sq.
m/gm. and 25-50A in size.

3,024,145   3/6/62

Nickerson
Monsanto Chemical Co.

Process of Bonding Glass Articles

11 Claims (9 Process, 2 Product)

7 claims - process for binding glass fibers in a bat.
1 claim   - product: glass fiber bat.
2 claims - process for bonding at least two separate glass articles.
1 claim   - product: a composite glass article.

The method of coating glass fibers in a bat and glass articles (as plates) with a film of a silica aquasol dissolved in a water soluble non-ionic organic polymer.

Silica aquasols are described - pH 8-10.5, SiO_2/Na_2O ratios noted, average particle size of colloidal silica, content of SiO_2.

Water-soluble, non-ionic, film-forming organic polymers - such as polyvinyl alcohol, methyl cellulose, starch, dextrin.

Organic binder may be removed by heating, leaving a colloidal silica bond (e.g. 350° - 650^OC. for composite plates).

3,025,588   3/20/62

Eilerman
Pittsburgh Plate Glass Co.

Glass Fiber Treatment

5 Claims (5 Methods, 0 Product), 1 Figure

The application of a sizing to glass filaments as they issue from a hot bushing, to form strands, which can be twisted, plied and woven to form a fabric, the sizing to be compatible with a resin for impregnation and curing

The size consists of an aqueous solution of a synthetic latex, zinc stearate, textile lubricant, wetting agent, as well as a coupling agent (in some cases).

3,026,200   3/20/62

Gregory
134 Woodworth Corporation

Method of Introducing Hard Phases into Metallic Matrices

6 Claims (6 Methods, 0 Product), 1 Figure

A mixture of metal powders is made - consisting of a matrix metal and a refractory oxide-forming metal. The compact is heated first in air to oxidize some of the surface matrix metal, which in later heatings will furnish oxygen which in turn will diffuse inwards to oxidize the refractory metal present.

Two examples give some of the numerous heating treatments required.

Examples shows Cu powder mixed with 0.5% Al powder and the treatment to get a dispersion of Al_2O_3 in the copper. A chart gives the creep rates vs. stress for Cu, unoxidized Cu-0.5% Al, and oxidized Cu - 0.5% Al.

The claims mention no specific matrix metals, but do mention the refractory oxide-forming metals - Si, Al, Mg, Be, Zr, Ti, Th, rare earths -- in the range 0.05-2.5% by weight.

Body of patent mentions matrix metals of ductile types, such as copper group, iron group, gold, platinum, etc.

3,030,183

4/17/62

Berry
duPont

Titanium Dioxide Fibers and Their Preparation

5 Claims (3 Product, 2 Process)

Rutile fibers (by x-ray diffraction patterns) are produced by passing a mixture of air and $TiCl_4$ over a melt of a metal halide.

Ti-Cl, -Br-, $-I_4$ may be used.

The metal halide may be alkali- Na, K, etc. or alkaline earth- as Ca, Ba with Cl, Br, I, --

Example gives a KCl melt at 800^oC, with air laden with $TiCl_4$ to form vertically grown fibers on the surface of the melt. On cooling, the fibers are washed with water to remove salts.

Claims are made for fibers with cross-section less than 25μ, and axial ratio of at least 10:1; also fibers less than 5μ, and at least 10:1.

Examples give fibers 5μ in diameter and 2-3 mm long.
Claims also mention feltable and felted fibers.
7 examples give directions for preparation of fibers in as many ways.

Anatase form of TiO_2 is found as a minor constituent in some cases.

3,031,322

Bugosh
duPont

Fibrous Boehmite Containing Compositions

5 Claims (5 Compositions, 0 Product)

The claims are for compositions using fibrous boehmite (described in USP 2,915,475 Bugosh, Dec. 1, 1959) with other ingredients to make paper, mats, insulating bats, flocculating agents for water suspensions, etc.

The other ingredients may be (1) anisodiametric inorganic materials - like asbestos or glass fibers, (2) materials containing Si, Al, Ti oxide minerals, ratios of max. to min. dimensions 5:1 to 100:1, these materials to have a length $2-3 \times 10^5$ that of the boehmite (50-1500 mμ), (3) a liquid carrier as water or organic liquids, and (4) flocculating agents, including colloidal silica, and organic binders for products, like resins, and fillers.

Example 9: Fibrous boehmite, 60 parts; asbestos fiber, 40 parts; borosilicate glass fibers, 500 parts; with calgon flocculating agent - made into inorganic paper.

3,031,417

Bruce
duPont

Preparation of Fibrous Alumina Monohydrate and Aquasols Thereof

6 Claims (6 Process, 0 Product)

Fibrils of alumina monohydrate (AlOOH) are prepared by putting into solution alum ($Al_2(SO_4)_3 \cdot 18H_2O$) with a soluble carbonate (such as sodium) in water to form a gel precipitate, which is washed. The filter cake is dispersed in water, mixed with acetic or formic acid, and heated in an autoclave to 140-180°C under its own developed pressure to form a translucent fluid solution containing about 6% Al_2O_3, the fibers present being about 100-1500 mμ in length. Closer ranges of length being 100-700 mμ (example - average 236 mμ) and surface area 200-400m^2/gm.

Stress in the body of the patent and in the claims is upon the use of a weak acid like acetic or formic.

13 examples, giving details of proportions of chemicals and procedures.

X-ray diffraction pattern of fibrils matches that of the mineral boehmite.

3,031,418

4/24/62

Bugosh
duPont

Chemically Modified Alumina Monohydrate, Dispersions Thereof and Processes for Their Preparation

6 Claims (5 Product, 1 Process), 6 Figures

The methods described are given in USP 2,915,475-Bugosh-Dec. 1, 1959, in detail, and the present patent refers to some modifications.

Colloidal chem. modified alumina hydrate is modified to contain appreciable amounts of anions, either inorganic or organic, without changing the "lines" found in certain x-ray diffraction patterns, these lines being given in detail in the claims. (The end product resembles boehmite in most details).

Al salts are put into acid solution, and the anions are added - being preferable sulfates, or sulfites, phosphates, vanadates, etc., or the organic oxalate, succinate, etc. Concentrations, pH, are adjusted, and the mixture subjected to an autoclave treatment. The gelatinous product is dried and may be purified. The dried powder product has large surface area (50-450 sq. meters per gm.), definite x-ray diffraction lines, and chemically an alumina monohydrate with water of adsorption (up to 20% which may be removed by further drying), and capable of being pressed into shapes.

There may also be present certain cations, like Li, Na, K, Ca, Mg, Cu, etc.

12 examples are given of preparations and methods, including the Al salts, the crystal modifiers, and methods of making.

The polyvalent anions (like sulfate) may be present in amount 3-30 equivalent % based on Al content.

3,039,849

6/19/62

Willcox
duPont

Aluminum Oxide Production

5 Claims (5 Process, 0 Product), 1 Figure

Microfibrous alumina is produced by passing a gas mixture of moist air, or moist oxygen containing gas, over an amalgam of Al-Hg, at 100-200°C. The amalgam in a sheet form permits the (hydrated) microfibers to grow vertically on the sheet to thicknesses of 2-10 mm, and the crop is harvested by scraping off. Al in the amalgam is 0.003-0.2%.

10 examples of variations in process with details of apparatus and methods, with compositions of gas reactants, flow rates, density, thermal conditions, etc.

When the fibrils are heated at 750-950°C, they lose water and Hg content. They are amorphous to diffraction by x-rays of electrons, have large surface area - 200 sq. m/gm; diameter $0.05-0.15\mu$, and $30-40\mu$ and 1-15 mm in length.

Some examples show use of microfibers in mixtures for compressing into plates, etc.

3,039,981   6/19/62

Shannon, Morgan and Sullivan
Owens-Corning Fiberglas Corp.

Aqueous Binder Composition of Phenol-Formaldehyde Condensate, Aluminum Sulfate and Mineral Oil, and Glass Fiber Coated Therewith

4 Claims (3 Product [Binder], 1 Product [glass fiber coated], 0 Process)

Aqueous dispersions of phenol-formaldehyde resins, with aluminum sulfate 0.25%-1%, with emulsified hydrocarbon mineral oil, applied as a binder to glass fiber.

3,041,131   6/26/62

Juras, Mazzucchelli, and Bugel
Union Carbide Corporation

Composite Plastic-Metal Fiber Articles and Method for Making Same

10 Claims (6 Product [Articles], 4 Method), 7 Figures

Surface layers of resin and metal fibers are built on porous and dense (organic) surfaces to have sharp contours, for molds and the like.

Metal fibers may be iron (steel wool), 0.0005-0.025" in diameter and 1/32-1/2" long. Resins may be epoxy or polyester, etc. Foams for porous strectures may be organic or inorganic.

3,041,663   7/3/62

Green
Pittsburgh Plate Glass Company

Method and Apparatus for Forming Fibers

5 Claims (3 Apparatus, 2 Method, 0 Product), 2 Figures

In the usual way of making glass fiber from a molten mass of glass, through a bushing, the fibers are carried to a rapidly rotating forming tube some distance away via various equipment located on a second floor.

Here the aim is to locate the forming tube close to the bushing. The claims refer to such improved mechanical equipment.

3,043,796   7/10/62

Novak and McCallum
The Commonwealth Engineering Company

Metallized Fibers for Decreasing Curing Time of Plastic Glass Fiber
Mixtures

3 Claims (3 Method, 0 Product), 9 Figures

The body of the patent describes various ways of gas–plating metal like
Ni, Fe, Al, Cu,from carbonyls and other sources, onto glass fibers,
thickness 1–10 u, and imbedding the fibers into a resin, thermoplastic
or thermosetting, to form a fiber reinforced laminate in which the
resin–curing time is decreased.

Another method describes the mixing of finely divided metals, such as Al
obtained from Al isobutyl by gas–plating technique onto a resin surface,
and mixing intimately the Al particles with the resin composition
(with 5-50% by weight of Al).

The three claims refer only to the latter method of Al particles in the
resin, heat curable epoxy resin, from Al isobutyl or other sources, to
effect a shortening of the heat–curing time.   There are 7 examples.

3,045,317   7/24/62

Shipman
J. P. Stevens and Company

Process for Producing Sized Glass Yarns and Cleaning Fabrics Woven
Therefrom

9 Claims (9 Process, 0 Product)

The application of a sizing made of highly hydrolyzed polyvinyl alcohol to
glass yarns, at a temperature of 105^{0}-115^{0} F in the bath, drying the yarn,
and then removing the thin film by "coronizing" – namely by passing the
yarn through an oven, maintained at 1000^{0}-1600^{0} F, to yield a product
substantially free of organic matter.

3,046,084   7/24/62

Veazie
Owens–Corning Fiberglas Corp.

Glass Surface Treatments

8 Claims (8 Method, 0 Product)

Glass marbles or other structures, used in making glass fibers, are first
treated with nitric acid to remove surface non–siliceous constituents.
Such treatment increases the yield of usable glass fiber, by decreasing
the down-time during a run.   It is postulated that the acid treatment de-

creases the occurrence of stones and seeds, which are formed from
devitrification products occurring at the surface of the unclean marbles.
Another theory postulates that the acid draws cations like Ca, Mg, Al, to
the marble surface, where they serve as fluxing media.

3,046,170   7/24/62

Toulmin, Jr.
Union Carbide Corp.

Laminates of Metal Plated Glass Fibers and Methods of Making Same

23 Claims (10 Method and Process, 13 Product), 4 Figures

To prevent fibers of glass from picking up moisture and thus making a
poor bond with a resin matrix, the invention calls for the immediate
coating of the fibers with a gaseous metal source, as they issue from a
molten.source. The claims call for the steps of (a) drawing glass fibers
from a molten bath, (b) plating with metal from a source (such as Fe, Ni,
Cr carbonyl), (c) weaving the glass fibers, (d) impregnating the woven
matt or shape with a polyester polymer resin to form a continuous resin
phase, (e) and heating the entire mass to form a structure and set the
resin.

The drawn fibers may be glass, siliceous glass , and silica glass.

Example: E glass fibers, Fe coating from iron pentacarbonyl (1 mil thick)
and polyester resin, etc.

3,047,383   7/31/62

Slayter
Owens-Corning Fiberglas Corp.

Polyphase Materials

1 Claim (1 Method, 0 Product)

The end product is a material in which glass fibers are distributed in a
discontinuous manner in a continuous phase of metal, the glass fibers
occupying 5-90% of the volume of the metal article.

The body of the patent describes the use of glass fibers with metal powders,
compressed under heat as in powder metal fashion, and gives compositions
of glasses and metals which may be used with them, softening points of
glasses and melting points of metals. The body also describes the mixing
of glass fibers with molten metals, and how to coat light glasses with a
metal layer to get them to sink.

The single claim describes only one method, which is also given in the
body, namely to mix glass powder with metal powder and press into a
shape, heating above sintering temperature of metal, but below softening
point of glass, and elongating the plastic mass to extend glass fibers into
a discontinuous phase in the continuous phase of the metal.

3,050,427

8/21/62

Slayter, Morgan, Morrison & Shannon
Owens-Corning Fiberglas Corp.

Fibrous Glass Product and Method of Manufacture

9 Claims (7 Method, 2 Product), 22 Figures

Intricate machinery is described for making by a continuous process, products and articles which contain mineral fiber (glass fiber in the title) and foamed resin binder.

Many binders are described as phenolic resin, etc. Foaming agents may be organic or inorganic. Fillers may be added to the product, such as polystyrene beads, glass beads, silica etc., to fill space, and to increase resistance to fire. 5 examples.

3,053,672

9/11/62

Labino
Johns-Manville Corporation

Glass Composition

8 Claims (8 Product [compositions], 0 Process)

Improved glass wool is claimed by the addition of small amounts of BaO (1-5%) and ZnO (1-5%) to conventional glass fiber compositions, the improvement being in light transmission, chemical durability (in body of patent).

Claims give range of compositions and specific compositions.

Range –		
SiO_2	46-54%	
B_2O_3	8-13	
Al_2O_3 (and Fe, Ti)	11-15	
$Na_2O + K_2O$	0.5-2.5	
CaO	15-22	
MgO	0-2	
BaO	1-5 (pref. 2)	
ZnO	1-5 (pref. 1, 3)	
CaF_2	0.1-2.5	

Softening point, liquids temperature, chemical durability given.

3,053,713   9/11/62

Juras
Union Carbide Corporation

Plastic Articles Reinforced with Preformed Precompressed Metal Fiber Elements.

13 Claims (7 Product, 6 Method), 10 Figures

In the past loose metal fibers were imbedded in organic plastics to form molds for production of case or molded plastic dies.

It is claimed here to use long metal fibers initially compressed into definite shapes, and to then fill the voids with numerous resins, etc.

12-60% of the volume of the object to be metal fibers. Numerous examples.

3,055,736   9/25/62

Bécue (France)
Pechiney Co., etc. (France)

Process for Manufacturing Aluminum Oxide

10 Claims (10 Process, 0 Product), 1 Figure

Al (or its alloys) is dissolved in Hg and/or Ga at 200-300°C in a non-oxidizing atmosphere to the extent of 0.01-4% Al by weight in the amalgam. The solution is cooled to 80-10°C (pref. 30°) in a hydrating atmosphere (such as moist O_2, air, water vapor) in a thin layer 2-50 (20) mm thick, wherein microcrystals of aluminum oxide are formed and float to the surface, from which they are scraped off.

The recovered hydrated aluminum oxide, which is amorphous with a bulk density of less than 0.01 g/cm^3 external spec. surface of 285 m^2/gm, particles - generally spherical, size about 10 millimicrons, (m μ), as described in body of patent.

Apparatus is described.

3,056,747   10/2/62

Arthur, Jr.
duPont

Process for the Production of Fibrous Alumina Monohydrate

3 Claims (3 Process, 0 Product), 2 Figures

Alumina sols made in various ways are treated in aqueous solution with a salt (or chemical) containing the sulfate radical. The suspension is then heated in the range 250-600° C under pressures of 1000-2200 atmospheres (in the example) for varying lengths of time. The aluminum monohydrate is formed, with the crystal structure of boehmite, as determined by x-ray diffraction, and the product is fibrous in nature.

The fibers contain the "sulfate" radical to the extent of 1.8% (also 1.5) to over 14% by weight. The discussion states that the presence of the sulfate causes the fibers to be longer and smaller in diameter.

Example: Fibers made at 500° C and at 3000 atmosphere contain 14% sulfate by weight, and they are 16 μ long and 0.025-0.04 μ in diameter.

Discussion talks about 1/2 time depolymerization, and dissociation constants, etc. to explain the limits in the sols -- so do the claims.

6 examples of processes.

3,059,311

10/23/62

Hockberg
duPont

Stable Non-Woven Batt of Polytetra-Flouroethylene Fibers

2 Claims (2 Process, 0 Product)

"Heat-retractable" fibers of polytetra-flouroethylene are heated above 450°F, carded, etc., to form a non-woven batt, and "needle-punched."

3,060,041

10/23/62

Loewenstein (England)
Microcell Ltd. (England)

Glass Compositions

11 Claims (10 Compositions, 1 Product [Fibers])

The purpose is to increase the modulus of elasticity of a base glass having a range of composition, by adding certain oxides.

Base Glasses	SiO_2	20-50% by wt.
	Al_2O_3	0-20
	CaO	10-25
	MgO	7-22

to which may be added

	LiO	0.1-2%
	MnO	0.5-5.0
	B_2O_3	1-10
	CoO	0.05-1

The additives to increase Young's modulus are:

	TiO_2	10-27%
(one or more)	ZrO_2	8.5-20
	HfO_2	8-15
(narrower	ThO_2	10-25
ranges are also given	Ta_2O_5	9.4-15
for glasses and for additives)	Nb_2O_5	5-20

Examples – 11 compositions, including one containing BaO 10% with Young's modulus for each, ranging from 14.8 to 18.0 $\times$ 10^6 psi.

A range of compositions is given in claims for fibers.

3,062,682   11/6/62

Morgan, Morrison & Shannon
Owens-Corning Fiberglas Corporation

Fibrous Glass Product and Method of Manufacture

5 Claims (5 Process and Method, 0 Product), 23 Figures

A matt of fiber, mineral fiber or glass fiber, is pierced by means of a group of hollow needles through which resin binder with catalyst and foaming agent can be made to penetrate the matt at many points, and cause binding and foaming at these points, to form a (fiber glass) structure.

Many thermosetting resins are named. Inorganic cementitious materials (like magnesium oxysulphate, etc.) are also mentioned in the body of the patent. Foaming agents may be organic or inorganic (as ammonium carbonate). Many examples and densities of products are given.

3,063,866   11/13/62

Mayer, Rickett and Stenemann
General Mills Inc.

Method of Forming Bismuth Whiskers

8 Claims (8 Method, 0 Product), 4 Figures

A glass plate is first covered with a thin layer of Mn by vapor deposition, and Bi is then vaporized on top of the Mn. The Bi whiskers are formed by heating the combination at 245-275° C.

The rate of formation of whiskers is increased by slightly oxidizing the Mn layer.

The Mn layer should have a thickness of 0.1-1.0μ, and the Bi layer should have a similar thickness.

The body of the patent talks of the properties of Bi, in terms of magneto-resistive effect, Hall effect, magnetic tape recording uses, piezo-resistive effect, etc.

<u>3,063,883</u>   11/13/62

Brissette
Union Carbide Corporation

Reinforced Resin Laminates

16 Claims (11 Product, 5 Method)

A "preform matt" suitable for use in the production of reinforced resin laminates is made by mixing staple length glass fibers with unoriented synthetic organic fibers (such as acrylonitrile - vinyl chloride copolymer), and heating the matt under light pressure to bond the synthetic fibers at their contacts without bonding to the glass fibers.

Numerous variations, compositions, etc. given.

<u>3,065,091</u>   11/20/62

Russell, Morgan and Scheffler
Owens-Corning Fiberglas Corporation

Crystalline Fibers

28 Claims (6 Body Compositions, 2 Melt Compositions, 20 Process)
4 Figures

Crystalline fibers of TiO_2, ZrO_2 and zircon are grown from a melt containing an alkali oxide, as Na_2O, K_2O, Li_2O, etc. with B_2O_3 in preferred proportions. The melts are cooled in stages, and at preferred rates in each stage, to initiate crystal growth and then to continue growing the same crystals.

Example: TiO_2, B_2O_3 and borax (in 1/1/1 mol proportions) melted at 2500° F, cooled at 30° F per minute to 2300°, then at 15° per minute to 2220° F, increased to 2240°, (near nucleation zone), and cooled at 2^O F per minute to 1800° F., then cooled to room temperature. Matrix washed away with hot water, leaving crystals of TiO_2, some 4" in length.

Mixed crystals can be obtained as ZrO_2 and zircon from a zircon/B_2O_3/borax melt. Also TiO_2 and ZrO_2 from a mixture of ingredients.

Crystals can also be obtained by heating a melt at a constant temperature permitting alkali and boron oxide to volatilize, thus concentrating the melt, and encouraging crystals to grow.

Additions of oxides of Fe, Cr, Cb, Ta can be made.

Claims call for a ratio of Na_2O/B_2O_3 -- 1/3 in several places, as preferred.

No cross-section dimensions of crystals given.

3,067,482   12/11/62

Hollowell
duPont

Sheet Material and Process of Making Same

9 Claims (9 Method, 0 Product)

"---Sheet material having a soft and drapy hand similar to that of soft leather suedes."

Non-woven material of staple (organic) fibers impregnated with solutions of synthetic polymers, and treated, etc.

3,069,277   12/18/62

Teja
Montecatini-Societa Generale etc. (Italy)

Aqueous Dispersions of High Molecular Weight Fibrils of Amorphous Silicates.

2 Claims (1 Method, 1 Product [Composition])

Alkaline silicates are treated to form high molecular weight, limear (in the molecular sense) polymers.

Example: An aqueous solution of Na or K silicate is mixed with an additive formed in many ways, which acts as a catalyst. The mixture is heated to evaporate water until at least 10% of initial water is removed, and the solids are at least 40% by weight. It is claimed that the alkaline silicates will have formed linear polymers.

The additives are many in the form of oxides of Si, Mg, Zn, B, etc. in amounts of say, 5% colloidal SiO_2, or 3% B_2O_3, etc.

Turbidity values and molocular weight numbers given in many cases.

3,073,679   1/15/63

Stone and Ruehrwein
Monsanto Chemical Co.

Production of Single Crystal Boron Phosphide.

2 Claims (2 Process)

Crude boron phosphide is heated in the presence of HCl (preferred), HBr, HI, in the temperature range 600–1500° C (preferred 800–1200°) and the gases moved to a zone of higher temperature 800–1800° C (preferred 1100–1500°) where crystals of boron phosphide are deposited. The higher temperature zone should be 200–500° C higher than the cooler initial zone. The process may be stationary or continuous.

Crystals are cubic, 9 on Mohs scale, density 2.94 g/cc

Single crystals measure in one dimension at least 0.1 mm.

3,075,831

1/29/63

Remeika
Bell Telephone Laboratories

Growth of Single Crystals of Corundum and Gallium Oxide

9 Claims (9 Method, 0 Product), 2 Figures

Crystals of Al_2O_3 and Ga_2O_3 are grown in a melt of PbO/B_2O_3 --100/1-10 by solution, and very slow cooling from maximum temperature to 900° C.

Rhombohedral crystals are obtained at 1300°C and hexagonal plates at 1250-1225°C.

Example: PbO 50 gm., B_2O_3 1 gm., Al_2O_3 6.8 gm. in covered Pt. crucible, heated at 1300°C for 8 hours, controlled cooling 2° per hour to 915°C. Remaining liquid poured out hot, and crystals separated from glass by dissolving the latter in dil. HNO_3. (Rhombohedral) crystals 1/2 cm. long. Yield 4.2 gm.

Example: PBO 50 gm., B_2O_3 4 gm., Al_2O_3 7 gm., same treatment as above, hexagonal plates 3 cm. long, yield 4.6 gm.

Colored crystals of corundum obtained by adding Cr_2O_3, Co_2O_3, Fe_2O_3 in amount 0.1-1.-3-5% to the melts, for maser uses.

In all cases, the higher B_2O_3 content (4 gms. above) give hexagonal plates and lower B_2O_3 (1 gm. above) give rhombohedral crystals.

Example: PbO 50 gm., B_2O_3 3 gm., Ga_2O_3 10 gm., treated as above gave crystals of Ga_2O_3 0.5 cm. long. Yield 5.5 gm.

Claims cover compositions only. Temperatures and rates of cooling found only in body of patent.

3,077,380

2/12/63

Wainer and Cunningham
Horizons Inc.

Preparation of Sapphire Fibers

5 Claims (5 Process, 0 Product)

Molten Al is maintained at 1370-1510° C in an atmosphere of H_2 gas containing a minor amount of water vapor (dew point -30 to -90°C), and in the presence of a refractory oxide (of heat of formation lower than that of alumina) like SiO_2, ZrO_2, TiO_2 and combinations including Al_2O_3, and alumina whiskers are formed. Oxygen for formation comes from two sources - water vapor and from the refractory oxide.

Two types of whiskers may be formed, depending upon conditions:

wool   --   0.1-7μ diam. and 1/4, 1/2-3" long

whiskers   --   80μ diam. and 1/4-3/4" long.

The speed of formation of whiskers can be increased by adding 25 mol % of Ti, Zr, Nb, Ta, Si to the molten Al, 8-17 fold (in examples).

Strength of fibers -- greater than 1 million psi tensile stength for fibers 0.02-0.04 mm. in diameter. Elsewhere strength is given in the range 1-3 million.

Fibers can be bent 360°.

3,078,564   2/26/63

Bourdeau
Alloyd Research Corp.

Bonding Process and Product Thereof

4 Claims (4 Process, 0 Product), 3 Figures

A thin layer of metal is deposited by vaporization onto the two surfaces of two objects which are to be joined, in a chamber where decomposable gases are available at temperatures 300-1000°C, and then pressing the two objects together while heated to form a diffusion bond.

Carbonyls, halogens etc. of many metals are claimed -- in particular chromium dicumene.

Deposited layer – 0.00001 to 0.001".

3,080,242   3/5/63

Berry
duPont

Fibrous Aluminum Borate and Its Preparation

10 Claims (3 Product [Fibers], 7 Method)

Two distinct methods are employed and described. The first involves a melt of boric oxide in which Al_2O_3 and/or $Al(OH)_3$ is dissolved and the fibers removed after slow cooling by water washing. The second method involves heating the same ingredients in a pressure vessel in water and extracting the fibers by washing. In both cases the fibrous product is Type A (determined by given x-ray diffraction lines and chemical analysis) with a formula:

$$(Al_2O_3)_{3 \pm 0.4} \cdot B_2O_3$$

These Type A fibers can be treated in several ways to reduce the B_2O_3 content to obtain Type B fibers, which are of the same physical dimensions but of different composition:

$$(Al_2O_3)_{9 \pm 0.4} \cdot B_2O_3$$

Distinct x-ray pattern and chemical composition of Type B fibers are also given.

Type A melts made in Pt. crucible at 1000-1600°C are cooled slowly to 900°C to permit fibers to grow in an otherwise glassy melt. Example: Melt made at 1400°C, cooled 10° per minute to 900°C, then more quickly to room temperature, washed in 2% aqueous NaOH solution; resulting fibers, Type A -

$$(Al_2O_3)_{3.0} \cdot B_2O_3$$

0.7 - 1.1 mm in length, cross-section 1 - 10μ. Example: B_2O_3, $Al(OH)_3$, and water in a pressure vessel at 750°C and 1500 atmospheres; resulting fibers are Type A.

Type B fibers are obtained by treating Type A fibers (1) in a hot furnace to volatize B_2O_3, for example, at 1750°C at atmosphere pressure; (2) by heating at 1400°C at low pressure, for example 0.1 mm Hg. pressure; (3) by heating in water at 750°C and 1500 atmospheres, for example,

$$(Al_2O_3)_{8.6} \cdot B_2O_3$$

One other method is described in which B_2O_3 plus Al_2O_3 are heated in a Pt. tube in the range 1450°C max. and cooler along the length to about 1250°C. This tube with molten charge is drawn slowly through a furnace, resulting in zone of crystal growth with both Type A and Type B, separately and mixed.

Claims call for fibers where length/width ratios are at least 10/1, maximum width of 25μ. Fibers are flexible and capable of being felted.

X-ray diffraction lines are given in detail for both Type A and Type B.

3,080,256   3/5/63

Bundy
Georgia Kaolin Co.

Coated Mineral Filler for Organic Polymers, and Method of Forming the Coated Filler

15 Claims (11 Method, 4 Product)

The method of forming a mineral material suitable as a filler for organic polymers, by making a water slurry of the mineral, and adding to it in prescribed manner a polyamine, an aqueous emulsion of an organic polymer, settling out the coated particles by adding acid, and drying the particles below the decomposition temperature of the organic polymer.

The minerals, including clay, should have a solubility less than 10^{-3} molar in water, exemplified by barites, sphalerite, rutile, kaolinite, mullite, graphite, (mentioned only in the body of the patent).

Clay is specifically mentioned in 7 claims.

3,082,051   3/19/63

Wainer and Beasley
Horizons Inc.

Fiber Forming Process

12 Claims (10 Process, 2 Product [ceramic fibers])

Solutions of metal organic salts are used separately and in combination to form sols, which are flowed in thin layers on glass plates, dried to form gels and remove liquid, thereby forming a thin solid sheet which cracks to form fibers, which may be heated further and sintered.

Metallic organic salts of such metals as Al, rare earths, Zr, Hf, Th, Nb, Ta, Cr, Mn, Fe, Co, Ni are used separately or in combinations and they are further mixed with other metallic organic salts from the group Zn, Be, Cu, Cd to form the sols used to make thin films, etc.

The thin film layers may be made either from acid sols or alkaline sols, with or without protective colloids. The solutions may be made in water or in organic solvents.

Specific claims are made for the formation of ZrO_2 fibers.

No dimensions are given in the claims for the size of the fibers, nor in the body of the patent.

Example: Zr carbonate, acetic acid, HCl -- in solution as the thin aqueous film.

3,082,099   3/19/63

Beasley and Johns
Horizons Inc.

Inorganic Fibers and Method of Preparation

9 Claims (4 Product [Fibers], 5 Method)

Fibers of rectangular cross section are made from metallic oxides to which SiO_2 has been added, using (aqueous) solution methods to lay down a thin layer of solution on a plate, drying the layer - causing it to shrink into fibers, and heating the fibers to higher temperatures, and sintering (600-1200°C).

Metallic oxides are from the group -- Al, rare earths, Zr, Hf, Th, Fe, Co, Mn, Ni, V, present in the amount 85-95%, remainder SiO_2. These oxides are introduced in solution form as metal salts of carboxylic acid, such as acetic acid (and others) - with metal salts of strong acids as HCl, in some cases - and with a silica sol, adjusting the pH to 1-2.

Examples show preparation of ZrO_2 fibers (2 methods), Al_2O_3 fibers, Fe_2O_3 fibers, a mixture of Fe plus Ni oxide fibers, Cr_2O_3 fibers, and compositions.

ZrO_2 fibers made with SiO_2 8.3% showed a tensile strength of 200,000–400,000 psi. ZrO_2 fibers made in the same way without silica showed strength of 50,000–100,000 psi, Al_2O_3 fibers with SiO_2 (9.5%) 1/w/th – 2-6"/up to $5\mu/0.5$-0.2μ.

3,082,143   3/19/63

Smith

Owens-Corning Fiberglas Corp.

Method of Forming a Substantially Rigid Laminated Fibrous Board

7 Claims (7 Method, 0 Product), 4 Figures

The method involves two separate pressing operations. The first pressing makes a mat of mineral fiber, including glass fiber under pressure and heat, using a thermosetting organic binder. A separately made textile layer or sheet (of glass fiber) is placed over the first mat, and pressed and heated with additional organic binder.

After the second pressing, the fibrous board returns to its earlier greater thickness –– and various claims are made for its superiority.

3,084,054   4/2/63

Tiede

A.E.C. (USA)

Glass Composition

37 Claims (37 Glass Compositions, 0 Process), 1 Figure

A triaxial diagram of the system SiO_2-Na_2O-U_3O_8 is shown, with dots representing compositions in an area, from which glass fibers can be made.

Claim 1 gives the range SiO_2 35-70% by wt., Na_2O 5-35%, U_3O_8 16-60%.

30 claims give 30 different compositions in this range, based evidently on chemical analysis. They include small amounts of Al_2O_3 0.02-0.07% and Fe_2O_3 0.02-0.03%.

1 claim includes CaO, 2 claims include CaO and MgO, 1 claim includes TiO_2 and ZrO_2, 1 claim includes K_2O, in all these cases in addition to the 3 main components.

1 claim substitutes K_2O for Na_2O in the 3 main component systems.

Melting temperatures 2600-2900°F. Fiber diameters and bushing temperatures given in many cases in the body of the patent.

36 examples of composition are covered exactly in the claims.

3,084,421   4/9/63

McDanels, Jech, Weeton & Petrasek
NASA (USA)

Reinforced Metallic Composites

4 Claims (3 Method, 1 Product)

Short lengths of strong metallic fibers are placed in a small diameter refractory tube and packed by vibration to maintain lengthwise position. The
assembly is heated in a vacuum or reducing atmosphere while molten metal
(ductile) is poured into the tube to fill all remaining space.

Examples are given of 5 mil W wires, 3/8" long, made into bars 1 1/2-3"
long of Cu. A table of data gives volume of Cu and of W, diameter, area,
break load and calculated ultimate tensile strength. E.g., 64.3% Cu,
35.7% — 120,000 psi.

3,085,876   4/16/63

Alexander and Yates
duPont

Process for Dispersing a Refractory Metal Oxide in Another Metal

6 Claims (6 Process, 0 Product)

The starting materials are all compounds, and solutions of them are made
in various ways, and they are co-precipitated. Drying is followed by
calcining in an oxygen–containing atmosphere (400°–1000°C) to increase
the particle size of the refractory oxide to 5-500 millimicrons (mμ). The
product is then heated in reducing atmosphere (hydrogen) to reduce the
other oxide to metal. Objects such as components for jet engines may be
made by compressing, and further sintering.

Examples: ZrO_2 in Ni metal. ZrO_2 in Mo metal.

            ThO_2 in Ni, Fe, Cr metal, ThO_2 in Ni.

Many matrix metals and refractory oxides are listed in the body of the
patent. Oxides and sulfides are included.

3,094,385   6/18/63

Brisbin and Heffernan
General Electric Co.

Method for the Preparation of Alumina Fibers

5 Claims (5 Process, 0 Product)

Single crystal fibers of hexagonal structure of Al_2O_3 are prepared from a
gaseous mixture of volatile aluminum compound with CO_2 and H_2.

Aluminum halide as $AlCl_3$, CO_2 and H_2— are introduced into a reaction chamber at least $1100^{\circ}C$ and at a reduced pressure of not more than 100 mm. Hg. and the temp. raised to approximately $1500^{\circ}C$ as the gases flow toward the exit end of the reaction chamber, where the crystal fibers deposit. Variations of the method are described.

4 example methods described, Example 1 giving dimensions of length, 0.2-2.5 mm; avg. diam. 0.05 mm.

3,095,642

7/2/63

Lockwood
Owens-Corning Fiberglas Corp.

Metal and Fiber Composite Materials and Methods of Producing

5 Claims (5 Process, 0 Product), 14 Figures

Glass fibers issuing from spinnerets (from a molten source) are arranged in various positions, such as in a row side by side, and the spacing infiltrated with molten metal, to form a composite sheet. Detailed descriptions of machinery are given.

The metals or alloys, with melting points 125°-$2000^{\circ}F$; may be Al, Pb, Zn, etc.

Other processes such as rolling and embossing may be applied to the sheets.

Uses - sheathing, telephone cable, etc.

3,096,144

7/2/63

Wainer and Mayer
Horizons Inc.

Method of Making Inorganic Fibers

6 Claims (6 Method, 0 Product), 4 Figures

A solution of a metal salt of a strong acid and a solution of an alkali metal of a weak acid are mixed, pH adjusted to 5-9, spread as a thin layer on a smooth surface, and dried ($90^{\circ}C$) to gel and to remove water. The thin film is removed and dried to $150^{\circ}C$ and then to 600°-$800^{\circ}C$ in an oxidizing atmosphere. The resulting fibers are cooled to room temperature and washed free of all water soluble salts. The fibers may be heated to higher temperatures as they are -- or they may be impregnated with a solution of an oxide-forming metal salt, and then dried and fired above $600^{\circ}C$. (The claims only cover compounds with the impregnated salt additions.)

Examples cover fibers of ZrO_2, TiO_2, Fe_xO_y, Al_2O_3, Cr_2O_3, CuO (?), U_xO_y, Al_2O_3 plus ZrO_2, barium zirconate, barium titanite, and modifications.

The fibers are rectangular in cross section, $1/w - > 100/1$; $1/th - > 1000/1$ and dimensions 2.5 mm-6"/5-25μ/0.2-1.5μ.

The claims are concerned with the methods of obtaining these filaments, including colored ones, firing above 600°C, and the formation of ferro-electric materials as titanates, zirconates, niobates - of the perovskite structure, and especially barium titanite.

3,098,723

7/23/63

Micks
The Rand Corporation

Novel Structural Composite Material

1 Claim (1 Product [Turbine Blade], 0 Process), 7 Figures

Objects are given both low temperature and high temperature strengths by making them with a composite material of fibers or filaments embedded in a matrix, the matrix being ductile in nature and the filaments are brittle in nature.

The matrix may be Nb, Mg, Zn, Al, Cr, Mo, Fe, Co, Ni, Cu, Re, Ag, Au, Ti and alloys (e.g. stainless steel).

The filaments may be of W, Be, graphite, and of non-metallic refractory substances as carbides, cemented carbides, nitrides, silicides, oxides, borides, and silicates.

Example 1. Al_2O_3 filaments with a coating of an "interlayer" like TiN, dipped into a molten bath of stainless steel to form a shape.

Example 2. W filaments covered with Mo powder and sintered.

Example 3. Graphite filaments sintered in powdered Nb.

The claim refers specifically to a turbine blade, made of ductile metal matrix with imbedded filaments of less than 10μ diameter, where the thermal expansion of the matrix is greater than that of the filaments.

3,099,548

7/30/63

Ducati
Plasmadyne Corporation

Method of Making Glass Fibers

1 Claim (1 Method, 0 Product), 9 Figures

Electric plasma-jet torches with high temperatures (2000°-5000°C) and high velocities are applied to the heating of vitreous materials in various processes and equipment to form small diameter threads or fibers (dimensions not given).

The body of the patent shows the use of such torches to form fibers from (1) a tapered conical bath with molten glass flowing from the tip (2) a rod or filament fed into the torches (3) a reservoir of molten glass fed from a tank (4) the surface of a molten body of glass (5) a pool of glass kept in circular motion through a vortex.

The single claim refers only to (5) above – wherein a refractory circular bath is rotated by motor; vitreous material is fed in at the bottom through an orifice which is off center and at an angle. The fibers issue downward from a central opening (of large diameter) through which the plasma issues from above.

The body of the patent says the vitreous substance may be glass, quartz, zirconia, and alumina.

3,102,835

9/3/63

White
Allen Industries Inc.

Fibrous Materials and Method of Making Same

10 Claims (10 Method, 0 Product), 2 Figures

An organic urethane polymer saturant is used to bind organic fibers together, to form an open-texture fibrous material.

8 examples, numerous chemical compositions, and methods of procedures given.

Compression load data given.

3,104,943

9/24/63

Berry
duPont

Spinnable Mullite Fibers and Their Preparation

Spinnable mullite fibers, some in the form of ribbons, are made by vapor reaction from mixtures of SiO_2, Al, and Al_2S_3 in the presence of H_2, diluted with Ar. The reactants are heated in a silica tube in the range of 800^0-1200^oC at 1 atmosphere or 0.5-5 atmospheres.

SiO_2, 2; Al, 1; Al_2S_3, 0.6-1 preferred mixtures.

Body of patent says F_2 0.1% is also desirable.

Example 1. 1000^oC for 100 hours, gave SiO_2 at one zone and mullite fibers at another. Mullite ribbons -1 cm $\times$ 4μ $\times$ 1μ.

Example 2. 8-10 mm $\times$ 1μ $\times$ 1μ.

H_2 gas needed 1% (1-5%), remainder an inert gas like Ar.

Other mixtures may be used to obtain Al_2S_3 as a gaseous product.

Claims indicate that the fibers are less than 5μ, and have lengths to diameter ratios of 100/1 and 1000/1.

Fibers can be worked into yarns, mats, etc., since they are flexible.

3,107,152 10/15/63

Ford and Mitchell
Union Carbide Corp.

Fibrous Graphite

10 Claims (4 Process, 6 Product), 2 Figures

Special heat treatments have been devised to convert cellulosic materials, especially rayon, into flexible graphitized products, such as single filaments, yarns or textiles.

The heat treatments consist of successive schedules in protective atmospheres as:

(1) 10°-50°C rise per hour in the range $\cdot$ 100°-400°C.

(2) up to 100°C rise per hour in the range 400° - 900°C.

(3) heating to about 3000°C until graphitization has occurred.

Three claims refer to flexible graphitized filaments and products made thereof, having a diameter 5–25 microns, tensile strength greater than 40,000 psi.

Specific electrical resistance from 1,800 - 5,500 micro–ohm–cms.

Properties. A table of properties gives tensile strength of monofilaments.

	Diam. in Microns	Tensile Strength in psi
By Invention	5–7	108,000 – 130,000
By Invention	20–25	48,000 – 53,000

	Diam. Microns	Tensile Strength in psi
Commercial)	146	28,000 – 33,000
Lamp Filaments)	205	18,500 – 22,000

Other data on specific resistance.

3,107,180 10/15/63

Diefendorf
General Electric   Company

Process for Deposition of Pyrolytic Graphite

6 Claims (6 Process, 0 Product), 2 Figures

Non–sooting carbon vapor is obtained by adding nitric oxide to the carbon vapor used in making pyrolytic graphite objects.

Nitric oxide added by volume -- 0.1-5%, preferably 1.5%.

Temperature: 1000°-2300°C.

3,108,888   10/29/63

Bugosh
duPont

Colloidal, Anisodiametric Transition Aluminas and Processes for Making
Them

18 Claims (3 colloidal alumina Product, 5 Porous Shapes, 3 Sintered
Bodies, 7 Process), 10 Figures

Aluminum hydrate (mono) is prepared by hydrolyzing basic aluminum di-
acetate and amm. sulfate in water at 160°C in an autoclave. The product
is called colloidal fibrous boehmite - "anisodiametric" in dimensions,
meaning unequal in diameters or axes. Fibers from 3-10 m μ in minimum dia. with axial ratios 3/1 -- 300/1, and maximum length ca. 1500
m μ.

Heating boehmite fibers at successive higher temperatures gives transition aluminas - gamma, kappa, eta, delta, theta, alpha.

The claims refer especially to the making and use of gamma alumina
made from colloidal anisodiametric boehmite, to make objects from (1)
boehmite alone, (2) boehmite plus other forms of alumina, (3) boehmite
(plus alumina) plus grain growth inhibitors like CoO, MgO, Cr_2O_3, NiO,
(4) boehmite (plus alumina) plus sintering promoters as MnO_2, Fe_2O_3,
CuO, TiO_2, (5) boehmite as a binder for numerous metal oxides.

Objects may be porous as sheets, catalyst carriers; and dense bodies up
to 98% theor. density by high firing.

37 examples are given in detail of methods and processes, including the
extrusion (example 24) of small diameter rods fired at 1600°C. to form
2 mil. dia. alpha Al_2O_3 fibers.

This is a very long patent, 38 columns, full of theory, examples, detailed
explanations, etc.

Some tensile strength data and densities given in the examples.

3,109,511   11/5/63

Slayter, Russell and Morgan
Owens-Corning Fiberglas Corp.

Muffler Liner

2 Claims (2 Product, 0 Process), 7 Figures

A lined muffling system for exhaust gases, consisting of a "liner" or a
"porous liner" between inlet and outlet openings.

The liner is made of highly refractory inorganic fibers of TiO_2, ZrO_2, or
zircon and an inorganic binder. Examples give ceramic sections 1/4 -
1 1/2" thick.

Example: TiO_2 fibers 87 gm., wood sawdust 20 gm., Montmorillonite 25 gm., and powdered soda-lime glass 13 gm., pressed and fired at 2500°F.

Also a layer of a mixture may be applied or sprayed onto a metal inner surface.

3,110,545   11/12/63

Beasley and Johns
Horizons Inc.

Inorganic Fibers and Preparation Thereof

20 Claims (14 Method, 6 Product [3 porous capillary containing filament, 3 nonporous ribbons])

A "colloid" is defined as a solid phase with particles finer than 0.2 to 0.0001 microns, in a liquid preferably water.

A colloid silica dispersion 20 microns thick on a smooth non-sticking surface gives fibers by rapid drying -- 10μ thick, 20–$40\ \mu$ wide, and 2-3" long. Rate of drying and pH of sol influences dimensions. Example - a $36\ \mu$ thick layer dried at 200°C gives fibers 6μ thick, 12μ wide and 3" long. High shrinkage aids in making fibers by induced strains. Such fibers contain submicroscopic capillaries as pores or channels. Sintering causes further shrinkage of about 1/3 in volume and closes capillaries in fibers.

Examples in body of patent and in claims cover fibers of SiO_2, Al_2O_3, Cr_2O_3, ZrO_2, ThO_2, and mixtures, and with the addition of B_2O_3.

Filaments with rectangular cross section can be prepared from a lyophobic sol by drying on a smooth surface, at 20-220°C, and sintering to close pores -- the film having a thickness 5-50μ. Ratio of l/w/th -- 1000/1-10/1.

Products may be produced as ribbons with porous capillaries -- l/w/th - 100/1-10/1.

There are specific claims for SiO_2 alone and with B_2O_3, and for ZrO_2.

5 examples of compositions and preparation given in detail.

3,110,571   11/12/63

Alexander
DuPont

Ceramic Material Bonded to Metal having Refractory Oxide Dispersed therein

8 Claims (4 Process, 4 Product [articles])

A "Bonder" is made by dispersing a finely divided refractory metal oxide in a molten metal (paying attention to free energy relations).

The refractory oxides may range from Al_2O_3 to Y_2O_3., also spinels and silicates which may range from silicates to zirconates and size range 5-500 mμ. (millimicrons), and in volume from 0.05 - 20%.

The molten metals may range from Al to Zr.

Examples: (1) Al 61%, Mg 33% plus colloidal ThO_2 5.8 by weight, for sealing Al metal parts to an alumina ceramic part. (2) Pb, Mg with BeO to bond Zn metal part to Pyrex, etc.

The bonder may be used in molten form (1) to bind itself to a solid metal plate, (2) to bind itself to a ceramic plate, (3) to bind two other materials together, as glass to glass, metal to glass, metal to ceramic oxide to make gas tight seals, (4) to bind a ceramic powder to form a cermet.

3,110,939   11/19/63

Lockwood
Owens-Corning Fiberglas Corp.

Apparatus and Method for the Preparation of Polyphase Materials

4 Claims (2 Method, 2 Apparatus, 0 Product), 2 Figures

Articles such as panels are made from a mixture of molten metal and glass fiber, the metals having lower melting points than the softening points of the glass.

The metals in examples are Pb, Zn, Al. The glass fibers may be fibers or flakes.

The molten metal and the glass fibers are poured from separate containers onto a screw, inclined at an angle in a container, and the increased viscosity of the mixture causes intimate mixing, and the formation of gobs which are raised by the screw - for collection and further processing.

3,114,197   12/17/63

DuBois and Roth
The Bendix Corp.

Brake Element having Metal Fiber Reinforcing

6 Claims (6 Product, 0 Process), 5 Figures

Friction elements made for aircraft brakes, by sintering metals alone, or ceramics alone, or mixtures of any two or three, with metal fibers preferably by shearing the metal.

Examples: Metals in powder form such as Cu, Fe, Mo, Monel
             Ceramics — mullite, MgO
             Graphite
             Sheared metal fibers - Stainless steel, 304, 1010 steel,
                carbon steel, Ni. Diam. less than 0.003 and less than 1/2"
                long.
             Compacts of MgO with metal fibers.
             By volume - metal powder 12-99%, non-metallic powder
                0-65%, metal fiber 1-35%.
             Friction test data by Dynamometer, modulus of rupture of
                compacts.

3,118,807   1/21/64

Holcomb
Johns-Manville Corp.

Bonded Fibrous Insulation

8 Claims (6 Product [lightweight blanket], 2 Product [liner for combustion chamber])

Lightweight, refractory, thermal insulating blankets resistant to temperatures of 2000°F and 2500°F are made by mixing refractory mineral fibers
with powdered glass bentonite and with a resin binder. Combustion liners
with such blankets backed up with stainless steel linings are also claimed.

The refractory fibers are not specified, but the body of the patent shows
aluminum silicate fibers as an example.

The glass may be powdered soda-lime-silica glass, or plate glass.

The resin may be thermosetting, e.g., phenol-formaldehyde.

Mixture-mineral fibers 80-85%, binder 15-20% by weight.

Binder contains 3 ingredients, bentonite clay 10-15 parts, glass powder
3-8 parts, resin 2-7 parts. There are other ratios given.

Initial heating cures resin, higher heating softens glass (and bentonite) to
give a bond. Heating in the combustion chamber to maximum temperature
will complete heating.

3,121,659   2/18/64

Amanzio

Apparatus for Producing Fiber Reinforced Cementitious Structure

2 Claims (2 Process, 0 Product), 7 Figures

The body of the patent describes improved means of making Portland
Cement and asbestos fiber mixtures into sheets, 8 feet by 4 feet by 1/8"-4"
thick, for structural purposes. The claims refer to fiber-reinforced
cementitious structures made in described special apparatus.

A wet slurry is fed into a container so that the fibers lay generally in one
direction for a certain length of time, then changing the flow pattern so
that the superimposed layer has the fibers laying in a transverse direction,
etc.

Drawings show the equipment.

3,125,404

3/17/64

Crawley
The Carborundum Company

Heat Resistant Fibrous Products and Method of Making the Same

10 Claims (10 Product, 0 Process), 5 Figures

Organic acrylic fibers when heated in an oxydizing atmosphere around 400°-600°F transform to a fairly stable strong heat-resistant state. The patent proposes to use such treated fibers to give strength to inorganic, ceramic fibers, for fire-proof clothing, pipe lagging, heat insulation, etc.

The cloth is made with the untreated yarn, and the final product is passed through an oven.

The body of the patent mentions examples of Dynel fiber, Orlon fiber, aluminum silicate fibers such as Fiberfrax, and asbestos. The claims refer to siliceous fibers, aluminum silicate, in major proportion of the siliceous fibers, 15-90% by weight of total blend of fabric, with modifications.

There are curves and data on temperature drop across insulating fabric, loss of tensile strength after keeping material at various temperatures up to 2000°F.

The key materials are Fiberfrax fibers and acrylic fibers.

3,125,428

3/17/64

Maczka

Method for Coating Silica Rods

2 Claims (2 Process, 0 Product), 6 Figures

A pyrolytic (carbon) coating is applied continuously to a hot silica fiber as it is drawn from a silica glass rod (or tube).

The moving gaseous hydrocarbon environment around the fiber may be acetylene, propane, methane, fed coaxially around the moving fiber.

The body of the patent states such coated fibers may be used for electrical resistors, and to eliminate creation of surface defects when bundles of fibers are made for light production purposes.

3,127,277

3/31/64

Tiede
Owens-Corning Fiberglas Corp.

Glass Composition

22 Claims (12 Compositions, 10 Product [Fibers])

The body of the patent says the object is to obtain glass fibers of high modulus of elasticity, in the order of $E = 15$-16×10^6. The claims only mention compositions of glasses and fibers made from them.

Two claims cover ranges of composition, 10 claims cover definite compositions given also as examples in the body, and 10 claims cover the glass fibers made from the definite compositions.

There is discussion of effect of BeO on modulus, of CeO, Fe_2O_3, etc. on properties of the molten glasses, liquids, crystallization, etc.

Numerous examples of compositions are given.

3,127,668   4/7/64

Troy
Illinois Institute of Technology Research Institute

High Strength-Variable Porosity Sintered Metal Fiber Articles and Method of Making the Same

11 Claims (7 Method, 4 Product), 9 Figures

Porous metal fiber articles are made in competition with powder metallurgy from fibers 0.002 to 2" long, (length/diam. ratio of about 10/1) by floating them in liquid, pouring the suspension onto a screen, porous mold, or continuous porous belt or screen, using suction to remove the liquid, compacting the article or sheet mechanically, and then sintering the object to make permanent junctions where the fibers cross each other. Filters may be made this way, with porosity of 50-95%. The claims refer to the product as a fiber metal skeleton.

The sintering may be autogenous - that is, high enough to cause the metal itself to adhere to itself, or a braze may be applied to the fibers, such as Cu.

Examples - Iron fibers - tensile strength, porosity compared with powder
               product impact strength also.
         Stainless steel type 430, coarse and fine fibers; tensile
               strength, porosity, permeability. Cold pressure
               in making 10-50 and 30-70 tons per sq. in., then
               sintered 2400°F in hydrogen.

3,129,105   4/14/64

Berry and Sowards
duPont

Fibrous Metal Titanates

7 Claims (5 Product [Fibers], 2 Process)

U.S. Patents 2,833,620 and 2,841,470 show how to make water-insoluble asbestos-like fibers of general formula $M_2O(TiO_2)_n$ where M is an alkali metal K, Na, Rb, Cs and n is an integer (2 to 7). The first by reacting alkali metal compound with TiO_2 in water at 400°C at 200 atm.; the second by heating an alkali metal chloride or fluoride at $1200^{\circ}C$ and dissolving in it a non-fibrous titanate of $M_2O(TiO_2)_{2n}$, whereby fibrous titanates form.

This patent shows how these fibers may be modified in composition by removing alkali and thereby increasing TiO_2 content, leaving the fibers still flexible, by adding also H_3O ions into the lattice. The treatment consists of subjecting fibers to acid solutions in water, pH 2-6.5 at temp. $25^{\circ}C$ up to critical temp. of water, and calcining the product at $300^{\circ}-700^{\circ}C$.

The general formula of acid-modified fibers is $M_xH_y(O(TiO_2)n)_z$, where x is 1-2, y is 1-15, z is 1/2 (x plus y), n is 6-7, M is Na, K, Rb, or Cs.

Examples: TiO_2 94.5%, K_2O 2.5%, H_2O 3% - formula $KH_7(O(TiO_2)_6)_4$
TiO_2 96%, Na_2O nd, H_2O nd.

Fibers lose flexibility above $1000^{\circ}C$, for use as high temperature insulation.

3,131,073   4/28/64

Long
Telecomputing Corporation

Ceramic Material and Method of Preparation

26 Claims (9 Process, 1 Eutectic Material, 16 Product)

A sintered article is made by mixing two ingredients (1) a refractory oxide eutectic and (2) a filter material in granular, flake or fiber form.

The eutectic is formed by mixing and sintering a metal pyrophosphate, of formula $X_2P_2O_7$, where X is Mn, Ti, Fe, Zr, Ni (especially Mn) with a refractory oxide such as Al_2O_3, ZrO_2, BeO, TiO_2, MgO, Cr_2O_3, ThO_2, HfO_2 -- their eutectic temperatures being given.

The filler oxides are the same as given above.

The claims give exact compositions of mixtures:

Example - Binder $Mn_2P_2O_7$ 74% by wt. plus Al_2O_3 26%

Filler Al_2O_3 26% of total batch wt.

Another claim calls for the addition of 10-20% $Na(PO_3)_3$ to lower melting point.

Modulus of rupture and modulus of elasticity at room temperature and 800° and $1000^{\circ}F$ given in several cases. Also thermal expansion.

Articles suggested: heat resistant cooking ware, radome bodies, brake shoe linings, and the like.

3,132,033   5/5/64

Tiede
U.S. Atomic Energy Commission

Fiberizable Glass Compositions

42 Claims (40 Glass Compositions, 2 Fibers made from Compositions)

The purpose is to get glass fibers with high radioactive material content, by adding ThO_2 or U_3O_8 to the compositions. All contain ThO_2, and some contain U_3O_8. The liquidus temperature must be within the range of commercial practice, for refractories and bushings, preferably not higher than $2600^\circ F$

The glasses are all silicates, containing alkalies as Na_2O and K_2O, and other oxides as Al_2O_3 (0.03 - 0.07%), Fe_2O_3 (0.01 - 0.04%), ZnO (one composition), SnO_2 (one composition), V_2O_5, F_2, ThO_2, U_3O_8, all compositions in weight %.

38 specific compositions are given as examples in the body of the patent, and all 38 form claims.

2 general claims are given:

	Claim 39	Claim 41
SiO_2	24 - 70%	24 - 56%
Na_2O	5 - 35	10 - 24
ThO_2	5 - 35	5 - 20
U_3O_8	--	sufficient quantity

2 fiber claims are based on Claims 39 and 41.

3,135,561   6/2/64

Kempthorne

Apparatus for Blowing and Spraying Light Weight Fibers
and Granulated Materials

2 Claims (2 Process, 0 Product), 9 Figures

A compact new machine for handling fibers and materials, designed with hopper, feed box, carding brush, "star gate," screw feed, blowers, etc.

The body of the patent mentions rock wool, glass wool, mineral wool, asbestos, perlite, vermiculite.

3,135,585

6/2/64

Dash

General Electric Company

Method of Growing Dislocation-Free Semiconductor Crystals

14 Claims (14 Method of which 6 refer to silicon), 6 Figures

The purpose is to grow single crystal masses of definite crystal orientation free of oxygen (usually coming from quartz glass crucibles) and free of dislocations.

A vertical round tube furnace contains the charge and the seed.

The main mass of crystals, used as a source, is a solid cylinder machined at the top to form a cone. The top of the cone is melted into a bead by means of a surrounding closely coupled induction heating coil.

A single crystal seed of definite crystal orientation is machined into a needle shape 0.010-0.020" diameter, 3/4" in length and one end touches the molten bead at the top of the cone. By control of speed of withdrawal of the needle, the heating of the cone and later of the main body of crystals, and movement of the induction coil, cylinder of single crystal can be built.

> Examples - Silicon: 1 cm. diameter, 12 cm. long.
>   Germanium: 6 mm. diameter, 10" long.

3,139,653

7/7/64

Orem

Secretary of Commerce, USA

Apparatus for the Growth of Preferentially Oriented Single Crystals of Metals

2 Claims (2 Apparatus, 0 Product), 6 Figures

Detailed description and drawings of a vertical wire wound furnace for the growing of a monocrystalline specimen (in vacuum if necessary), having the matrix (i.e. the charge) mold cavity connected vertically with a seed cavity just below it. All pertinent parts of the inner mold structures are marked and notched, for alignment and for indexing of axes of the seed crystal (determined elsewhere by X-ray diffraction methods).

The vertical axis of the seed cavity can be canted, and the whole inner crucible plus contents can be rotated.

The matrix mold and the seed receptacle are moved vertically to melt the matrix (polycrystalline) and to cause it to grow a monocrystal from the orientation of the seed, which is not melted.

The claims do not mention any specific material, although the text refers to metals.

This patent is a division of 3,060,065 - Orem - October 23, 1962.

3,141,786   7/21/64

Bugosh
duPont

Fibrous Boehmite Alumina Molding Powder and Processes for Molding
them into Dense, Alpha Alumina Articles

6 Claims (3 Product, 3 Process), 6 Figures

Bugosh makes use of a pulverulent fibrous alumina monohydrate having
the boehmite crystal lattice, the alumina fibrils having a surface area of
200 - 400 sq. meters per gram, and an average length of 25 - 1500 millimicrons (Bugosh US Patent 2,915,475).

Molding powders are made by mixing the fibrous alumina with a grain
growth inhibitor, such as nickel oxide, magnesium oxide, chromium oxide,
cobalt oxide, magnesium fluoride or their decomposition compounds (as
acetates) in the proportion of 0.5 to 5% by weight of oxide Al_2O_3.

The water mixtures are dried, pulverized, dry-pressed, and then hot-
pressed in graphite molds, at 1000 - 6000 psi, at 1600°C - 1800°C, for 5 -
60 minutes, to give highly dense products, with average grain size 3 - 6
microns.

> Transverse strength, 80,000 psi; compressive strength, 250,000 psi;
> rockwell hardness, 92.5; coefficient of thermal expansion, 8×10^{-6}
> per 1°C in the range of 25° - 700°C. Chemical analysis given.

The product is suitable for cutting tools, nozzles, dies.

3,141,809   7/21/64

DiMaio and Miller
Johns-Manville Fiber Glass Inc.

Mineral Fiber Laminate and Method of Making Same

10 Claims (6 Process, 4 Product), 5 Figures

An unwoven porous laminate, which is relatively rigid, consisting of
layers of parallel individual glass fibers, of varying density, dependent on
void content; density of first layer 1/2 - 4 lbs. per cu. ft., density of second layer 6 - 16 lbs. per cu. ft.; using thermosetting resins (15 - 25% by
wt.), and means of compressing and curing the product.

The glass fiber shall have a diam. 0.00010 - 0.00015 inch.

Mineral wool laminates are similarly made.

3,142,551

7/28/64

Von Wranau

Furnace for Manufacture of Glass Fibers

3 Claims (3 Furnace Design), 11 Figures

Details of a furnace of the continuous tank type, for the production of glass fiber, containing in one unit the melting chamber, refining chambers, with spinneret chambers placed between the forehearths, in various positions to maintain temperature uniformly, save fuel, etc.

3,142,598

7/28/64

Rosen
Pacific Plastics Company

Method of Making Resin-Impregnated Glass Fiber Automobile Leaf Springs

6 Claims (6 Process, 0 Product), 2 Figures

Method of making non-metallic leaf springs for automobile vehicles.

Unidirectional glass fiber plastic reinforced roving is used, employing 70% glass fiber and 30% resin (poly epoxide types), wound and cured in an oven at 120°F. Pressures are applied.

A spring 1/2" thick, 2 3/4" wide and 43" long is described.

3,142,869

8/4/64

Gould, Hahn, Coleman, Simmers
Johns-Manville Corp.

Process and Apparatus for Opening and Cleaning Fibrous Material

10 Claims (8 Apparatus, 2 Process), 2 Figures

Apparatus for opening, cleaning and felting a fibrous material and removing unwanted particles -- using ducts, screens, aeration chambers, agitators, etc.

The body of the patent mentions mineral wool with unwanted shot-like particles.

3,143,405

8/4/64

Wong
Owens-Corning Fiberglas Corp.

Method of Producing Polyamide Coated Glass Fibers

5 Claims (5 Method, 0 Product), 2 Figures

Coating glass fibers issuing molten from a group of orifices at speeds greater than 4000 ft. per minute by passing them through two baths in tandem, to form a polyamide coating (1 - 25% by wt.).

One bath contains a dicarboxylic acid halide, adipyl chloride, the second bath a polyamine such as hexamethylene diamine.

Tensile strength of a fiber -- 28 gms. breaking load - equal to 439,000 psi.

3,144,687   8/18/64

Skalko and Owens
Owens-Corning Fiberglas Corp.

Method for Forming and Processing Textile Fibers and Filamentary Materials.

11 Claims (11 Method, 0 Product), 4 Figures

The main purpose of the detailed equipment described is to weave together continuously two different types of filaments, to form a "linear composite" product or structure.

One filament may be a mineral filament such as glass, issuing from a series of orifices, and the second filament may be cotton, wool or resin.

Numerous details but no property data are given.

3,145,981   8/25/64

Messler

Furnaces for the Production of Mineral, Especially Basalt Wool

2 Claims (2 Design), 2 Figures

A furnace for melting basalt wool compositions, which require a higher temperature than the usual glass and mineral wool compositions, plus a trough and flame blast passage are described.

3,147,085   9/1/64

Gatti
General Electric Company

Apparatus for Growing Whiskers

4 Claims (4 Apparatus Design), 2 Figures

Metal or metal oxide whiskers are the product of an apparatus which consists of an enclosure to contain the desired atmosphere and contains three rotatable drums. Two of the drums are of equal size and are the surfaces on which the whiskers are nucleated and grown. A third and larger drum is mounted nearly concentrically with one of the first described drums. It passes first through a boat of the desired raw material (a metal) and then between the peripheries (but not touching) the other two drums, all of which at this point pass through a furnace which provides sufficient heat to vaporize the metal carried by the supply drum and causes this metal to vaporize and deposit whiskers on the previously described drums from which the whiskers are later scrapped off. Alumina whiskers are mentioned in the text as an example.

3,147,159

9/1/64

Lowe
Norton Company

Hexagonal Silicon Carbide Crystals Produced from an Elemental Silicon Vapor Deposited onto a Carbon Plate

12 Claims (12 Process), 4 Figures

Single crystal silicon carbide platelets of electronic quality are produced by heating elemental silicon in a carbon furnace and a protective atmosphere of nitrogen plus hydrogen, carbon monoxide or an inert gas at temperatures of 2300°C to 2500°C.

3,148,027

9/8/64

Richmond
Laporte Titanium Ltd. (England)

Vapor Phase Process for Producing Metal Oxides

7 Claims (7 Process), 3 Figures

Titanium dioxide particles are produced by reacting titanium tetrachloride in an oxidizing atmosphere at 750°C to 1500°C and in a fluidized bed of titanium dioxide particles.

3,149,910

9/22/64

Tauber, et al
U.S. of America

Method of Growing Single Crystals

1 Claim (1 Method)

Single crystals of $BaZn_2Fe_{16}O_{27}$ for electronic applications are obtained by melting a mixture of 52.1 to 68.5% Bi_2O_3, 6.8 to 10.3% BaO, 3.6 to 5.5% of ZnO, and 21.1 to 32.1% Fe_2O_3 (all percents are by weight) in a platinum crucible in the temperature range of 1250°C to 1300°C, then cooling at 2.5°C per hour to 1050°C, then to room temperature, and leaching the matrix away in hot dilute nitric acid for 8-25 hours.

3,150,925   9/29/64

Gambino
United States of America

Method of Growing Single Crystals

2 Claims (2 Method)

Magnetic single crystals of $Ba_2Me_2Fe_{12}O_{22}$ and others of higher Fe and O content in which the Me_2 represents Mg, Zn, Co, Ni, or Fe^{+2} or combinations are produced by dissolving BaO, Fe_2O_3 and the MeO constituents in Na_2CO_3 at about 1300°C then cooling at a rate of 1-5°C per hour, particularly at temperatures in the range of 1000°C to 1050°C. The crystals are then recovered by treating the solidified melt with hot dilute nitric acid solution.

3,152,006   10/6/64

Basche
High Temperature Materials, Inc.

Boron Nitride Coating and a Process of Producing the Same

8 Claims (4 Process, 4 Product), 1 Figure

A pyrolytic process of producing boron nitride coatings on a substrate at high temperatures from ammonia and boron trichloride is described. The process is carried out at temperatures between 1450°C and 2300°C at pressures below 50 mm of mercury.

3,152,878   10/13/64

Levecquie, et al
Saint Bogain (France)

Manufacture of Fibers, Particularly Glass Fibers

4 Claims (4 Apparatus Design), 5 Figures

A centrifuge apparatus for producing glass fibers is described.

3,152,992   10/13/64

Heinrich, et al
North American Phillips Co.

Delayed Addition of Phosphorous to Aluminum Melt in the Process of Forming Aluminum Phosphide Crystals

6 Claims (6 Method)

A method of preparing aluminum phosphide crystals in an aluminum melt is described where the aluminum is first heated to 1000°C then on to 1300°C to 1600°C when an oxygen free atmosphere of hydrogen and phosphorous plus doping agents as desired is provided. The text provides four examples and indications that times of about 50 minutes at 1500°C and pressures of 1500 torr are sufficient to produce 1 x 5 x 8 mm crystals. A method of leaching the matrix and cleaning the resulting crystals is also described in the text and is a quite standard chemical procedure.

3,154,463

10/27/64

Kjell-Berger and Grane
Rockwool Aktiebalaget (Sweden)

Mineral Wool

3 Claims (3 Product)

A mineral wool insulating material comprised of air and fibers having a range of diameters of 1-4 microns is described.

3,155,475

11/3/64

Ashman
A.D. Little Inc.

Process for Drawing Fiber through a Supernatant Liquid

12 Claims (12 Process), 2 Figures

A process for drawing glass fibers from a pool of heated glass upwards through a layer of materials such as KCl, RbCl, or molten metal is described. The supernatant liquid must be immisible and non-reactive with the glass below it. The text describes the process as being useful for drawing fibers of glass which do not have a viscosity-temperature relation that would easily permit them to be drawn through bushings or by other conventional methods. The supernatant liquid is later washed or etched off.

3,155,476

11/3/64

Drummond
Pittsburgh Plate Glass Co.

Apparatus for Producing Glass Fibers

4 Claims (4 Apparatus Design), 11 Figures

An improved means of cooling the bushings through which glass fibers are drawn is provided by troughs or wires which are liquid (such as water) cooled and to which radiant heat from the bushings is transmitted.

11/17/64

3,157,541

Heywang, et al
Siemens (Germany)

Precipitating Highly Pure Compact Silicon Carbide upon Carriers

15 Claims (15 Method), 2 Figures

A highly pure or doped gaseous silicon compound and hydrocarbon substance are passed over a pure silicon substrate heated at 1150°C to 1430°C
to make semiconductor products. See also U.S. Patent 3,160,476.

11/17/64

3,157,722

Moore
Plessey Co., Ltd. (England)

Method of Making Reinforced Refractory Bodies

4 Claims (4 Process), 2 Figures

A flame spraying process is described in which matrix material and fibrous
reinforcements are simultaneously deposited onto a mandrel. Metal wires
and alpha alumina composite materials are particularly described.

12/1/64

3,159,475

Chen, et al
Johns-Manville Corp.

Apparatus for Forming Fibers

4 Claims (4 Apparatus Design), 2 Figures

An improved 3 roll apparatus and associated feed distributor mechanism
are described primarily for producing higher quality mineral wool type
materials.

12/8/64

3,160,476

Sirtl
Siemens (Germany)

Process of Producing Compact Boron, particularly in Monocrystalline
Form.

15 Claims (15 Process), 6 Figures

The desired boron product is prepared by vapor depositing (from a halogen
gas form of boron) onto an especially well cleaned and prepared boron
surface. The essential feature of the process is to use a (gaseous) hydro-

gen-halogen compound in a quantity at which the equilibrium temperature for boron redissolution by the reaction gases is at most 200°C below the temperature for depositing boron. This results in a more nearly perfect boron crystal with respect to both dislocations and major protuberances. (Also see U.S. Patent 3,157,541).

3,161,473

12/15/64

Pultz
Corning Glass Works

Method of Making Beta-Silicon Carbide Fibers

4 Claims (4 Method), 2 Figures

Carbon and silica in a range of molar ratios of 1:1 to 3.5"1 are heated to 1375°C-1550°C in a reaction chamber. The chamber is then evacuated (to at least 300μ pressure) and an atmosphere of nitrogen and hydrogen (ammonia) is provided at an initial total pressure of 400-700 mm. The reactions are permitted to occur for 1-50 hours. The text describes the product of this process as whiskers of up to 3" in length by 1-5 microns in diameter for use in reinforcing other materials.

3,168,423

2/2/65

Krieglstein and Reiss
Siemens (Germany)

Method of Producing Monocrystalline Wafers from the Vaporous Phase with Alternative Cooling and Intermediate Cooling Steps

8 Claims (8 Method), 3 Figures

Leaf shaped monocrystals of semiconductor materials such as of germanium, silicon and A^{III} and B^{V} compounds are grown from the gas phase. The material is first heated to above the vaporization temperature, then is cooled in steps of 8° to 50°C per minute. 9 examples are given.

IX. BIBLIOGRAPHY

A. Introduction

This bibliography is arranged in two forms to assist the reader in locating references of interest:

(1) An alphabetical list by author of 550 references to publications concerned with ceramic and graphite fibers (Section IX-B). Entries suggested as first reading for persons desiring a review of a given field are indicated by an asterisk (*). The coded numbers in the right-hand column of the list are a cross-reference to the categorized tabulation (Section IX-C).

(2) A tabulation of the references by primary categories, such as whiskers, fused silica and glass fibers, composites, etc. (Section IX-C). Coded numbers are used so that the reader can locate a given subject and its related references. For example, group W1 contains 60 references describing the measured properties of whiskers.

Since the subject categories were selected on a broad basis, references overlap, and therefore, the categorized portion is liberally cross-indexed.

An explanation of the code is:

W - Whiskers

      W1 - Measured Properties

      W2 - Factors Affecting Measured Properties
            W2a - Effect of Size and/or Shape
            W2b - Effect of Structural Perfection and/or Composition
            W2c - Effect of Surface Treatments and/or coatings
            W2d - Effect of Environment

W3 – Growth Techniques
   W3a – Vapor Deposition
   W3b – High Temperature Reduction of Metallic Salts
   W3c – Hydrogen Reduction of Halides
   W3d – Electrolysis
   W3e – Controlled Solidification of Eutectic Alloys
   W3f – From Solution
   W3g – Spontaneous
   W3h – Proper
   W3i – Vapor-Liquid-Solid
   W3j – Pressure
   W3k – Cleavage

W4 – Test Methods and Equipment

W5 – General Review and/or Theoretical Strength

F – Fused Silica and Glass Fibers

F1 – Measured Properties

F2 – Factors Affecting Measured Properties
   F2a – Effect of Size and/or Shape
   F2b – Effect of Structural Perfection and/or Composition
   F2c – Effect of Surface Treatment and/or Coatings
   F2d – Effect of Forming Methods
   F2e – Effect of Environment

F3 – Manufacturing Processes
   F3a – Filaments
   F3b – Fabrics, Tapes, etc.

F4 – Test Methods and Equipment
   F4a – Filaments
   F4b – Fabrics, Tapes, etc.

F5 – General Review and/or Theoretical Strength

C – Composites

C1 – Whisker Reinforcement
   C1a – Metal Whisker – Metal Matrix
   C1b – Metal Whisker – Inorganic Matrix
   C1c – Metal Whisker – Organic Matrix
   C1d – Ceramic Whisker – Metal Matrix
   C1e – Ceramic Whisker – Inorganic Matrix
   C1f – Ceramic Whisker – Organic Matrix

C2 – Fiber Reinforcement
    C2a – Glass Reinforced Plastics
    C2b – Carbon/Graphite Reinforced Plastics
    C2c – Reinforced Metal
    C2d – Reinforced Inorganics

C3 – Test Methods

C4 – Interfacial Studies
    C4a – Whisker Composites
    C4b – Fiber Composites

C5 – General Review and/or Strength Theory

A – Applications

A1 – Whiskers
    A1a – Reinforcement
    A1b – Thermal Insulation
    A1c – Fillers
    A1d – Papers
    A1e – Others

A2 – Fibers
    A2a – Reinforcement
    A2b – Thermal Insulation
    A2c – Fillers
    A2d – Fabrics and Papers
    A2e – Others

CF – Composite and Polycrystalline Fibers

CF1 – Carbon or Graphite Fibers;
    Properties, Uses and Forming Methods

CF2 – Polycrystalline;
    Properties, Uses and Forming Methods

CF3 – Composite,
    Properties, Uses and Forming Methods

CF4 – Metal Filaments

R – Related Studies with Bulk Materials and/or Theoretical Reviews

R1 – Graphite

R2 – Sapphire

R3 – Glass

R4 - Plastics

R5 - Metals

R6 - Theoretical Reviews
  R6a - Crystals
  R6b - Filaments

R7 - Crystal Growth

B. Alphabetical list of publications by author

Asterisk (*) in lefthand column denotes suggested first reading.

Coded numbers in righthand column refer to categorized tabulation in Section IX-C.

* 1. Abbott, H.M., "Composite Materials: An Annotated Bibliography," Lockheed Missiles and Space Co., Sunnyvale, Calif., SB-62-58, February 1963.   C5/1

* 2. Ibid., "Supplement No. 1," SB-63-49, June 1963.   C5/2

3. Abbott, R.L., "Evaluation of National Research Corporation Glass Monofilaments," Naval Air Eng. Center, Phila., Pa., Rpt. AML-1884, February 1964.   F2c/1

4. Accountius, O.E., "Whiskered Metals Reach for 1,000,000 psi," Machine Design, 35 (11) pp. 194-199, May 9, 1963.   W5/1
C5/3

5. Adams, J.J. and Sterry, J.P., "High Temperature Fibrous Insulation," H.I. Thompson Fiber Glass Co., Gardena, Calif., AFML, WPAFB, ML TDR 64-156, October 1964.   CF3/1

6. Ibid., "Zirconia Fibrous Insulations," ASD-TDR-63-725, December 1963.   CF3/2

7. Ainslie, N.G., et al, "Devitrification Kinetics of Fused Silica," G.E. Co., Research Lab., Schenectady, N.Y., Report 61-RL-2640M (Revised), March 1961.   F2c/2

8. Aleksandrov, L.N. and Kogan, A.N., "Investigation of the Strength of Needle-Like Tungsten Crystals," Fiz. Tverd Tela, 6 (1) p. 307, 1964. (in Russian)   W1/1

9. Amelinckx, S., "Dislocations in Alkali Halide Whiskers," Growth and Perfection of Crystals, edited by R.H. Doremus, B.W. Roberts, and D. Turnbull, John Wiley & Sons, Inc., N.Y., p. 139, 1958.   W2b/1

10. Amelinckx, S., "On Whisker Growth Shapes," Phil. Mag., pp. 425-428, May 3, 1958.   W2a/1

* 11. Amelinckx, S. and Delavignette, P., "Electron Optical Study of Basal Dislocations in Graphite," J. Appl. Phys., 31 (12) pp. 2126-2135, 1960.   R1/1

12. Anderegg, F.O., "Strength of Glass Fiber," Ind. & Eng. Chem., 31 (3) pp. 290-298, 1939.   F1/1
F2/1

13. Anderson, J.A., "Metal Whiskers," NASA - Goddard Space Flight Center, Greenbelt, Maryland, Goddard Summer Workshop, Program in Measurements and Simulation of the Space Environment, Summer 1963.   W2d/1
W3/1

14. Anderson, O. L. , "Cooling Time of Strong Glass Fi-   F2d/1
     bers," J. Appl. Phys. , 29 (1) pp. 9-12, 1958.

15. Andreyeuskaya, G. D. , "Soviet Trends in Oriented   C2a/1
     Glass Fiber Reinforced Plastics," Aerospace Infor-
     mation Division, Washington, D. C. , AID Report No.
     62-173, May 1962.

16. Anon. , "Glass Gets Tough," Chem. Week, 94 (15) pp.   F1/2
     59-60, April 11, 1964.

17. Anon. , "Growing Whiskers for Strength," The Thiokol   W1/2
     Magazine, 3 (1) pp. 28-29, 1964.

18. Anon. , "New Sinews for Rugged Service," Chem.   F5/1
     Week, 95 (25) pp. 33-34, December 19, 1964.

* 19. Anon. , "The Promise of Composites," Mat. in Des.   C5/4
        Eng. , 58 (3) pp. 80-126, 1963.

20. Anon. , "Space Age Fiber Goal: Strength at 2000°F, "   F2d/2
     Chem. Week, 89 (15) pp. 61-62, April 29,1961.

21. Anon. , "Stronger Metals with Silicon Nitride Whisk-   C1d/1
     ers," New Sci. , 19 (351) p. 291, August 8, 1963.

22. Anon. , "Structural Ceramic Fibers of Virtually Any   CF2/1
     Oxide," Mat. in Des. Eng. , 52 (1) p. 5, July 1960.

23. Anon. , "Whiskers Go Commercial," Chem. Week, 96   C1/1
     (7) pp. 65-68, February 13, 1965.

* 24. Arledter, H. F. and Knowles, S. E. , Synthetic Fibers   A1/1
        in Papermaking, edited by O. A. Battista, Interscience   A2/1
        Publishers, New York, pp. 185-243, 1964.   CF4/1

25. Arnold, S. M. and Koonce, S. E. , "Filamentary   W3g/1
     Growths on Metals at Elevated Temperatures," J.
     Appl. Phys. , 27 (8) pp. 962-963, 1956.

26. Arridge, R. G. C. , et al, "Metal Coated Fibers and   F2c/3
     Fiber-Reinforced Metals," J. Sci. Instr. , 41 pp. 259-   C2c/1
     261, May 1964.

27. Arridge, R. G. C. and Prior, K. , "Cooling Time of   F2d/3
     Silica Fibers," Nature, 203 (4943) pp. 386-387, July
     1964.

28. Austerman, S. B. , "Growth of BeO Single Crystals, "   W3a/1
     J. Am. Cer. Soc. , 46 (1) pp. 6-10, 1963.

29. Austerman, S.B., "Role of Si, Al, and Other Impu-   W2b/2
rities in BeO Crystal Growth," Atomic International,   W3a/2
Canoga Park, California, Atomic Energy Commis-
sion Contract AT (11-1)-GEN-8, Report Number
NAA-SR-8235, July 15, 1963.

* 30. Bacon, R., "Growth, Structure and Properties of   W2/1
Graphite Whiskers," J. Appl. Phys. 31 (2) pp. 284-   W3/2
290, 1960.

31. Bacon, R. and Tang, M.M., "Carbonization of Cellu-   CF1/1
lose Fibers - II. Physical Property Study," Carbon,
2 (3) pp. 221-225, December 1964.

32. Baker, G.S., "Angular Bends in Whiskers," ACTA   W2a/2
Met 5 pp. 353-357, July 1957.

33. Baker, W.S. and Kaswell, E.R., "Handbook of Fi-   W5/2
brous Materials," McGraw-Hill Tech. Writing Service
and Fabric Res. Labs. Inc., Dedham, Massachusetts,
ASD, WPAFB, AF 33(616)-7504, WADD TR 60-584,
Pt. II, October 1961.

34. Barber, D.J., "Electron Microscopy and Diffraction   W4/1
of Al_2O_3 Whiskers," Phil. Mag., 10 (103) pp. 75-94,
July 1964.

35. Barish, L., et al, "Mechanical and Thermal Degrada-   F2/2
tion Mechanisms of Quartz Fibers and the Develop-   A2d/1
ment of Experimental Quartz Fabrics with Improved
Finishes," Fabrics Research Labs., Inc., Dedham,
Mass., AFML, WPAFB, Cont. AF 33(616)-7557,
ASD-TDR-63-802, September 1963.

36. Barnes, R.E., comp., "Silane Coupling Agents, Bib-   F2c/4
liography for 1953-1962," Armed Services Tech. Inf.
Agency, Alexandria, Virginia, September 1962.

37. Bartenev, G.M., "Flawless Glass Fibres," presented   F1/3
to Symposium on Physics of Noncrystalline Solids,   F2d/4
sponsored by Int. Union Pure and Applied Phys.,
Delft, Holland, July 6-10, 1964.

38. Baskey, R.H., "Fabrication of Core Materials from   F2c/5
Aluminum-Coated, Fuel Bearing Fiberglas," Clevite
Corporation, presented at Winter Meeting, Am. Nu-
clear Society, Chicago, Illinois, November 9, 1961.

39. Baskey, R.H., "Fiber Reinforcement of Metallic and   C1a/1
Non-Metallic Composites," Clevite Corp., Cleveland,
Ohio, AFML, contract AF 33(657)-7139, Int. Rpt.
Vol. 1, February 1962.

40. Ibid., Vol. 2, May 1962.   C1a/2

41. Ibid., Vol. 3, August 1962.   C1a/3

42. Ibid., Vol. 4, December 1962.   C1a/4

43. Ibid., Final Rpt., ASD-TDR-63-619, July 1963.   C1a/5

44. Bateson, S., "Critical Study of the Optical and Me-   F2/3
     chanical Properties of Glass Fibers," J. Appl. Phys.,
     29 (1) pp. 13-21, 1958.

45. Bateson, S., "A Note on the Structure of Glass Fi-   F2b/1
     bers,"J. Soc. Glass Tech., 37 pp. 302T-305T, 1953.

46. Bell, J.E., "Effect of Glass Fiber Geometry on Com-   F2a/1
     posite Material Strength," ARS Journal, 31 (9) pp.
     1260-1265, 1961.

47. Bernal, J.D., et al, "A Discussion of Non-Materials,"   F2/4
     Proc. Royal Soc. A., 282 pp. 1-154, 1964.   F5/2

48. Bershtein, V.A. and Glikman, L.A., "On a Rapid   C2a/2
     Method for Determining the Fatigue Strength of Fi-   C3/1
     ber Glass Reinforced Plastics," Industrial Labora-
     tory 30, pp. 274-277, Sept. 1964.

49. Bevis, R.E. and Thomas, G.L., "Glass Fiber Bun-   F1/4
     dles, Theoretical vs. Actual Tensile Strengths,"
     Proc. 19th Annual Mtg. Soc. Plastics Ind., 17-D, pp.
     1-6, February 1964.

50. Bigot, J., "Preparation and Study of Iron Oxide Whisk-   W3a/3
     ers," Mem. Sci. Rev. Met., LX (7-8) pp. 541-550,
     1963. (in French).

51. Bigot, J. and Talbot-Besnard, S., "Origin and Growth   W3a/4
     of Iron Sesquioxide Filaments and Platelets," Compt.
     Rand, 255, pp. 1927-1929, October 15, 1962. (in
     French)

52. Bokshtein, S.Z., et al, "Tensile Testing of Filamen-   W4/2
     tary Crystals of Copper Nickel and Cobalt to Failure,"
     Soviet Physics - Solid State, 4 (7) pp. 1272-1277,
     1963. (in English)

53. Bokshtein, S.Z. and Svetlov, I.L., "Determination of   W2a/3
     the Shape and Dimensions of Transverse Cross Sec-   W4/3
     tions of Threadlike Crystals," Translated from
     Zavodskaya Laboratoriya, 28, (5) pp. 595-596, May
     1962.

54. Boller, K. H., "Effect of Pre-Cyclic Stresses on Fatigue Life of Plastic Laminates Reinforced with Unwoven Fibers," For. Prdts. Lab., USDA, Madison, Wisc., AFML, WPAFB, cont. AF 33(657)-63358, ML TDR 64-168, Sept. 1964.   C2a/3 C3/2

55. Boller, K. H., "Fatigue Properties of Plastic Laminates Reinforced with Unwoven Glass Fibers," For. Prdts, USDA, Madison, Wisc., AFML, WPAFB, cont. AF 33(657)-63358 ASD-TDR-62-464, Mar. 1962.   C2a/4 C3/2

56. Boller, K. H., "Strength Properties of Reinforced Plastic Laminates at Elevated Temperatures," For. Prdts. Lab., USDA, Madison, Wisc., AFML, WP-AFB, cont. AF 33(657)-63358, ML TDR 64-167, Aug. 1964.   C2a/5 C3/2

* 57. Bradstreet, S. E., "Principles Affecting High Strength to Density Composites with Fibers or Flakes," AFML, WPAFB, Dayton, Ohio, ML TDR 64-85, May 1964.   C5/5

58. Brenner, S. S., "The Case for Whisker Reinforced Metals," J. Metals, 14 (11) pp. 809-811, 1962.   C1d/2

* 59. Brenner, S. S., "Factors Influencing the Strength of Whiskers," presented to ASM Seminar on Fiber Composite Materials, Phila., Pa., October 1964.   W2/2 W5/4

60. Brenner, S. S., "Growth and Properties of Whiskers," Science 128 (3324), pp. 569-575, 1958.   W3/3

* 61. Brenner, S. S., "Mechanical Behavior of Sapphire Whiskers at Elevated Temperature," J. Appl. Phys., 33 (1) pp. 33-39, 1962.   W1/3 W2/3

62. Brenner, S. S., "Metal Whiskers," Sci. American, 203 (7) pp. 65-72, 1960.   W5/3

63. Brenner, S. S., "Tensile Strength of Whiskers," J. Appl. Phys., 27 (12) pp. 1484-1491, 1956.   W4/4

64. Brookfield, K. J. and Pickthall, D., "Some Further Studies on the Effect of Glass-Resin Type and Cure on the Strength of Laminates," SPE Trans., 2 (4) pp. 332-338, Oct. 1962.   C2a/6

65. Brossy, J. F. and Provance, J. D., "Development of High Modulus Fibers from Heat Resistant Materials," Houze Glass Co., Point Marion, Pa., AF 33(657)-8904, WADC TR-58-285, Part II, March 1960.   F1/5 F3a/1

66. Brown, R.G., "Fundamentals of Filament Winding,"   C2a/7
    Missiles and Rockets, 13, pp. 27-29, Aug. 12, 1963.

67. Bryant, P.J., "Mechanism of Lubrication of Graphite   W1/4
    Single Crystals," Midwest Res. Inst., Kansas City,
    Mo., ASD, WPAFB, AF 33(616)-6277, WADD TR 60-
    529, Part I, October 1960.

68. Ibid., WADD TR 60-529, Part II, June 1962.   W1/5

69. Bryant, P.J., et al, "On Graphite Whiskers," J.   W2b/3
    Appl. Phys., 30 (11) p. 1839, 1959.

70. Burke, J.E., editor, Progress in Ceramic Science,   F5/3
    Volume I, Pergamon Press, New York, 232 pp., 1961.

71. Ibid., Volume III, 262 pp., 1963.   F5/4

72. Bushong, R.M., et al, "Pyrolytic Coating of Carbon   W2c/1
    Filaments," Union Carbide Corp., Lawrenceburg,
    Tenn., AFML, WPAFB, Contract AF 33(657)-11297,
    Qtrly. Prgr. Rpt., November 10, 1963.

73. Bushong, R.M., et al, "Research and Development on   W1/6
    Advanced Graphite Materials, Vol. XLII, Summary   W2/4
    Tech. Rpt.," Union Carbide Corp., Lawrenceburg,   W3/4
    Tenn., AFML, WPAFB, AF 33(616)-6915, WADD TR   W4/5
    61-72, Vol. XLII, pp. 76, 89-104, 174-184, January   R1/2
    1964.

74. Butler, J.B., "An Investigation of Glass Tensile   F1/6
    Strength Variations Caused by Induced Mechanical   F2c/6
    Damage," G.E. Co., Missile & Space Division, King
    of Prussia, Pa., Report 63SD291, August 1963.

75. Butler, J.B., "A Statistical Evaluation Technique Ap-   F1/7
    plied to Glass Filament Yarn Tensile Strength," G.E.
    Co., Missile & Space Division, King of Prussia, Pa.,
    Report 62SD798, October 1962.

76. Cabrera, N. and Price, P.B., Growth and Perfection   W1/7
    of Crystals, edited by R.H. Doremus, B.W. Roberts,   W2/5
    and D. Turnbull, John Wiley and Sons, Inc., New   W3/5
    York, p.212, 1958   W3h/1
                                                               W4/6

77. Calow, C.A. and Begley, B., "A Study of the Han-   W1/8
    dling, the Mechanical Properties and Crystalline   W2/6
    Structure of Silicon Nitride Whiskers," United Kingdom
    Atomic Energy Auth., Aldermaston, Berkshire, Rpt.
    AWRE 0-70/64, September 1964.

78. Cameron, N. G. , "The Effect of Heating on the Room   F1/8
     Temperature Strength of Pristine E-Glass Rods,"   F2e/1
     Univ. of Ill. , Urbana, Ill. , Naval Res. Lab. , Wash-
     ington, D. C. , NONR 2947 (02) (X), T&A. M. Rpt. 252,
     November 1963.

* 79. Cameron, N. G. , "An Introduction to the Factors In-   F2/5
     fluencing the Strength of Glass," Univ. of Illinois,   F5/5
     Urbana, Illinois, Naval Research Lab. , Washington,
     D. C. , Proj. 62R05 19A, Tech. Memo 158, March
     1961.

80. Campbell, W. B. , "Feasibility of Forming Refractory   W1/9
     Fibers by a Continuous Process," Lexington Labs.   W3a/5
     Inc. , Cambridge, Mass. , AMRA cont. DA-19-020-
     AMC-0068 (X), Qtrly. Prgr. Rpt. 1, June 1963.

81. Ibid. , Prgr. Rpt. 3, November 1963.   W1/10

82. Ibid. , Prgr. Rpt. 4, April 1964.   W1/11

83. Canter, J. M. and Heller, R. B. , "Experimental Meth-   W3/6
     od for Observing Growth of Graphite Whiskers," NASA
     Goddard Space Flight Center, Greenbelt, Md. , God-
     dard Summer Workshop Program in Measurement and
     Simulation of the Space Environment, N 64-28219, 1963.

84. Capps, W. and Blackburn, D. H. , "The Development   F1/9
     of Glass Fibers Having High Young's Modulus of   F2/6
     Elasticity," National Bureau of Standards, Washing-   F2c/7
     ton, D. C. , NBS Rpt. 5188, April 1957.   F2d/5
                                                                 F3a/2

85. Carroll, M. T. and Pritchard, D. J. , "Evaluation of   A2a/1
     Metal Fiber Composite," General Dynamics, Ft.
     Worth, Texas, AFML, WPAFB, AF 33(657)-7248,
     Rpt. No. ERR-FW-121, Dec. 1961.

* 86. Carroll-Porczynski, C. Z. , <u>Advanced Materials,</u>   A1/2
     Chemical Publishing Co. , 286 pp. , 1962.   A2/2

87. Cech, R. E. , "On Whisker-Forming Reactions," G. E.   W3/7
     Co. , Research Laboratory, Schenectady, N. Y. , Rpt.
     59-RL-2316M, November 1959.

88. Charles, R. J. and Hillig, W. B. , "The Kinetics of   F2e/2
     Glass Failure by Stress Corrosion," G. E. Co. , Re-   F4a/1
     search Lab. , Schenectady, N. Y. , Rpt. 61-RL-2790M,
     July 1961.

89. Chase, G. H. and Phillips, C. J., "Elastic Modulii   F1/10
and Tensile Strength of Glasses and Other Ceramic   F2b/2
Materials," Rutgers Univ., New Brunswick, N. J.,
Progress Rpt. 4, June 1964.

90. Clausen, E. M., et al, "Synthesis of Fiber Reinforced   C2c/2
Inorganic Laminates," Illinois Univ., Urbana, Ill.
AFML, WPAFB, AF 33(616)-6283, WADD TR 60-299,
Pt. 2, Final Rpt., June 1962.

91. Clausen, E. M. and Reutter, J. W., "Preparation   W3/8
Techniques for Growth of Single Crystals of Non-
Metallic Materials," G. E. Co., Research Laboratory, Schenectady, N. Y., AFML, WPAFB, AF
49(638)-1247, Final Rpt., June 1964.

92. Cobb, E. S., Jr., "High Strength Glass Fiber Web-   A2d/2
bings, Tapes, and Ribbons for High Temperature
Pressure Packaged Decelerators," Owens-Corning
Fiberglas Corp., Ashton, R. I., AFML, WPAFB,
AF 33(616)-7441, ASD-TDR-62-518, July 1962.

93. Cobb, E. S., Jr., "High Temperature Resistant Flex-   F1/11
ible Fibrous Structural Materials of New Glass Com-   F2/7
position," Owens-Corning Fiberglas Corp., Ashton,
R. I., AFML, WPAFB, AF 33(616)-7441, ML TDR
64-10, December 1963.

* 94. Coleman, B. D., "Statistics and Time Dependence of   R6b/1
Mechanical Breakdown in Fibres," J. Appl. Phys.,
$\underline{29}$ (6) pp. 968-983, 1958.

* 95. Coleman, B. D., "On the Strength of Classical Fibres   R6b/2
and Fibre Bundles," J. Mechs. and Phys. Solids, $\underline{7}$
pp. 60-70, 1958.

* 96. Coleman, R. V., "The Growth and Properties of   W3/9
Whiskers," Metallurgical Reviews, $\underline{9}$ (35) pp. 261-
304, 1964.

97. Coleman, R. V., "Observations of Dislocation in Iron   W2b/4
Whiskers," J. Appl. Phys. $\underline{29}$ (10) pp. 1487-1492,
1958.

98. Coleman, R. V. and Sears, G. W., "Growth of Zinc   W3a/6
Whiskers," ACTA Met., $\underline{5}$, pp. 131-136, March 1957.   W3c/1

* 99. "Conference on Structural Plastics, Adhesives and   C2a/8
Filament Wound Composites. Vol. I: Organic and In-
organic Resins and Adhesives, Vol. II: Filament
Wound Composites, Vol. III: Reinforcements,"

* 99.   symposium sponsored by Plastics and Composites   C2a/8
(cont) Branch, NML, Mat. Central, WPAFB, Dec. 1962.

100.   Conrad, H., "Mechanical Behavior of Sapphire,"   R2/1
       Aerospace Corporation, El Segundo, Calif., Rpt. No.
       ATN-64 (9236)-14, Mar. 1964.

101.   Cooney, J.J., "Development of Improved Resin Sys-   C2a/9
       tems for Filament Wound Structures," Aerojet-
       General Corp., Sacramento, California, BuWeps,
       cont. NOw-63-0627-c, Task VI, Final Rpt., Aug.
       1964.

102.   Coplan, M.J., "Flexible Low Permeability Metal and   F3b/2
       Ceramic Cloths," Fabric Res. Labs. Inc., Dedham,
       Mass., WADD, ARDC, contract AF 33(616)-7222, Bi-
       monthly Prog. Rpt. 1, Aug. 1960.

103.   Coplan, M.J., et al, "Conversion of High Modulus   F3b/3
       Materials into Flexible Fabric Structures," Fabric
       Res. Labs. Inc., Dedham, Mass., ASD, WPAFB,
       AF 33(616)-7222, ASD-TDR-62-851, Oct. 1962.

104.   Coplan, M.J., et al, "Flexible Fibrous Structural   F3b/1
       Materials," Fabric Res. Labs. Inc., Dedham, Mass.,
       AFML-WPAFB, contract AF 33(657)-10541, ML-
       TDR-64-102, Feb. 1964.

105.   Cornish, R.L., et al, "An Investigation of Material   C2a/10
       Parameters Influencing Creep and Fatigue Life in Fil-
       ament Wound Laminates," IITRI, Chicago, Ill., Bu-
       Ships cont. NOw-86461, Qtrly. Prgr. Rpt. 5, May
       1963.

106.   Cornish, R.H. and Chaney, R.M., "Glass Fiber   F2a/2
       Strength Enhancement Through Bundle Drawing Oper-   F2d/6
       ations," IITRI, Chicago, Ill., BuWeps   contract   F3a/3
       N600(19)-58450, Qtrly. Prgr. Rpt. 4, July 1963.

107.   Coskren, R.J. and Chu, C.C., "Investigation of the   F4b/1
       High Speed Impact Behavior of Fibrous Materials,
       Part II: Impact Characteristics of Parachute Materi-
       als," Fabric Res. Labs. Inc., Dedham, Mass., ASD,
       WPAFB, ASD-TDR-60-511, Part II, February 1962.

* 108. Cottrell, A.H., Dislocation and Plastic Flow in Crys-   R6a/1
        tals, Oxford Press, New York, 223 pp., 1958.

* 109. Cox, H. L., "The Elasticity and Strength of Paper and   R6b/3
        Other Fibrous Materials," Brit. J. Appl. Phys., 3
        p. 72, 1952.

110. Cox, J.E., et al, "Exploratory Investigation of Glass–   CF3/3
     Metal Composite Fibers," United Aircraft Corp.,
     East Hartford, Conn., BuWeps cont. NOw-64-0389-f,
     Bi-Monthly Prgr. Rpt. No. 1, July 13, 1964.

111. Ibid., Prgr. Rpt. No. 2, September 1964.   CF3/4

112. Ibid., Prgr. Rpt. No. 3, November 1964.   CF3/5

113. Cratchley, D., "Factors Affecting the UTS of a Metal/   C2c/4
     Metal-Fibre Reinforced System," Powd. Met., 11,   C3/3
     pp. 59-72, 1963.

* 114. Cratchley, D. and Baker, A.A., "The Tensile   C2c/3
     Strength of a Silica Fibre Reinforced Aluminum Al-   C3/3
     loy," Metallurgia, 69, pp. 153-159, April 1964.   C5/6

115. Cunningham, A.L., "Mechanism of Growth and Phys-   W3/10
     ical Properties of Refractory Oxide Fibers," Horizons   W5/5
     Inc., Cleveland, Ohio, Office Naval Rsch., NOw-
     2619(00), Final Rpt., April 1960.

* 116. Daniels, H.E., "The Statistical Theory of the Strength   R6b/4
     of Threads," Proc. Royal Soc., London, 183A p. 405,
     1945.

117. Davidson, D.A., "The Mechanical Behavior of Fab-   F4b/2
     rics Subjected to Biaxial Stress: Canal-Model Theory
     of the Plain Weave," General American Transporta-
     tion Company, AFML, WPAFB, contract AF 33(657)-
     8479, ML-TDR-64-239, July 1964.

118. Davies, G.J., "On the Strength and Fracture Charac-   W1/12
     teristics of Intermetallic Fibers," Phil. Mag., 9 (102)   W3e/1
     pp. 953-964, June 1964.

119. Davies, L.G., et al, "A Study of High Modulus, High   CF3/6
     Strength Filament Materials by Deposition Tech-
     niques," General Tech. Corp., Alexandria, Va., Bu-
     Weps cont. NOw-64-0176-c, 1st Bi-Monthly Rpt.,
     January 1964.

120. Ibid., 2nd Bi-Monthly Rpt., March 1964.   CF3/7

121. Ibid., 3rd Bi-Monthly Rpt., May 1964.   CF3/8

122. Ibid., 4th Bi-Monthly Rpt., July 1964.   CF3/9

123. Ibid., 5th Bi-Monthly Rpt., September 1964.   CF3/10

124. Ibid., Final Report, January 1965.   CF3/11

125. Davies, T. J. and Evans, P. E., "Aluminum Nitride   W3a/7
      Whiskers," Nature, 197 (4867) p. 587, 1963.

126. Davis, J. W., "Standards for Filament Wound Plas-   C2a/11
      tics," SPE Journal, 20 (7) pp. 601-604, July 1964.

127. Davis, R. M. and Milewski, J. C., "High Tempera-   C2c/5
      ture Composite Structure," Martin Company, Balti-
      more, Md., ASD-TDR-62-418, Sept. 1962.

128. Dawn, F. S. and Ross, J. H., "Investigation of the   CF1/2
      Thermal Behavior of Graphite and Carbon-Based Fi-
      brous Materials," ASD, WPAFB, Dayton, Ohio, ASD-
      TDR-62-782, Oct. 1962.

* 129. Dietz, A. G. H., "Fibrous Composite Materials," Int.   C5/7
        Sci. and Tech., (32) pp. 58-62, 64, 66-69, August
        1964.

130. Dikina, L. S. and Shpunt, A. A., "Strength of Cleavage   W1/13
      Whiskers," Soviet Physics–Solid State, 4, pp. 405-
      406, 1962. (in English)

131. Dix, E. H., Jr., "Aluminum Alloys for Elevated Tem-   R5/1
      perature Service," Symp. on Structures for Thermal
      Flight, ASME Aviation Conf., Los Angeles, Calif.,
      paper 56-AV-8, March 14-16, 1956.

132. Dixon, T. E. and Kane, R. A., "Growth of Manganeous   W3c/2
      Oxide Whiskers," J. Appl. Phys., 34 (9) pp. 2774-
      2775, 1963.

133. Douglas, R. W., et al, "A Study of Configuration   R3/1
      Changes in Glass by Means of Density Measurements,"
      J. Soc. Glass Tech., 32 pp. 309-339, 1948.

* 134. Dow, N. F., "Study of Stresses Near a Discontinuity in   C5/8
        a Filament-Reinforced Composite Metal," G. E. Co.,
        Missile and Space Div., King of Prussia, Pa., U.S.
        Navy BuWeps cont. NOw-60-0465-d, August 1963.

135. Downey, T. F., "Development and Evaluation of Im-   C2a/12
      proved Resins for Filament Wound Plastics," U.S.
      Polymeric Chem. Inc., Santa Ana, Calif., Proj.
      208LR107, Qtrly. Rpt. No. 2, May 1964.

136. Drum, C. M. and Mitchell, J. W., "Electron Micro-   W2b/5
      scopic Examination of Role of Axial Dislocations in
      Growth of AlN Whiskers," Appl. Phys. Letters, 4 (9)
      pp. 164-165, May 1964.

137. Drummond, W.W. and Cash, B.A., "Development of   F3a/4
      Textile Type Vitreous Silica Yarns," Bjorksten Res.
      Labs. Inc., Madison, Wisc., ASD, WPAFB, con-
      tract AF 33(616)-6255, WADC TR 59-699, March 1960.

138. Duft, B., et al, "Potential of Filament Wound Com-   C2a/13
      posites," Narmco Res. & Dev., San Diego, Calif.,
      BuWeps cont. NOw-61-0623-c (FBM), Monthly Prgr.
      Rpt. No. 2, May 1961.

139. Dukes, W.A., "The Endurance of Textile Materials   R6b/5
      Under Constant Load, A Review," Expl. Res. and
      Dev. Est., Waltham Abbey, England, ERDE Survey
      No. 1/5/62, AD 290-678, August 1962.

140. Dunn, S.A. and Roth, W.D., "High Viscosity Re-   F5/6
      fractory Fibers," Bjorksten Res. Labs. Inc., Madi-
      son, Wisc., BuWeps NOrd-19100, Summary Tech.
      Rpt., February 1962.

141. Eakins, W.J., "The Degree of Bonding Between Cou-   F2c/9
      pling Agents and Glass Filaments: A Study," SPE
      Trans., $\underline{1}$ (4) pp. 234-244, Oct. 1961.

142. Eakins, W.J., "Glass/Resin Interface: Patent Survey,   C4b/1
      Patent List and General Bibliography," DeBell &
      Richardson Inc., Hazardville, Conn., Plastics Tech.
      Eval. Cntr., Picatinny Arsenal, Dover, N.J., Plastic
      Rpt. 18, Sept. 1964.

* 143. Eakins, W.J., "The Interfaces: The Critical Region   C4b/2
      of a Composite Material," DeBell & Richardson Inc.,
      Hazardville, Conn., USAMC, Picatinny Arsenal,
      cont. DA-28-017-AMC-343(A), A State-of-the-Art
      Evaluation, Nov. 1963.

144. Eakins, W.J., "Materials Study. High Strength, High   C2a/14
      Modulus, Filament Wound, Deep Submergence Struc-
      tures," DeBell and Richardson Inc., Hazardville,
      Conn., BuShips cont. NObs-84672, Prgr. Rpt. No.
      3, June 1962.

145. Eakins, W.J., "The Relative Stability of Cation Im-   F2c/8
      purities and Silica Fiber Surfaces," SPE Trans., $\underline{2}$ (4)
      pp. 354-362, October 1962.

* 146. Eakins, W.J., "A Study of the Finish Mechanism: F2c/10
      Glass to Finish, Finish Thickness and Finish to
      Resin," SPE Trans., $\underline{1}$ (4) pp. 214-223, Oct. 1961.

147. Eakins, W.J. and Humphrey, R.A., "Studies of Hol-   F2a/3
      low Multipartitioned Ceramic Structures," DeBell &

147. Richardson Inc., Hazardville, Conn., NASA-Hqs.,   F2a/3
(cont) contract NASW-672, NASA CR-142, December 1964.

148. Edmunds, W.M., "Research on Protection of High   F2c/11
Temperature Glass Fibers Against Flexuring and   F3a/5
Abrasion," Owens-Corning Fiberglas Corp., Grans-
ville, Ohio, AFML, WPAFB, AF 33(616)-7441, Final
Rpt., Nov. 1962.

149. Edwards, P.L. and Happel, R.J., Jr., "Beryllium   W2b/6
Oxide Whiskers and Platelets," J. Appl. Phys., 33 (3)   W3a/8
pp. 943-948, 1962.

150. Ibid., "Alumina Whisker Growth on a Single Crystal   W2b/7
Alumina Substrate," pp. 826-827.   W3a/8

151. Edwards, P.L. and Svager, A., "Spiral Copper   W2a/4
Whiskers," J. Appl. Phys. 35 (2) pp. 421, 1964.

152. Ehlers, S., "Woven Fabrics: Their Properties and   F3b/4
Uses," Mat. in Des. Eng., 59 (5) pp. 88-91, May
1964.

153. Eischen, G., "Effect of Adsorbed Gases on the Tor-   F2c/12
sional Elasticity of Glass Fibers," Compt. Rend., 248
pp. 3160-3162, 1959. (in French)

154. Eischen, G., "The Influence of Adsorbed Gas on the   F2c/13
Rupture Strength of Glass Fibers," Compt. Rend.,
250 pp. 2194-2196, 1960. (in French)

155. Ellis, R.B., "Investigation of Techniques and Materi-   CF2/2
als for the Formation of High Temperature (1500°) Inorganic Fibers," Southern Res. Inst., Birmingham,
Ala., ASD, WPAFB, AF 33(616)-6955, ASD-TDR-61-
134, June 1961.

156. Epstein, G., "Filament Winding Puts Strength Where   C2a/15
Needed," Mat. in Des. Eng., 59 (2) p. 90, 1964.

157. Epting, J. L., Jr., "Silicone Finished Glass Fiber   A2d/3
Tapes and Webbing for Decelerators," ASD, WPAFB,
ASD-TDR-63-808, Sept. 1963.

158. Erickson, P.W., "Glass Fiber Surface Treatments: F2c/14
Theories and Navy Research," Naval Ord. Lab.,
White Oak, Md., NOL TR 63-253, Nov. 1963.

159. Erickson, P.W., et al, "Chemical and Physical Ef-   C4b/4
fects of Glass Surfaces Upon Laminating Resins,"
Naval Ord. Lab., White Oak, Md., NOL TR 63-267,
Jan. 1964.

160. Erickson, P.W. and Volpe, A.A., "Chemical and   C4b/3
      Physical Effects of Glass Surfaces Upon an Epoxide
      Polymer System," Naval Ord. Lab., White Oak, Md.,
      NOL TR 63-10, Mar. 1963.

* 161. Ernsberger, F.M., Progress in Ceramic Science,   R3/2
      Vol. III, edited by J. H. Burke, The Macmillan Co.,
      New York, p. 57, 1963.

162. Fahland, F.W., "Survey on Metal Coatings of Glass   F2c/15
      Filaments," U.S. Naval Res. Lab., Tech. Memo.
      No. 237, June 1963.

163. Farmer, R.W., "Air Aging of Inorganic Fibrous In-   CF2/3
      sulations," AFML, WPAFB, Dayton, Ohio, ASD-
      TDR-64-137, June 1964.

164. Fechek, F. and Hennessey, M., "Boron Fiber Rein-   CF3/12
      forced Structural Composites," paper presented AIAA
      Launch and Space Vehicle Shell Structure Conference,
      Palm Springs, Calif., April 1-3, 1963.

165. Fellows, B.T., "Polycrystalline Ceramic Fiber Re-   CF2/4
      inforcements for High Temperature Structural Com-
      posites," H.I. Thompson Fiber Glass Co., Gardena,
      Calif., ASD, WPAFB, AF 33(616)-8080, Qtrly.
      Prgr. Rpt. 7, January 1963.

166. Ibid., ASD-TDR-62-1108, May 1963.   CF2/5

167. Ibid., Qtrly. Prgr. Rpt. 9, June 1963.   CF2/6

168. Ibid., Qtrly. Prgr. Rpt. 11, December 1963.   CF2/7

169. Ibid., Qtrly. Prgr. Rpt. 12, March 1964.   CF2/8

170. Feltzin, J., "Research and Development of High   R4/1
      Temperature Stable Inorganic Resins and Elastomers,"
      Aerojet-General Corp., Azuza, Calif., Dept. of the
      Army cont. DA-04-495-ORD-3075,   Final Report
      2365, Aug. 1962.

171. Fiedler, W.S., et al, "Quartz Fibers in High Tem-   F1/12
      perature Resistant Materials," Bjorksten Res. Labs.
      Inc., Madison, Wis., AFOSR, Wash. D.C. contr.
      AF 18(603)-42, Rpt. TR 58-91, May 1958.

172. Florentine, R.A., et al, "Magnesium Oxide in Fiber   CF2/9
      Form by Extrusion Methods," Penn-Salt Chemicals
      Corp., Phila., Pa., Naval Res. Lab. cont. Nonr
      2687(00), Qtrly. Prgr. Rpt., August 1960.

173. Forcht, B.A. and Rudick, M.J., "Reinforced Car-   C2a/16
      bonaceous Materials," Astronautics, 6 (4) pp. 32-33,   C2b/1
      1961.

174. Ford, J.A., et al, "Analytical and Experimental In-   W3e/2
      vestigations of Fracture Mechanisms of Controlled
      Polyphase Alloys," United Aircraft Corp., U.S. Navy
      cont. N600(19)-59361, Final Rpt., November 1963.

175. Fracture, edited by B.L. Averbach, D.K. Felbeck,   R6/1
      G.T. Hahn, and D.A. Thomas, Technology Press and
      John Wiley and Sons, Inc., New York, 646pp., 1959.

176. Franks, J., "Growth of Whiskers in the Solid Phase,"   W3j/1
      ACTA Met., 6 pp. 103-109, February 1958.

177. Frazer, A.H. and Reed, T.A., "Research on Aro-   R4/2
      matic Polymers for Thermally-Stable Fibers and
      Films," E.I. duPont de Nemours & Co., Textile Fi-
      bers Dept., ASD, WPAFB, AF 33(616)-8253, Qtrly
      Prgr. Rpt. April 1963.

178. Freeston, W.D. and Gardella, J.W., "Phase II Re-   F3b/5
      port on Process Development for Stranding and Plying
      of Metallic Yarns," Fabric Res. Labs. Inc., Dedham,
      Mass., AFML, WPAFB, AF 33(657)-11624, Proj. Nr.
      8-134, Jan. 1964.

179. Fridman, V.Ya. and Shpunt, A.A., "Investigation of   W1/14
      the Strength of Lithium Fluoride Crystal Fragments,"
      Soviet Physics-Solid State, 5 (3) pp. 575-579, 1963.
      (in English)

180. Fridman, V.Ya. and Shpunt, A.A., "Tensile Tests on   W4/7
      Crystal Fragments (Cleavage Whiskers)," Soviet
      Physics-Solid State, 5 (3) pp. 570-575, 1963. (in
      English)

181. Garber, R.I. and Gindin, I.A., "Physics of the   R6a/2
      Strength of Crystalline Materials," Soviet Physics-
      Uspekhi, 3 (70) pp. 41-77, 1960. (in English)

182. Gates, L.E., et al, "Development of Ceramic Fibers   F1/13
      for Reinforcement in Composite Materials," Hughes   F3a/6
      Aircraft Co., Culver City, Calif., NASA contract
      NAS 8-50, NASA Rpt. CR-50793, April 1961.

183. Gates, L.E. and Lent, W.E., "Development of Re-   F3b/6
      fractory Fabrics," Hughes Aircraft Co., Culver City,
      Calif., NASA-Huntsville, NAS8-50, Final Summary
      Report, April 1963.

184. Gatti, A., et al, "The Synthesis of B_4C Filaments,"   W1/15
     G.E. Co., Missile and Space Div., King of Prussia,   W3a/9
     Pa., NASA-Hqs., NASw-670, 1st Qtrly. Rpt., Octo-   W4/8
     ber 1963.

185. Ibid., 2nd Qtrly. Rpt., January 1964.   W1/16

186. Ibid., 3rd Qtrly. Rpt., April 1964.   W1/17
                                                                    W1/18
187. Ibid., Final Rpt., September 1964.   W3A/9
                                                                    W4/8
188. Gatti, A. and Higgins, J., "Research on High   CF3/13
     Strength, High Modulus, Low-Density Continuous Fil-
     aments of Boron Carbide," G.E. Co., Missile and
     Space Div., King of Prussia, Pa., cont. AF 33(615)-
     1644, 2nd Qtrly Prgr. Rpt., February 1965.

189. Geminov, U.N. and Kop'yev, I.M., "The Strength of   CF4/2
     Fine Metallic Threads," Joint Publications Res.
     Service, OTS, U.S. Dept. Commerce, TPRS 16567,
     Dec. 1962.

190. Gilman, J.J., "High Strength Materials of the Future,"   R6/2
     G.E. Co., Research Laboratory, Schenectady, N.Y.,
     Rpt. No. 60-RL-2579M, December 1960.

* 191. Girard, E.H., "Continuous Filament Ceramic Fibers,   F1/14
     Part II, "Carborundum Co., Niagara Falls, N.Y.,   F3a/7
     ASD, WPAFB, AF 33(616)-6246, WADD TR-60-244,
     Part II, February 1961.

* 192. Glasser, J.H. and Sump, C.H., "Report on Evalua-   C5/9
     tion of Manufacturing Methods for Fibrous Composite
     Materials," survey conducted for MPB, Mfg. Tech.
     Div., AFML, WPAFB, July 10, 1964.

193. Glorioso, S.V., "Preliminary Evaluation of Glass Fi-   C2a/17
     ber Reinforced Phenolic Laminate," Gen. Dynamics,   C3/4
     Ft. Worth, Texas, AFML, WPAFB cont. AF 33(657)-
     7248, Rpt. No. FTDM 1968, Feb. 1962.

194. Goldberg, R.S., et al, "Micromechanics of Fiber-   C2a/18
     Reinforced Composites," Rocketdyne Corp., Canoga   C5/10
     Park, Calif., cont. AF 33(615)-1627, Qtrly. Rpt.
     R-5908-1, Oct. 1964.

195. Gordon, J.E., Growth and Perfection of Crystals,   W1/19
     edited by R.H. Doremus, B.W. Roberts, and D.   W2/7
     Turnbull, John Wiley and Sons, Inc., New York, p.   W4/9
     219, 1958.

196. Gordon, J.E., "Whiskers," Endeavour, <u>23</u> (88) pp. 8-12, 1964.   W5/6

197. Gordon, J.E., et al, "On the Strength and Structure of Glass," Proc. Royal Soc., London, A249, pp. 65-72, 1959.   R3/3

198. Gordon, J.E. and Menter, J.W., "Mechanical Properties of Whiskers and Thin Films," Nature, <u>182</u> (4631) pp. 296-299, 1958.   W1/20   W2/7

199. Gorsuch, P.D., "On the Crystal Perfection of Iron Whiskers," G.E. Co., Research Laboratory, Schenectady, N.Y., 58-RL-2113, 1958.   W5/7

200. Gorton, C.A. and McMahon, C.C., "Ultra-Fine, High Temperature, High Strength Metallic Fibers," Hoskins Mfg. Co., Detroit, Mich., AFML, WPAFB, cont. AF 33(616)-8366, ML TDR 64-48, Feb. 1964.   CF4/3

201. Grand d'Hauteville, E., "Textile Glass Products for Reinforced Plastic Laminates," CIBA Review, pp. 34-37, 1963/1965.   C2a/19   C5/11

* 202. Griffith, A.A., "Phenomena of Rupture and Flow in Solids," Trans. Royal Soc., London, A221, pp. 163-169, 1920.   R6a/3

203. Griffith, J.R. and Whisenhunt, F.S., Jr., "Filament Winding Plastics, Part I: Molecular Structure and Tensile Properties," Naval Res. Lab., Washington, D.C., NRL Rpt. 6047, Mar. 1964.   C2a/20

204. Grove, C.S., et al, "Developments in Fiber Technology," Ind. and Eng. Chem., <u>54</u>, (8) pp. 55-57, 1962.   F3a/8   F4a/2

* 205. <u>Growth and Perfection of Crystals</u>, edited by R. H. Doremus, B.W. Roberts, and D. Turnbull, John Wiley & Sons, Inc., New York, 275 pp., 1958.   W5/8

206. Grunfest, I.J. and Dow, N.F., "New Shapes for Glass Fibers," Astronautics, <u>6</u> (4) pp. 34-35, 1961.   F2a/4   C2a/21

207. Gutfreund, K. and Weber, H.S., "Interaction of Organic Monomers and Water with Fiberglass," SPE Trans., <u>1</u> (4) pp. 191-198, Oct. 1961.   C4b/5

208. Guzeev, V.V. and Malinski, Iv.M., "Device for Measuring Stress Relaxation in Fibers," Industrial Laboratory, <u>29</u> (3) pp. 1529-1530, April 1964. (in English)   W1/21   W4/10

209. Hamilton, D.R., "Interferometric Determination of   W2b/8
      Twist and Polytype in Silicon Carbide Whiskers,"
      J. Appl. Phys., 31 (1) pp. 112-116, 1960.

210. Hammond, M.L. and Pavitz, S.F., "Influence of En-   F2e/3
      vironment on Brittle Fracture of Silica," J. Am.
      Cer. Soc., 46 (7) pp. 329-332, 1963.

211. Hargreaves, C.M., "On the Growth of Sapphire Mi-   W3a/10
      crocrystals," J. Appl. Phys., 32 (5) pp. 936-938,
      1961.

212. Harris, R.S., "Investigation of Glass-Metal Compos-   C2c/6
      ite Materials," Owens-Corning Fiberglas Corp.,
      Newark, Ohio, Naval Res. Lab. cont. NOrd 15764,
      Qtrly. Rpt., July 1960.

213. Hasselman, D.P.H. and Batha, H.D., "Strength of   W1/22
      Single Crystal Silicon Carbides," Appl. Phys. Letters,   W2/8
      2 (6) pp. 111-113, March 1963.

214. Hawthorne, A.T., et al, "Development of Fabric Base   F3b/7
      Materials for Space Applications," Westinghouse De-
      fense & Space Center, Baltimore, Md., AFML, WP-
      AFB, AF 33(616)-8259, ML-TDR-64-20, Jan. 1964.

215. Henning, R.G., "Growth Parameters of Tin Metal   W3j/2
      Whiskers Produced by Pressure," Master's Thesis,
      Univ. of Utah, August 1963.

* 216. Herring, C. and Galt, J.K., "Elastic and Plastic   W1/23
       Properties of Very Small Metal Specimens," Phys.   W2/9
       Rev., 85 (6) pp. 1060-1061, March 1952.   W3b/1

217. Hertzberg, R.W., "Effect of Growth Rate and Cr   W3e/3
      Whisker Orientation on Mechanical Behavior of a Uni-
      directional by Solidified Cu-Cr Eutectic Alloy," Trans.
      ASM, 57 (2) pp. 434-441, 1964.

218. Hillig, W.B., "The Factors Affecting the Strength of   R3/5
      Bulk Fused Silica," G.E. Co., Research Laboratory,
      Schenectady, N.Y., Rpt. 61-RL-2777M, July 1961.

* 219. Hillig, W.B., "The Sources of Weakness and the Ulti-   R6/4
       mate Strength of Brittle Amorphous Solids," G.E.
       Co., Research Laboratory, Schenectady, N.Y., Rpt.
       No. 62-RL-2896M, February 1962.

220. Hillig, W.B., "The Strength of Bulk Fused Quartz,"   R3/4
      G.E. Co., Research Laboratory, Schenectady, N.Y.,
      Rpt. No. 60-RL-2449M, Aug. 1960.

221. Hillig, W.B. and Charles, R.J., "Concerning Dislocations in Glass," G.E. Co., Research Laboratory,
Schenectady, N.Y., Rpt. 60-RL-2535M, Oct. 1960.   R3/6

* 222. Hillig, W.B. and Charles, R.J., "Surface, Stress-
Dependent Reactions, and Strength," G.E. Co., Re-   R6/3
search Laboratory, Schenectady, N.Y., Rpt. No. 64-
RL-3756M, October 1964.

223. Hoffman, G.A., "The Exploitation of the Strength of
Whiskers," The Rand Corp., Santa Monica, Calif.,   W5/9
Rpt. P-1294, March 1958.

224. Holland, L., The Properties of Glass Surfaces, John
Wiley & Sons Inc., New York, 546 pp., 1964.   R3/7

225. Hollinger, D.L., "Influence of Stress Corrosion on
Strength of Glass Fibers," G.E. Co., AETD, Even-   F1/15
dale, Ohio, Naval Res. Lab. contract NOnr 4486(00)   F2e/4
(X), 1st Bi-monthly Prgr. Rpt., May 1964.

226. Hollinger, D.L., et al, "High Strength Glass Fibers
Development Program," G.E. Co., FPLD, Evendale,   F1/19
Ohio, Dept. of Navy contract NOw 61-0641-c (FBM),   F2e/5
Final Rpt., May 1963.

227. Hollinger, D.L., et al, "Influence of Stress Corrosion
on Strength of Glass Fibers," G.E. Co., AETD,   F1/16
Evendale, Ohio, Naval Res. Lab. contract NOnr   F2e/5
4486(00)(X), 2nd Bi-monthly Prgr. Rpt., July 1964.

228. Ibid, 3rd Bi-monthly Prgr. Rpt., September 1964.   F1/17
F2e/5

229. Ibid, 4th Bi-monthly Prgr. Rpt., November 1964.   F1/18
F2e/5

230. Holloway, D.G., "The Strength of Glass Fibres,"
Phil. Mag., 4 pp. 1101-1106, 1959.   F1/20

231. Holloway, D.G. and Schlapp, D.M., "The Strength of
Glass Fibres Drawn in Vacuum," Trans. Brit. Cer.   F1/21
Soc., 59 pp. 424-431, 1960.   F2e/b

232. Hough, R.L., "Continuous Pyrolytic Graphite Composite Filaments," Proc. 5th Annual Structures and   CF1/3
Materials Conference AIAA, pp. 471-479, 1964.   CF3/14

233. Hough, R.L., "Refractory Reinforcements for Ablative Plastics. Part I: Synthesis and Reaction   CF2/10

233.   Mechanism of Fibrous Zirconium Nitride," AFML,   CF2/10
(cont) WPAFB, Dayton, Ohio, Proj. 7340, ASD-TDR-62-
        260, Pt. I, June 1962.

234.   Ibid., "Part II: Synthesis of Zirconium Nitride and   CF2/11
        ZrO$_2$ Flakes," October 1963.

235.   Ibid., "Part III: Pyrolytic Boride Reinforcing Agents," CF2/12
        August 1964.

236.   Ibid., "Part IV: Synthesis Apparatus for Continuous   CF2/13
        Filamentous Reinforcements," December 1963.

237.   Howard, J.S., Jr., "When to Design for Filament   C2a/22
        Winding," Prod. Eng., 35 (22) pp. 102-107, Oct. 26,
        1964.

238.   Hull, E.H. and Malloy, G.T., "The Strength of Dia-   R6a/4
        mond," G.E. Co., Research Laboratory, Schenecta-
        dy, N.Y., Rpt. No. 64-RL-3824C, November 1964.

239.   Hulse, C.O., "Formation and Strength of MgO   W1/24
        Whiskers," J. Am. Cer. Soc., 44 (11) pp. 572-   W2/10
        575, 1961.   W3a/11

240.   Islinger, J., et al, "Mechanism of Reinforcement of   C2a/23
        Fiber-Reinforced Structural Plastics and Composites,"
        IITRI, Chicago, Ill., ASD, WPAFB, AF 33(616)-5983,
        WADC TR 59-600, Part II, June 1960.

241.   Jaccodine, R.J. and Kline, R.K., "Whisker Growth   W3a/12
        from Quartz," Nature, 189 (2) p. 298, 1961.

242.   Jeszenski, B. and Hartmann, E., "Growth and Mechan-   W1/25
        ical Properties of NaCl Whiskers," Soviet Physics-   W3f/1
        Crystallography, 1 (3) pp. 341-344, 1962. (in English)

243.   Jobaris, J., "Optimum Filament Diameter," Narmco   F2a/5
        Res. & Dev., San Diego, Calif., Navy BuShips, con-
        tract NO1s 86347, 4th Qtrly, Prog. Rpt., Jan 1964.

244.   Johnson, D.E., et al, "Candidate Materials for High   A2/4
        Temperature Fabrics," A.D. Little Inc., Cambridge,   CF4/5
        Mass., WADC, WPAFB, AF 33(616)-5880, WADC TR
        59-155, Sept. 1959.

245.   Johnson, D.E. and Newton, E.H., "Metal Filaments   A2/3
        for High Temperature Fabrics," A.D. Little, Inc.,   CF4/4
        Cambridge, Mass., AFML, WPAFB, AF 33(657)-
        10539, ASD-TDR-62-180, Part II, Feb. 1963.

246. Johnson, E.R. and Amick, J.A., "Formation of Single   W3a/13
      Crystal Silicon Fibers," J. Appl. Phys., 25 (9) pp.
      1204-1205, 1954.

247. Johnson, R.C., et al, "Synthesis and Some Properties   W1/26
      of Fibrous Silicon Nitrides," Bureau of Mines, Norris,   W3a/14
      Tenn. Rpt. BM-RI-6467, 1964.

248. Keller, E., "Whiskers as Reinforcement Materials,"   C1/2
      Ind. and Eng. Chem., 56 (4) pp. 9-10, 1964.

* 249. Kelly, A., "Fiber-Reinforced Metals," Sci. Am., 212   C1/3
        (2) pp. 28-37, 1965.   C5/12

250. Kelly, A., "The Promise of Fiber Strengthened Mate-   C1/4
      rials," Discovery, 24 (9) pp. 22-27, 1963.   C5/12

251. Kelsey, R.H., "Reinforcement of Nickel Chromium   Cld/3
      Alloys with Sapphire Whiskers," Horizons Inc., Cleve-
      land, Ohio, BuWeps cont. NOw-63-0138-c, Final
      Report, October 1963.

252. Kerper, M.J., et al, "Properties of Glasses at Elevat-   R3/8
      ed Temperatures," National Bureau of Standards,
      Washington, D.C., ASD, WPAFB, AF 33(657)-62-362,
      WADC TR 56-645, Part VIII, Mar. 1963.

253. Kestigan, M. and Tombs, N.C., "Growth of Single   R7/1
      Crystals of Electronic Materials," Sperry Eng. Rev.,
      16 (3) pp. 2-9, Fall 1963.

* 254. Kies, J.A., "The Strength of Glass," U.S.N. Res.   R3/9
        Lab., Washington, D.C., NRL Rpt. 5098, April 1958.

255. Kies, J.A., "The Strength of Glass Fibers and the   F1/22
      Failure of Filament Wound Pressure Vessels," U.S.N.   F2/8
      Res. Lab., Washington, D.C., NRL Rpt. 6034,   F5/7
      February 1964.

256. King, H.A., "Reinforced Plastics for Aerospace and   C5/14
      Hydrospace," Research and Dev., 15 (12) pp. 17-20,
      1964.

257. King, H.A., "Reinforced Plastics in Aerospace and   C5/13
      Hydrospace in the Year 2000," Proc. Soc. Plastic Ind.,
      19th Annual Mtg., 16-A, pp. 1-12, Feb. 1964.

258. Kirschner, H.P. and Knoll, P., "Silicon Carbide   W1/27
      Whiskers," J. Am. Cer. Soc., 46 (6) pp. 299-300,   W3c/3
      1963.

259. Kiselev, B.A., "Glass Fiber Reinforced Plastics,"   C5/15
      U.S. Dept. Comm., Washington, D.C., Joint Publi-
      cations Res. Ser. Rpt. 19389, May 1963. (in English)

260. Klassen-Neklyudova, M.V. and Rozhanskii, V.N.,   R6a/5
      "Fundamental Problems in the Physics of the Strength
      and Plasticity of Crystals," Soviet Physics-
      Crystallography, $\underline{1}$ (4) pp. 403-409, 1963. (in English)

* 261. Kliman, M.I., "Formation of Alumina Whiskers,"   W1/28
      Watertown Arsenal Laboratories, Watertown, Mass.,   W3a/15
      D/A Proj. 59332007, Rpt. No. WAL TR 371/52,
      November 1962.

262. Kliman, M.I., "Formation and Properties of Alumina   W1/29
      Fiber," Watertown Arsenal Laboratories, Watertown,   W3a/15
      Mass., D/A Proj. 59332007, Rpt. No. WAL TR 371/
      50, November 1962.

263. Kliman, M.I., "Impact Strength of Alumina Fiber-   C1e/1
      Ceramic Composites," Watertown Arsenal Labs.,
      Watertown, Mass., D/A Proj. 59332007, Rpt. WAL
      TR 371/51, November 1962.

264. Kliman, M.I., "Transverse Rupture Strength of Alu-   W2/11
      mina Fiber-Ceramic Composites," Watertown Arse-
      nal Laboratories, Watertown, Mass., D.A. Proj.
      59332007, Rpt. WAL TR 371/53, November 1962.

265. Knippenberg, W.F., et al, "Crystal Growth of Silicon   W3a/16
      Carbide," Philips Tech. Rev., $\underline{24}$ (6) pp. 181-183,
      1962/1963.

266. Koch, P.A., "Glass Filaments and Textiles," CIBA   F3a/9
      Review, pp. 19-33, 1963/1965.   F3b/8
                                                                F5/8

267. Kostyuk, V.E., et al, "Strength of Metal Whisker   W1/30
      Crystals Containing Impurities," Soviet Physics-Solid   W3c/4
      State, $\underline{5}$ (11) pp. 2241-2245, May 1964. (in English)

268. Kress, E., "Whiskers--- But Not on Your Face,"   W5/10
      Space Digest, pp. 70-72, 1960.

269. Krock, R.H. and Kelsey, R.H., "Whiskers - Their   W5/11
      Promise and Problems," Ind. Res., $\underline{1}$ (2) pp. 46-57,   C5/16
      1965.

270. Kroenke, W.J., "Flexible Glass Fibers," The B.F.   F1/26
      Goodrich Co., Akron, Ohio, AFML, WPAFB, con-   F2/9
      tract AF 33(657)-8905, ML-TDR 64-119, Mar. 1964.   F3a/10

271. Kroenke, W.J., "Linear Structured Glass Fibers," B.F. Goodrich Co., Akron, Ohio, AFML, WPAFB, contract AF 33(657)-8905, Qtrly. Prog. Rpt., February 1964.

F1/23
F2/9
F3a/10

272. Ibid., Qtrly. Prog. Rpt., December 1964.

F1/24

273. Kroenke, W.J., et al, "Structured Glasses Patterned After Asbestos," The B.F. Goodrich Co., Akron, Ohio, AFML, WPAFB, Contract AF 33(657)-8905, RTD-TDR-63-4043, Part I, April 1963.

F1/25
F2/9
F3a/10

274. Lachman, W.L. and Sterry, J.P., "Ceramic Fibers," Chem. Eng. Prog., 58 (10) pp. 37-41, 1962.

CF2/14

275. Laird, J.A., et al., "Glass Surface Chemistry for Glass Fiber Reinforced Plastics," A.O. Smith Corp., Milwaukee, Wisc., BuWeps cont. NOw-62-0679-c (FBM), Final Summary Report, June 1963.

C4b/6

276. Lambertson, W.A., et al, "Continuous Filament Ceramic Fibers," Carborundum Co., Niagara Falls, N.Y., ASD WPAFB, AF 33(616)-6246, WADD TR 60-244, June 1960.

F1/27
F3a/11

277. Lasday, A.H., "Development of High Modulus Fibers from Heat Resistant Materials," Hauze Glass Co., Point Marion, Pa., ASD, WPAFB, AF 33(616)-5263, WADC TR 58-285, Oct. 1958.

F1/28
F3a/12

278. Lauchner, J.H., et al, "High Temperature Inorganic Structural Composite Materials," Miss. State Univ., State College, Miss., AFML, WPAFB, AF 33(616)-7765, ASD-TDR-62-202, Pt. I, Nov. 1962.

C2d/1

279. Ibid., ASD-TDR-62-202, Part II, Jan. 1963.

C2d/2

280. Laudise, R.A., "Growing Oxide Crystals," Bell Labs. Record, 40 (7) pp. 244-250, 1962.

R7/2

* 281. Lemkey, F.D. and Kraft, R.W., "Tensile Testing Technique for Submicron Specimens," Rev. Sci. Inst., 33 (8) pp. 846-849, 1962.

W4/11

282. Lewis, A., et al, "Research to Obtain High Strength Continuous Filaments from Type 29-A Glass Formulations," Aerojet-General Corp., Azuza, Calif., AFML, WPAFB, AF 33(657)-8904, RTD-TDR-63-4241, Dec. 1963.

F1/29
F2/10
F3a/13

283. Linares, R.C., "Growth of Refractory Oxide Single   R7/3
      Crystals," J. Appl. Phys., 33 (5) pp. 1747-1749,
      1962.

284. Lindsay, E.M., et al, "Glass Reinforcements for Fil-   F1/31
      ament Wound Composites," Owens-Corning Fiberglas
      Corp., Toledo, Ohio, AFML, WPAFB, contract AF
      33(657)-9623, Int. Eng. Prg. Rpt., Aug. 1963.

* 285. Lindsay, E.M. and Rood, J.C., "Glass Reinforce-   F1/30
       ments for Filament Wound Composites," Owens-   F5/9
       Corning Fiberglas Corp., Toledo, Ohio, AFML,
       WPAFB, AF 33(657)-9623, Final Tech. Eng. Rpt.
       TR-64-8-104, Dec. 1963.

286. Linz, A., et al, "Growth of Crystals by Flame Fusion,"   R7/4
      MIT, Cambridge, Mass., AFML, WPAFB, AF 33(616)-
      8353 and AF 19(628)-395. Rpt. MIT-TR-185, Decem-
      ber 1963.

287. Lockhart, R.J., "Experimental Research on Filamen-   C1e/2
      tized Ceramic Radome Materials," Horizons Inc.,
      Cleveland, Ohio, AFML, WPAFB, cont. AF 33(615)-
      1167, Int. Eng. Prgr. Rpt., September 1964.

288. Loewenstein, K.L., "Studies in the Composition and   R3/10
      Structure of Glasses Possessing High Young's Modulus
      Part 1: The Composition of High Young's Modulus
      Glasses and the Function of Individual Ions in the Glass
      Structure," Phys. and Chem. Glasses, 2 (3) pp. 69-82,
      June 1961.

289. Ibid., "Part 2: The Effects of Changes in the Configu-   R3/11
      ration Temperature," Phys. and Chem. Glasses, 2 (4)
      pp. 119-125, Aug. 1961.

290. Luborsky, F.E., et al, "Crystallographic Orientation   W1/32
      and Oxidation of Sub-Micron Whiskers of Iron, Iron   W3a/18
      Cobalt, and Cobalt," J. Appl. Phys., 34 (9) pp. 2905-
      2909, 1963.

291. Luborsky, F.E., and Morelock, C.R., "Magnetization   W1/31
      Reversal in Cobalt Whiskers," G.E. Co., Research   W3a/17
      Laboratory, Schenectady, N.Y., Rpt. 64-RL-3787M,   W3c/5
      October 1964.

* 292. Lynch, C.T., et al, "Growth and Analysis of Alumina   W3/11
       Whiskers," ASD, WPAFB, Dayton, Ohio, Proj. 7350,   W5/12
       ASD-TDR-62-272, May 1962.

293. Machlan, G.R., "The Development of Fibrous Glasses
Having High Elastic Moduli,"Owens-Corning Fiberglas
Corp., Newark, Ohio, WADC, WPAFB, AF 33(616)-
2422, WADC TR 55-290, Nov. 1955.

F1/32
F2/11
F3a/14
F4a/3

* 294. Machlin, E.S., "Non-Metallic Fibrous Reinforced
Metal Composites," Mat. Res. Corp., Yonkers, N.Y.,
BuWeps NOw-61-0209-c, Status Report, Sept. 1961.

C5/17

295. Mallinder, F.P. and Proctor, B.A., "Elastic Con-
stants of Fused Silica as a Function of Large Tensile
Strain," Phys. & Chem. Glasses, 5 (4) pp. 91-103,
Aug. 1964.

F1/33
F2/12
F5/10

296. Marco, D.M., "New and Improved Materials for Ex-
pandable Structures," Goodyear Aircraft Corp.,
Akron, Ohio, AFML, WPAFB, AF 33(616)-7865, ASD-
TDR-62-542, Part I, June 1962.

C2a/26

297. Margo, G., "Exploratory Investigation of Inorganic
Fiber Reinforced Inorganic Laminates," Goodyear
Aircraft Corp., Akron, Ohio, ASD, WPAFB, AF
33(616)-5251, WADC TR 58-298, Part I, Oct. 1958.

C2d/3

298. Ibid., Part II, June 1959.

C2d/4

299. Markali, J., Mechanical Properties of Engineering
Ceramics, edited by W.W. Kriegel and J. Palmour,
III, Interscience Publishers, New York, p. 93, 1961.

W1/33
W3/12

* 300. Marsh, D.M., "Micro Tensile Testing Machine,"
J. Sci. Inst., 38 (6) pp. 229-234, 1961.

W4/13

301. Marsh, D.M., "Tensile Testing Machine for Microscopic Specimens," J. Sci. Inst., 36 (4) pp. 165-169,
1959.

W4/12

302. Matta, J.A. and Outwater, J.O., "The Nature, Ori-
gin and Effects of Internal Stresses in Reinforced
Plastic Laminates," SPE Trans., 2 (4) pp. 314-320,
Oct. 1962.

C2a/27

303. Mazdiyasni, K.S. and Lynch, C.T., "An Approach to
the Preparation of Powders, Fibers and Films of
Ultra-High Purity Ceramics," AFML, WPAFB, Dayton, Ohio, ML TDR 64-269, September 1964.

CF2/15

304. Mazur, J. and Rafalowicz, J., "Hypothesis on the
Ionic Mechanism of the Growth of Whiskers Obtained
by the Reduction of Metal Halides," Brit. J. Appl.
Phys., 12 pp. 569-571, October 1961.

W3c/6

305. McDaniels, D.L., et al, "Stress-Strain Behavior of   C1a/6
      Tungsten-Fiber Reinforced Copper Composites,"   C3/5
      NASA-Lewis Res. Center, Cleveland, Ohio, NASA TN
      D-1881, October 1963. .

306. McMullen, P., "The Tensile and Flexural Strengths of   C2a/24
      Some Glass Reinforced Plastics Tested at Various   C3/6
      Rates of Loading," Royal Aircraft Est., Farnborough,
      England, Tech. Note CPM 16, April 1963.

307. McMurdie, H.F., editor, "Research on Crystal   R7/5
      Growth and Characterization at the National Bureau of
      Standards, January to June 1964," National Bureau of
      Standards, Washington, D.C., NBS Tech. Note 251,
      October 1964.

308. McNeil, D.W., et al, "Mechanism of Water Absorp-   C2a/25
      tion in Glass Reinforced Plastics," Battelle Mem.   C4b/7
      Inst., Columbus, Ohio, BuShips Cont. NObs-86871,
      Qtrly, Prog. Rpt. No. 3, April 1963.

309. Mehan, R.L., Feingold, E., and Gatti, A., "Evalua-   W1/34
      tion of Sapphire Wool and Its Incorporation into Com-   W4/14
      posites of High Strength," G.E. Co., Missile and   C1d/4
      Space Div., King of Prussia, Pa., AFML, WPAFB,
      AF 33(615)-1696, 3rd Qtrly, Rpt., February 1965.

310. Merolo, S.A., "Properties and Characteristics of   C2a/28
      Ablative Plastic Chars. Part I: Organic-Fiber Rein-
      forced Plastics," AFML, WPAFB, Dayton, Ohio,
      ASD-TDR-62-1028, Feb. 1963.

* 311. Metcalfe, A.G. and Schmitz, G.K., "Effect of Length   F1/34
        on the Strength of Glass Fibres," ASTM Preprint No.   F2a/6
        87, presented at 67th Annual Mtg., June 1964.   F4a/4

312. Mettes, D.G., "Glass Fibers in Aerospace," pre-   F5/11
      sented to SPE, New York City Meeting, May 1964.   C2a/29

313. Milewski, J.V. and Shyne, J., "Whiskers Make Rein-   C1f/2
      forced Plastics Better Than Metals," paper presented
      Soc. Plastics Industry, Chicago Mtg., February 1965.

314. Milewski, J.V. and Shyne, J., "Whiskers - The Ulti-   C1f/1
      mate Reinforcement for Plastics," SPE 20th Annual
      Conf., 10, January 1964.

315. Morelock, C.R., "Pyrolytic Graphite Coated Fused   F2c/16
      Silica," G.E. Co., Research Lab., Schenectady,
      N.Y., Rpt. 61-RL-2672M, Mar. 1961.

316. Morelock, C.R., "Sub-Micron Whiskers by Vapor Deposition," ACTA Met., 10 pp. 161-167, 1962.   W3a/19

317. Morelock, C.R. and Sears, G.W., "Growth of Whisk-ers by Chemical Reaction," J. Chem. Phys., 34 pp. 1008-1009, 1961.   W3a/20

318. Morley, J.G., "Reinforcing Metals with Fibers," New Scientist, (322) pp. 122-125, Jan. 17, 1963.   F1/35
F2/13
F2c/17
C2c/7

319. Morley, J.G., "Strength of Glass Fibres," Nature, 184 (4698) p. 1560, Nov. 1959.   F1/36
F2c/17

* 320. Morley, J.G., et al, "Strength of Fused Silica," Phys. and Chem. Glasses, 5 (1) pp. 1-10, Feb. 1964.   F1/37
F2/13
F3a/15
F4a/5
F5/12

321. Morris, W.G. and Ogilvie, R.E., "Kossel Studies of Iron Whiskers," MIT, Cambridge, Mass., AFML, WPAFB, cont. AF 33(657)-8906, RTD-TDR-63-4198, January 1964.   W4/15

322. Morrison, A.R., et al, "High Modulus, High Tem-perature Laminates with Fibers and Flakes," Owens-Corning Fiberglas Co., Granville, Ohio, AFML, WPAFB, AF 33(616)-5802, WADD TR-60-24, Suppl. 4, Feb. 1962.   F1/38
F2/14
C2a/30

323. Mould, R.E., "Cross-Bending Tests of Glass Fibers and the Limiting Strength of Glass," J. App. Phys., 29 (11) pp. 1263-1264, 1958.   F1/39
F4a/6

324. Muller, E.W. and Nishikawa, O., "Field Ion Micros-copy of Iron Whiskers," Penn. State U., Univ. Park, Pa., AFML, WPAFB, AF 33(616)-6397, Final Rpt., November 1963.   W4/16

325. Murgatroyd, J.B., "The Delayed Elastic Effect in Glass Fibers and the Constitution of Glass in Fiber Form," J. Soc. Glass Tech., 37 pp. 291-300, 1953.   F1/40
F2b/3

326. Murgatroyd, J.B., "The Strength of Glass Fibers. Part I: Elastic Properties," J. Soc. Glass Tech., 28 (130) pp. 368-387, 1944.   F1/41
F2b/3
F2d/7

327. Ibid., "The Strength of Glass Fibers. Part II: The Effect of Heat Treatment on Strength," pp. 388-405T.   F1/42
F2b/3
F2e/7

328. Murphy, E.A. and O'Rourke, R.G., "Fabrication of Ultrafine Beryllium Wire," The Brush Beryllium Co., Cleveland, Ohio, BuWeps cont. NOw-63-0137-c, TR 318-240, Aug. 1963.   CF4/6

329. Murphy, W.K. and Sears, G.W., "Iron Whisker Resonators," Appl. Phys. Letters, 3 (4) pp. 55-57, Aug. 1963.   A1e/1

330. Myers, N.C., et al, "Investigation of Structural Problems with Filament Wound Deep Submersibles," H.I. Thompson Fiber Glass Co., Gardena, Calif., BuShips cont. NObs-88351, 3rd Qtrly. Rpt., Oct. 1963.   C2a/31

331. Nabarro, F.R.N. and Jackson, P.J., Growth and Perfection of Crystals, edited by R.H. Doremus, B.W. Roberts and D. Trumbull, John Wiley & Sons, Inc., New York, p. 13, 1958.   W3/13 W5/13

332. Nadgornyi, E.M., "Perfection and Strength of Crystals," FTD Rpt., Contemp. Probl. of Phys. Met., N64-16606, July 1963. (in English)   W5/16

* 333. Nadgornyi, E.M., "The Properties of Whiskers," Soviet Physics-Uspekhi, 5 (3) pp. 462-477, 1962. (in English)   W5/14

* 334. Nadgornyi, E.M., et al, "Filamentary Crystals with Almost the Theoretical Strength of Perfect Crystals," Soviet Physics-Uspekhi, 67 (2) pp. 282-304, 1959. (in English)   W5/15

335. Nannichi, Y., "Formation of Silicon Whiskers on a Subliminating Surface," Appl. Phys. Letters, 3 (8) pp. 139-142, October 1963.   W3/14

336. Newkirk, J.B. and Sears, G.W., "Growth of Potassium Halide Crystals from Aqueous Solution," ACTA Met., 3 (1) pp. 110-111, 1955.   W3f/2

337. Newton, E.H. and Johnson, D.E., "Fine Metal Filaments for High Temperature Applications," A.D. Little, Inc., Cambridge, Mass., AFML, WPAFB AF 33(657)-10539, ML TDR 64-92, Feb. 1964.   CF4/7

338. Nieto, M.M. and Russell, A.M., "Growth of Whiskers Due to Solid-to-Solid Phase Transformation in Zr," J. Appl. Phys., 35 (2) p. 461, 1964.   W3/15

339. Nikolayov, V. Yu and Kasatochkin, V.I., "Fibrous   CF1/4
      Carbon," AFSC, WPAFB, FTD Rpt. TT63-330, May
      1963. (in English)

340. Norris, L.F. and Parravano, G., "Sintering of Zinc   W3g/2
      Oxide," J. Am. Cer. Soc., 46 (9) pp. 449-452, 1963.

341. Otto, W.H., "Compaction Effects in Glass Fibers,"   F1/43
      J. Am. Cer. Soc., 44 (2) pp. 68-72, 1961.

* 342. Otto, W.H., "Relationship of Tensile Strength of Glass   F1/44
        Fibers to Diameter," J. Am. Cer. Soc., 38 (3) pp.   F2a/7
        122-124, 1955.

343. Otto, W.H., et al, "Potential of Filament Wound Com-   F1/45
      posites," Narmco Res. & Dev., San Diego, Calif.,   C2a/32
      BuWeps cont. NOw-61-0623-C (FBM), Qtrly. Rpt.,   A2a/2
      May 1962.

344. Otto, W.H. and Vidanoff, R.B., "Silica Fiber Forming   F1/46
      and Core Sheath Composite Fiber Development,"   F3a/16
      Narmco Ind. Inc., San Diego, Calif., BuWeps, cont.   CF3/15
      N600(19)59607, Qtrly. Prgr. Rpt. 1, April 1963.

345. Ibid., Final Summary Rpt., Jan. 1964.   F1/47
                                                               F3a/16

346. Outwater, J.O. and Ozaltin, O., "Surface Effects of   F2c/18
      Various Environments and of Thermosetting Resins on
      the Strength of Glass," SPE Trans., 2 (4) pp. 321-325,
      Oct. 1962.

347. Papalegis, F.E. and Bourdeau, R.G., "Pyrolytic Re-   W1/35
      inforcements for Ablative Plastic Composites," High   W3/17
      Temp. Mtls., Inc., Boston, Mass., AFML WPAFB,   W4/17
      cont. AF 33(657)-11094, ML TDR 64-201, July 1964.   CF1/5

348. Papalegis, F.E. and Bourdeau, R.G., "Pyrolytic Re-   W3/16
      inforcing Agents for Ablative Erosion-Resistant Com-
      posites," High Temp. Mtls. Inc., Boston, Mass.,
      ASD, WPAFB, cont. AF 33(657)-8651, ASD-TDR-63-
      403, May 1963.

349. Papalegis, F.E. and Bourdeau, R.G., "Research on   W1/36
      New and Improved Thermally Protective Fiber Rein-   W3/17
      forced Composite Materials," High Temp. Mtls. Inc.,   CF1/5
      Boston, Mass., ASD WPAFB, AF 33(657)-8651,
      August 1962.

350. Ibid., Qtrly. Prgr. Rpt., November 1962.   W1/37
                                                               CF1/5

351. Ibid., Qtrly. Prgr. Rpt., February 1963.   W1/38
                                                               CF1/5

352. Papkov, V.S. and Berezhkova, G.V., "Production of   W2a/5
      Filamentary Al$_2$O$_3$ Crystals," Kristallografiya 9 (3)   W3a/21
      pp. 442-443, 1963. (in Russian)

353. Parikh, N.M., "Fiber-Reinforced Metals and Alloys,"   C1a/7
      IITRI, Chicago, Illinois, BuWeps cont. NOw-64-   C1d/5
      0066-c, Bi-Monthly Rpt., February 1964.

                                                                  C1a/8
354. Ibid., April 1964.   C1d/5

                                                                  C1a/9
355. Ibid., June 1964   C1d/5

                                                                  C1a/10
356. Ibid., August 1964.   C1d/5

357. Parikh, N.M. and Adamski, R., "Fiber Reinforced   C1a/11
      Metals and Alloys," IITRI, Chicago, Illinois, BuWeps   C1d/5
      cont. NOw-64-0066-c, IITRI-B6020-5, October 1964.

* 358. Parratt, N.J., "Defects in Glass Fibers and Their Ef-   F2/15
        fect on the Strength of Plastic Mouldings," Rubber and   C2a/33
        Plastics Age, 41 (3) pp. 263-266, March 1960.

359. Parratt, N.J., "Reinforcing Effects of Silicon Nitride   C1d/6
      Whiskers in Silver and Resin Matrices," Powder Met-   C1f/3
      allurgy, 7 (14) pp. 152-167, 1964.

* 360 Patrick, R.L., "Physical Techniques Used in Studying   C4b/9
       Interfacial Phenomena," SPE Trans., 1 (4) pp. 181-
       190, Oct. 1961.

361. Patrick, R.L. and Port, E.A., "Interfacial Inter-   C4b/8
      action in Composite Structures," Alpha Res. & Dev.
      Inc., Blue Island, Ill., BuWeps cont. NOw-64-0413-c,
      Qtrly. Prgr. Rpt. 2, Sept. 1964.

* 362. Pearson, G.L., et al, "Deformation and Fracture of   W4/18
        Small Silicon Crystals," ACTA Met., 5 (4) pp. 181-
        191, 1957.

363. Peterson, G.P., "Optimum Filament Wound Compos-   C2a/34
      ites," ASD, WPAFB, Dayton, Ohio, WADD TN 61-50,
      June 1961.

364. Petker, I., et al, "Development of High Strength Im-   F1/48
      pregnated Roving for Filament Winding," Aerojet-   F3a/17
      General Corp., Sacramento, Calif., U.S. Navy Spl.   C2a/35
      Proj. cont. NOw-63-0627-C (FBM), Final Rpt.,
      July 1964.

365. Phillips, C.F., Glass: Its Industrial Applications,   R3/12
      Reinhold Publishing Corp., New York, 252pp., 1960.

366.   Pierce, C.M., "The Growth of Metal and Metallic   W3/18
       Oxide Whiskers: An Annotated Bibliography," Lock-   W5/17
       heed Missile and Space Co., Sunnyvale, Calif., SB
       63-68, May 1964.

367.   Pieser, H.S., editor, "Research on Crystal Growth   R7/6
       and Characterization at the National Bureau of Stand-
       ards, July to December 1963," National Bureau of
       Standards, Washington, D.C., NBS Tech. Note 236,
       April 1964.

368.   Pitt, C.H. and Henning, R.G., "Pressure Induced   W3j/3
       Growth of Metal Whiskers," J. Appl. Phys., 35 (2)
       pp. 459-460, 1964.

369.   Pladek, O.J., "Reinforced Polycarbonates," Rein-   C2a/36
       forced Plastics, 4 (1) pp. 28-29, Jan./Feb. 1965.   C5/18

* 370. Polanyi, Von M., "The Nature of Fracture," Z. Phys., R6a/6
       7 pp. 323-327, 1921.

371.   Popper, P. and Ruddlesden, S.N., "The Preparation,   W1/39
       Properties, and Structure of Silicon Nitride," Trans. W3/19
       Brit. Cer. Soc., 60 (9) pp. 615-619, 1961.

* 372. Poremba, F.J., comp., "Mechanical Properties of   W5/18
       Whiskers (Filamentary Crystals) 1957-1964," U.S.
       Steel Corp., Monroeville, Pa., Metals/Materials Div.,
       Bibliography No. 64-4, October 1964.

* 373. Price, P.B., "The Mechanical Properties of Whisk-   W1/40
       ers," Ph.D. Thesis, U. of Virginia, 1958.   W5/19

374.   Price, P.B., et al, "On the Growth and Properties of W1/41
       Electrolytic Whiskers," ACTA Met., 6, pp. 524-531,   W3d/1
       August 1958.

375.   "Properties of Selected Commercial Glasses," Bul-   R3/13
       letin B-83, Corning Glass Works, Corning, New York.

376.   Prosen, S.P., "Some Properties of NOL Rings of   C2b/2
       Carbon-Base Fibers and Epoxy Resin," Naval Ord-
       nance Lab., White Oak, Md., NOL TR 64-183, Dec.
       4, 1964.

377.   Prosen, S.P., et al, "Interlaminar Shear Properties   C2a/37
       of Filament Wound Composite Materials for Deep
       Submergence," Naval Ord. Lab., White Oak, Md.,
       Proc. 19th Annual Mtg. Soc. Plastics Ind., 9-D, pp.
       1-8, Feb. 1964.

378. Provance, J.D., "Glass Fiber Properties of a   F1/50
      Magnesia-Alumina-Silica System," presented at Fall   F2/16
      Mtg., Glass Div., Am. Cer. Soc., Bedford Springs,   F3a/18
      Oct. 1963.   F4a/7

                                                            F1/49
379. Provance, J.D., "Stronger Glass Fibers Lighten Aero- F2/16
      space Vehicles," Cer. Ind., 82 (4) pp. 109-111, 132, 142, F3a/18
                                                            F4a/7
      1964.   F5/13

380. Provance, J. and Kelly, L.W., "Optimizing Prop-   F1/51
      erties of Hauze 29-A Glass Fibers," Hauze Glass   F2/16
      Corp., Point Marion, Pa., AFML, WPAFB, AF   F3a/18
      33(657)-8904, 1st Qtrly, Prgr. Rpt., Sept. 1962.   F4a/7

                                                            F1/52
381. Ibid., 2nd Qtrly, Rpt., Dec. 1962.   F3a/18
                                                            F4a/7

382. Pullam, G.R., "Ceramics," Space Aeronautics R&D   R1/3
      Handbook, 42 (4) pp. 156-158, Sept. 1964.   R2/2

383. Pustovalov, V.V., "Change in Thermal Conductivity of R3/14
      Quartz Glass in the Process of Crystallization,"
      Steklo i Keramika, 5 pp. 28-30, May 1960, trans. by
      A.J. Peat, G.E. Co., Res. Lab., Schenectady, N.Y.,
      Rpt. 60-RL-2525M, Sept. 1960.

384. Rambauske, W. and Gruenzel, R.R., "Micro-X-Ray   W4/19
      Diffraction of Whiskers," Dayton U., Ohio Res. Inst.,
      AFML, WPAFB, AF 33(616)-8416, July 1963.

385. Raynes, B.C., "Studies of the Reinforcement of Met-   C1d/7
      als with Ultra High Strength Fibers (Whiskers)," Ho-
      rizons, Inc., Cleveland, Ohio, BuWeps cont. NOw-
      61-0207-c, Final Report, November 1961.

386. Raynes, B.C. and Kelsey, R.H., "Studies of the Re-   C1d/8
      inforcement of Metals with Ultra High Strength Fibers
      (Whiskers)," Horizons Inc., Cleveland, Ohio, BuWeps
      cont. NOw-62-0235-c, Final Report, October 1962.

387. Ibid., Int. Rpt. No. 2, January 1963.   C1d/9

388. Regester, R.F., "The Strength and Perfection of   W1/42
      Aluminum Oxide Whiskers," Masters Thesis, U. of   W5/20
      Penna., 1963.

389. Revere, A., "Electron Microscope Studies of the Sur- W4/21
      face Structure of Whiskers Grown from Iron Chlo-
      ride," Althea Revere, Vineyard Haven, Mass.,
      AFML, WPAFB, AF 33(657)-8554, ASD-TDR-63-4136,
      November 1963.

390. Revere, A., "Replication Techniques for Electron Microscopic Studies of Iron Whisker Surface," Althea Revere, Vineyard Haven, Mass., AFML, WPAFB, AF 33(616)-6604. WADD TR 60-287, March 1961.   W4/20

391. Riley, M.W., "Reinforced Plastics Get Stronger," Mat. in Des. Eng., 56 (2) pp. 99-103, Aug. 1962.   C2a/39

392. Riley, M.W., et al, "Filament Wound Reinforced Plastics: State of the Art," Mat. in Des. Eng., 52 (2) pp. 127-146, Aug. 1960.   C2a/38 C5/19

393. Roberts, D.A., "Physical and Mechanical Properties of Some High-Strength Fine Wires," Battelle Memorial Institute, DMIC Memo 80, January 1961.   CF4/8

394. Roberts, E.C. and Lawrence, S.C., "Whisker-on-Whisker Growth," ACTA Met., 5 p. 335, June 1957.   W2b/9

395. Roberts, R.W. and McElligott, P.E., "Formation of Silicon Whiskers," G.E. Co., Research Lab., Schenectady, N.Y., Rpt. 63-RL-3427C, September 1963.   W3a/22

396. Rollins, F.R., Jr., "Use of Graphite Whiskers in a Study of the Atmosphere Dependence of Graphite Friction," J. Appl. Phys., 32 (4) pp. 1454-1458, 1961.   W2d/2

397. Romstad, K., "Investigation of Methods for Evaluating Unwoven Glass-Fiber-Reinforced Plastic Laminates in Flexure," For. Prdts. Lab., Madison, Wisc., FPL Rpt. 024, Feb. 1964.   C2a/40 C3/7

398. Romstad, K., "Methods for Evaluating Tensile and Compressive Properties of Plastic Laminates Reinforced with Unwoven Glass Fibers," For. Prdts. Lab., USDA, Madison, Wisc., Rpt. No. FPL-052, Aug. 1964.   C2a/41 C3/7

* 399. Rosato, D.V., "Non-Woven Fibers in Reinforcing Plastics," Ind. & Eng. Chem., 54 (8) pp. 30-38, 1962.   C2a/42

400. Rosen, B.W., "Effect of Glass Form on Strength," G.E. Co., Missile and Space Div., King of Prussia, Pa., BuWeps cont. NOw-63-0674-C, Rpt. R64SD18, October 1963.   F2a/8 C2a/43

401. Ibid., "Hollow Glass Fiber-Reinforced Laminates," Final Rpt., Sept. 1964.   F2a/9 C2a/43

402. Rosen, B.W. and Ketler, A.E., Jr., "Hollow Glass Fiber-Reinforced Plastics," G.E. Co., Missile and   F2a/10 C2a/43

402.   Space Div., King of Prussia, Pa., BuWeps cont.   F2a/10
(cont)   NOw-61-0613-d, Final Rpt., May 1963.   C2a/43

403.   Roskos, T.G., et al, "Interfacial Properties of an   C2a/44
Amine-Modified Silane and Its Preformed Polymer at   C4b/10
E-Glass Surfaces, SPE Trans., 2 (4) pp. 326-331
Oct. 1962.

404.   Ross, J.H., "Needed: Textiles for the Space Age,"   A2a/3
Mat. on Des. Eng., 51 (6) pp. 138-140, June 1960.

405.   Ross, J.H. and Little, C.O., "Thermal Stability of   A2a/4
Flexible Fibrous Materials," AFML, WPAFB, Dayton,   A2b/1
Ohio, Rpt. RTD-TDR 63-4031, Oct. 1964.   A2d/4

406.   Ruane, A.G., comp., "Fiber Reinforcement in Metals   C5/20
and Materials, A Literature Search," G.E. Co., Res.
Lab., Schenectady, N.Y., Aug. 1964.

* 407.   Ruth, V. and Hirth, J.P., "The Kinetics of Diffusion   W3/20
Controlled Whisker Growth," Met. Eng. Dept., Ohio
State Univ., Columbus, Ohio, Office of Naval Res.,
cont. Nonr 495 (26), Tech. Rpt. 4, March 1964.

* 408.   Sadowsky, M.A., et al, "Effect of Couple Stresses on   C5/23
Force Transfer Between Embedded Microfibers,"
Watervliet Arsenal, Watervliet, N.Y., Rpt. WVT RR
6407, June 1964.

409.   Sadowsky, M.A. and Jussain, M.A., "Alpha-Clas-   C5/22
sification of Microfilms, Part IV," Watervliet Arsenal, Watervliet, N.Y., Rpt. WVT RR 6317, October
1963.

410.   Sadowsky, M.A. and Hussain, M.A., "Thermal Stress   C5/21
Discontinuities in Microfibers," Watervliet Arsenal,
Watervliet, N.Y. WVT RR 6401, AD 601-057, April
1964.

* 411.   Salkind, M.J., "Candidate Materials for Whisker   W5/22
Composites," Watervliet Arsenal, Watervliet, N.Y.,   C5/24
Rpt. WVT RR 6411, May 1964.

* 412.   Salkind, M.J., "Whiskers and Whisker Strengthened   W5/21
Composite Materials," Watervliet Arsenal, Water-   C5/24
vliet, N.Y., Rpt. WVT RR 6315, October 1963.

413.   Sandulova, A.V., et al, "Preparation and Some Prop-   W1/43
erties of Whisker and Needle-Shaped Single Crystals   W3/21
of Germanium, Silicon and Their Solid Solutions,"
Soviet Physics-Solid State, 5 (9) pp. 1883-1888,
March 1964. (in English)

414. Saunders, R.D. , "Filament Winding of Thick Lam-   C2a/45
inates for Deep Submergence," Proc. 19th Annual
Meeting, Soc. Plastics Ind. , Sect. 9-B, pp. 1-5, Feb.
1964.

415. Scala, E. , "The Design and Performance of Fibers   C5/25
and Composites," Cornell University, Materials Science Center, Ithaca, New York, Rpt. 286, November
1964.

416. Schmidt, D.L. and Hawkins, H.T. , "Pyrolyzed Rayon   C2b/3
Fiber Reinforced Plastics," AFML, WPAFB, Dayton,   A2a/5
Ohio, ML TDR 64-47, AD 438-892, April 1964.

417. Schmidt, D.L. and Jones, W.C. , "Carbon-Base   C2b/4
Fiber Reinforced Plastics," AFML, WPAFB, Day-   A2a/6
ton, Ohio. ASD-TDR-62-635, Aug. 1962.   A2c/1
  CF1/6

418. Schmidt, D.L. and Jones, W.C. , "Carbon-Base   C2b/5
Fiber-Reinforced Plastics," ASD, AFSC, WPAFB,   A2a/6
Chem. Eng. Prog. , 58 (10) pp. 42-50, 1962.   A2c/1
  CF1/6

419. Schmidt, D.L. and Jones,W.C. , "Carbon Fiber Plas-   C2b/6
tic Composites," SPE Journal, 20 (2) pp. 162-169,   A2a/6
Feb. 1964.   A2c/1
  CF1/6

420. Schmitz, G.K. , "Exploration and Evaluation of New   F1/53
Glasses in Fiber Form," Int'l. Harvester Corp. ,   F2a/11
Solar Div. , San Diego, Calif. , U.S. Naval Res. Lab.   F4a/8
contract NOnr-3654(00) (X) A2, Bi-monthly Prog. Rpt.
2, May 1964.

421. Schuerch, H. , "Boron Filament Composite Materials   C2c/8
for Space Structures. Part I: Compressive Strength of   CF3/16
Boron-Metal Composites," Astro Research Corp. ,
Santa Barbara, Calif. , cont. NASw-652, Rpt. ARC-
R-168, Nov. 1964.

422. Schulman, S. , "Elevated Temperature Behavior of   CF1/7
Fibers," ASD, WPAFB, Dayton, Ohio, Rpt. WADD   CF2/16
TN 60-298, August 1961.

423. Ibid. , Rpt. ASD-TDR-63-62, April 1963.   CF1/8

424. Schulman, S. and Epting, J. , "ASurvey of Test Pro-   W4/22
cedures for Evaluating Short Brittle Fibers," ASD,
WPAFB, Dayton, Ohio, ASD-TDR-63-92, May 1963.

* 425. Schuster, D. M. and Scala, E., "The Mechanical In-   C4a/1
         teraction of Sapphire Whiskers with a Birefringent
         Matrix," Trans. AIME, 230, pp. 1635-1640, Dec.
         1964.

  426.   Schwartz, H., "Tensile Strength of Ceramic Staple   F4a/9
         Fibers," WADC, WPAFB, Dayton, Ohio, WORT TM-
         56-104, Feb. 1957.

  427.   Sears, G. W., "Effect of Poisons on Crystal Growth,"   W2b/10
         J. Chem. Phys., 29 (5) pp. 1045-1048, 1958.

  428.   Sears, G. W., "Growth of Anthracene Whiskers by Va-   W3a/23
         por Deposition," J. Chem. Phys., 39 (11) pp. 2846-
         2847, 1963.

  429.   Sears, G. W., "A Mechanism of Whisker Growth,"   W3/23
         ACTA Met., 3 (7) pp. 367-369, 1955.

  430.   Sears, G. W., "Mercury Whiskers," ACTA Met., 1   W3/22
         (8) pp. 457-458, 1953.

  431.   Sears, G. W., "Ultimate Strength of Crystals," J.   W4/23
         Chem. Phys., 36 (3) p. 862, February 11, 1962.

  432.   Sears, G. W. and DeVries, R. C., "Growth of Alumina   W3a/24
         Crystals by Vapor Deposition," G. E. Co., Research
         Lab., Schenectady, N. Y., Rpt. 60-RL-2377M, March
         1960.

  433.   Shaffer, P. T. B. and Batha, H. D., "A Study of Funda-   W1/44
         mental Mechanical Properties of Ceramic Single Crys-   W2/12
         tals," Carborundum Co., Niagara Falls, N. Y., Bu-   W3/24
         Weps, N600 (19) 59749, BiMonthly Prgr. Rpt. No. 3,   W4/24
         July 1963.

  434.   Shand, E. B., "Experimental Study of Fracture of   R3/15
         Glass: I, The Fracture Process," J. Am. Cer. Soc.,
         37 (2) pp. 52-60, 1954.

  435.   Ibid., "II, Experimental Data," J. Am. Cer. Soc.,   R3/16
         37 (12) pp. 559-572, 1954.

  436.   Sharkitt, R. L. et al, "High Temperature Resistant   W1/45
         Beryllia Fiber-Reinforced Structural Composites,"   W2/13
         National Beryllia Corp., AFML, WPAFB, AF 33-   W3/25
         (616)-8066, ASD-TDR-62-632, Part II, June 1963.   C1d/10

  437.   Sherby, O. D., "Cambridge Conference on the   W1/46
         Strength of Metal Whiskers and Thin Films," Office   W2/14
         of Naval Res., Washington, D. C., ONRL Rpt. 34-58,
         April 1958.

* 438. Shindo, A., "Studies on Graphite Fibers," Japanese   CF1/9
Gov't. Ind. Res. Inst., Osaka, Japan, Rpt. #317,
December 1961.

439. Shoffner, J.E., et al, "Ceramic Matrix-Ceramic   C1e/3
Fiber Composite Systems," General Dynamics/Con-
vair, San Diego, Calif., AFML, WPAFB, cont. AF
33(657)-8926, Rpt. No. 8926-101, October 1959.

440. Sicka, R.W., et al, "Reinforcement of Nickel Chro-   C1d/11
mium Alloys with Sapphire Whiskers," Horizons Inc.,
Cleveland, Ohio, BuWeps cont. NOw-64-0125-c,
Final Report, October 1964.

441. Siefert, R.F., "Hollow Fibers," Reinforced Plastics,   F2a/12
$\underline{4}$ (1) pp. 22-23, Jan/Feb. 1965.

442. Silverman, A., "High Temperature Glass Fiber Pro-   F3a/19
duction," Trans. Soc. Glass Tech., $\underline{XL}$, pp. 413-428,
1956.

443. Sines, G., "Filamentary Crystals Grown from the   W3g/3
Solid Metals," J. Phys. Soc. Japan, $\underline{15}$ (10) pp. 1199-
1210, 1960.

* 444. Slayter, G., "Strength of Glass," Bull. Am. Cer.   F5/14
Soc., $\underline{31}$ (8) pp. 276-278, 1952.

445. Slayter, G., "Two Phase Materials," Sci. Am., $\underline{206}$ F5/15
(1) pp. 124-134, 1962.

446. Smalley, A.K., Atomics Int'l, Canoga Park, Calif.,   C1d/12
private communication, February 1965.

447. Smith, H.G. and Rundle, R.E., "X-Ray Investigation   W4/25
of Perfection in Tin Whiskers," J. Appl. Phys., $\underline{29}$
(4) pp. 679-683, 1958.

448. Sollenberger, N.J., "Investigation of Glass Fibers as   A2a/7
Reinforcement for Prestressing Concrete," Dept. Civ-
il Eng., Princeton U., Dept. Navy cont. NBy 8996,
Final Rpt., July 1963.

449. Soltis, P.J., "Anisotropic Mechanical Behavior in   W1/47
Sapphire (Al_2O_3) Whiskers," Naval Air Eng. Center,   W2/15
Phila., Pa., AML Rpt. 1831, April 1964.   W4/26

450. Soltis, P.J., "Investigation of Horizons Inc. Zirconia   CF2/17
Fibers," Naval Air Eng. Cntr., Phila., Pa., AML
Rpt. 1549, AD 404 358L, November 1962.

451. Soltis, P.J., "Study of Size Effect in Fine Beryllium   CF4/9
Wires," Naval Air Eng. Cntr., Phila., AML Rpt.
1909, AD 438 478, March 1964.

452. Souttar, H.S., "Demonstration of a Method of Making   CF3/17
Capillary Filaments," Phys. Soc. Proc., London, <u>24</u>,
pp. 166-167, 1912.

453. Spencer, S.B. and Jackson, W.O., "A New Fiber for   CF2/18
Reinforced Plastic Aerospace Applications," presented
at Regional Meeting of SPE, Garden City, Long Island,
May 13-14, 1964.

454. Spretnak, J.W., "A Summary of the Theory of Frac-   R5/2
ture in Metals," Battelle Mem. Inst., BMIC Rpt. 157,
August 1961.

455. Stambler, I., " Whisker Research Promises Ultra-   W5/23
Strong Materials," Space/Aeronautics, <u>34</u> (5) pp. 48-
52, 1960.

456. Statton, W.O. and Hoffman, L.C., "Structure in Vit-   F4a/10
reous Silicate Fibers as Shown by Small-Angle Scat-
tering of X-Rays," J. Appl. Phys., <u>31</u> (2) pp. 404-
409, 1960.

457. Stephens, D.L. and Alford, W.J., "Dislocation Struc-   R2/3
tures in Single Crystal Alumina," J. Am. Cer. Soc.,
<u>47</u> (2) pp. 81-84, 1964.

458. Sterman, S. and Bradley, H.B., "A New Interpreta-   C4b/11
tion of the Glass-Coupling Agent Surface Through Use
of Electron Microscopy," SPE Trans., <u>1</u> (4) pp. 224-
233, October 1961.

459. Sterry, J.P., "New Refractory for Ablation, Thermal   CF2/19
Insulation," Mat. in Des. Eng., <u>56</u> (4) pp. 12-13,
1962.

460. Stevens, D.W., et al, "Potential of Filament Wound   C2a/46
Composites," Narmco Res. & Dev., San Diego, Calif.
BuWeps cont. NOw-61-0623-3 (FBN), Final Summary
Rpt., May 1963.

461. Stirling, F.J., "Frozen Strains in Glass Fibers," J.   F2d/8
Soc. Glass Tech., <u>39</u> (188) pp. 134-144, June 1955.

462. Stites, J. and Sterry, J.P., "Non-Vitreous, Non-   CF2/20
Metallic Fiber Synthesis," H.I. Thompson Fiber
Glass Co., Gardena, Calif., ASD, WPAFB, AF 33-
(616)-8080, 6th Qtrly. Rpt., October 1962.

463. Stover, E.R., "Structure of Whiskers in Pyrolytic   CF1/10
      Graphite," G.E. Co., Res. Lab., Schenectady, N.Y.,
      Rpt. 64-RL-3609M, March 1964.

* 464. Stowell, E.Z. and Liu, Tien-Shih, "Parametric Stud-   C5/26
      ies of Metal Fiber Reinforced Ceramic Composite
      Materials," Southwest Res. Inst., San Antonio, Texas,
      BuWeps cont. NOas-60-6077-c, Int. Rpt. No. 2, May
      1960.

465. Strelkov, P.G. and Shpunt, A.A., "Dependence of the   W2a/6
      Strength of Cleavage Whiskers on Their Dimensions,"   W4/27
      Soviet Physics-Solid State $\underline{4}$ (8), February 1963. (in
      English)

466. Sutton, W.H., "Development of Composite Structural   C1d/23
      Materials for Space Vehicle Applications," ARS Jour.,
      $\underline{32}$ (4) pp. 593-600, 1962.

467. Sutton, W.H., "Investigation of Bonding in Oxide-   C4a/2
      Fiber (Whisker) Reinforced Metals," G.E. Co., Mis-
      sile and Space Div., King of Prussia, Pa., U.S.
      Army Materials Research Agency, AMRA CR 6301/1,
      November 1962.

468. Ibid., AMRA CR 6301/2, January 1963.   C4a/3

469. Ibid., AMRA CR 6301/3, March 1963.   C4a/4

470. Ibid., AMRA CR 6301/4, June 1963.   C4a/5

471. Ibid., AMRA CR 6301/5, September 1963.   C4a/6

472. Ibid., AMRA CR 6301/6, December 1963.   C4a/7

473. Ibid., AMRA CR 6301/7, March 1964.   C4a/8

474. Ibid., AMRA CR 6301/8, June 1964.   C4a/9

475. Sutton, W.H., "Investigation of Metal-Whisker Com-   C1d/24
      posites," G.E. Co., Missile and Space Div., King of
      Prussia, Pa., Rpt. R62SD65, June 1962, presented
      at the 64th Annual Mtg. Am. Cer. Soc., May 1962.

476. Sutton, W.H., et al, "Development of Composite   C1d/13
      Structural Materials for High Temperature Applica-   C4a/10
      tions," G.E. Co., Missile and Space Div., King of
      Prussia, Pa., U.S. Navy BuWeps cont. NOw-60-
      0465-d, Prgr. Rpt. 1, July 1960.

477. Ibid., Prgr. Rpt. 2, October 1960.   C1d/14

478.   Ibid., Prgr. Rpt. 3, January 1961.   Cld/15

479.   Ibid., Prgr. Rpt. 4, April 1961.   Cld/16

480.   Ibid., Prgr. Rpt. 5, August 1961.   Cld/17

481.   Ibid., Prgr. Rpt. 6, November 1961.   Cld/18

482.   Ibid., Prgr. Rpt. 7, February 1962.   Cld/19

483.   Ibid., Prgr. Rpt. 8, May 1962.   Cld/20

484.   Ibid., Prgr. Rpt. 9, August 1962.   Cld/21

485.   Ibid., Prgr, Rpt. 10, November 1962.   W1/48
                                                                   Cld/22
                                                                   C4a/11

486.   Ibid., Prgr. Rpt. 11, February 1963.   W1/49

487.   Ibid., Prgr. Rpt. 12, May 1963.   W1/50

488.   Ibid., Prgr. Rpt. 13, August 1963.   W1/51

489.   Ibid., Prgr. Rpt. 14, November 1963.   W1/52

490.   Ibid., Prgr. Rpt. 15, February 1964.   W1/53

491.   Ibid., Prgr. Rpt. 16, May 1964.   W1/54

492.   Ibid., Prgr. Rpt. 17, September 1964.   W1/55

* 493.   Sutton, W.H., et al, "Whisker Reinforced Plastics for   Clf/4
         Space Applications," SPE Jour., 20 (11) pp. 1203-
         1209, 1964.

* 494.   Sutton, W.H. and Chorne, J., "Development of High-   Cld/26
         Strength, Heat-Resistant Alloys by Whisker Reinforce-
         ment," Metals Eng. Qtrly., 3 (1) pp. 44-51, February
         1963.

* 495.   Sutton, W.H. and Chorne, J., "Potential of Oxide-   Cld/25
         Fiber Reinforced Metals," G.E. Co., Missile and   C5/27
         Space Div., King of Prussia, Pa., Rpt. R65SD2, Jan-
         uary 1965.

496.   Swica, J., et al, "Metal Fiber Reinforced Ceramics,"   C2d/5
         Alfred Univ., Alfred, N.Y., AFML, WPAFB, AF 33-
         (616)5298, WADC TR 58-452, Pt. II, Jan. 1960.

497. Synthetic Fibers in Papermaking, edited by O.A.   R6b/6
Battista, Interscience Publishers, New York, 340pp.,
1964.

498. Talley, C.P., et al, "Boron Reinforcements for   CF3/18
Structural Composites," Texaco Exp. Inc., Richmond,
Va., AFML, WPAFB, AF 33(616)-8067, Summary
Rpt. 1, May 1962.

499. Ibid., "Part II," ASD-TDR-62-257, April 1963.   CF3/19

500. Tang, M.M. and Bacon, R., "Carbonization of Cel-   CF1/11
lulose Fibers - I. Low Temperature Pyrolysis," Car-
bon, 2 (3) pp. 211-220, December 1964.

501. "Tensile Stress-Strain Properties of Fibers," E.I.   F4a/11
duPont de Nemours, Inc., Wilmington, Delaware,
Tech. Information Booklet, Bulletin X-82.

* 502. Thomas, W.F., "An Investigation of the Factors Like-   F1/54
ly to Affect the Strengths and Properties of Glass   F2/17
Fibers," Phys, and Chem. Glasses, 1 (1) pp. 4-18,   F3a/20
Feb. 1960.   F4a/12
                                                                        F5/16

503. Tighe, N.J., "Fused Quartz Fibers: A Survey of Prop-   F1/55
erties, Applications and Production Methods,"National   F5/17
Bureau of Standards, Washington, D.C., NBS Circular   A2/5
569, Jan. 1956.

504. Tomashot, R.C., "AF-994: A Superior Glass Fiber   F1/56
Reinforcement for Structural Composites," AFML,   F3a/21
WPAFB, Dayton, Ohio, ASD-TDR-63-81, March 1963.   F4a/13

505. Treuting, R.G., "Torsional Strain and the Screw Dis-   W1/56
location in Whisker Crystals," ACTA Met., 5, pp.   W4/28
173-175, March 1957.

* 506. Tyson, W.R. and Davies, G.J., "A Photoelastic   C5/28
Study of the Shear Stresses Associated with the Trans-
fer of Stress During Fibre Reinforcement," Brit. J.
Appl. Phys., 16 (3) pp. 199-206, 1965.

507. Vahldiek, F.W., et al, "Determination of Impurities   W2b/11
in Aluminum Oxide Whiskers," Anal. Chem., 34 (12)
pp. 1667-1668, November 1962.

508. Vanderbilt, B.M. and Jarvzelski, J.J., "The Bonding   C4b/12
of Fillers to Thermosetting Resins," SPE Trans., 2
(4) pp. 363-366, October 1962.

509. Veazie, F.M. and Brady, W.C., "New High Perform-   F1/57
      ance Vitreous Reinforcements," Conference on
      Structural Plastic Adhesives and Filament Wound
      Composites, ASD-TDR-63-396, pp. 310-314, April
      1963.

510. Venables, J.D., "Cleavage Whiskers," J. Appl.   W3k/1
      Phys., 31 (8) pp. 1503-1504, 1960.

511. Vidanoff, R. and Bradford, D., "Silica Fiber/Core   F3a/22
      Sheath Fibers," Narmco Res. & Dev., San Diego,   CF3/20
      Calif., BuWeps N600(19)61810, Qtrly. Prgr. Rpt. 1,
      May 1964.

512. Villers, R., "New Muscle for Metals," United Air-   We3/4
      craft Corp., Bee Hive Qtrly., 34 (2) pp. 30-32,
      Spring 1964.

513. Vondracek, C.H., et al, "Fiber Reinforced Boron   C2d/6
      Phosphate Structural Composites," Westinghouse
      Electric Corp., Pittsburgh, Pa., AFML, WPAFB,
      Dayton, Ohio, AF 33(657)-7587, ML TDR 64-60, Jan.
      1964.

514. Vondracek, C.H., et al, "Synthesis and Formulation   C2d/7
      of Inorganic Bonded-Inorganic Fiber Reinforced Non-
      Metallic Structural Materials," Westinghouse Electric
      Corp., Pittsburgh, Pa., AFML, WPAFB, Dayton,
      Ohio, AF 33(657)-7587, Qtrly, Prgr. Rpt., June 1963.

515. Vuillard, G., "Boron Carbide Crystals," Compt,   R6a/7
      Rend., 257 (25) pp. 3927-3929, 1963. (in French)

516. Vukasovich, M.S., "Reinforcement of Nickel Chro-   Cld/27
      mium Alloys with Sapphire Whiskers," Horizons Inc.,
      Cleveland, Ohio, BuWeps cont. NOw-64-0125-c, Int.
      Rpt. No. 1, March 1964.

517. Ibid., Int. Rpt. No. 2, June 1964.   Cld/28

518. Wagner, H.J., "Review of Recent Developments: C5/29
      Fiber-Reinforced Metals," Battelle Mem. Inst.,
      Columbus, Ohio, BMIC Bulletin, March 6, 1964.

519. Ibid., December 11, 1964.   C5/30

520. Wagner, R.S., et al, "Study of the Filamentary   W3a/25
      Growth of Silicon Crystals from the Vapor," J. Appl.
      Phys., 35 (10) pp. 2993-3000, 1964.

* 521. Wagner, R.S. and Ellis, W.C., "The Vapor-Liquid-   W3i/2
      Solid Mechanism of Crystal Growth and Its Applica-

521.   tion to Silicon," to be published in Trans. AIME, $\underline{233}$,
(cont.) 1965.

522.   Wagner, R.S. and Ellis, W.C., "Vapor-Liquid-Solid   W3i/1
         Mechanism of Single Crystal Growth," Appl. Phys.
         Letters, $\underline{4}$ (5) pp. 89-90, 1964.

523.   Waugh, J.A., et al, "The Development of Fibrous   F1/58
         Glasses Having High Elastic Modulii," Owens-Corning   F2/18
         Fiberglas Corp., Granville, Ohio, AFML, WPAFB,
         AF 33(616)-2422, WADC TR 55-290, Part II, May
         1958.

524.   Webb, W.W., "Dislocation Structure and the Form-   W1/57
         ation and Strength of Sodium Chloride Whiskers," J.   W2b/12
         Appl. Phys., $\underline{31}$ (1) pp. 194-206, 1960.

525.   Webb, W.W., et al, "Dislocations in Whiskers,"   W2b/13
         Phys. Rev., $\underline{108}$ (2) pp. 498-499, October 15, 1957.

526.   Webb, W.W. and Bertolone, N.P., "Novel Mechanism   W3f/3
         for Mass Transport During Whisker Growth of Cesium
         Chloride from Aqueous Films," J. Appl. Phys., $\underline{31}$
         (1) pp. 207-209, 1960.

527.   Webb, W.W. and Forgeng, W.D., "Growth and Defect   W2b/14
         Structure of Sapphire Microcrystals," J. Appl. Phys.,   W3a/26
         $\underline{28}$ (12) pp. 1449-1454, 1957.

* 528.   Webb, W.W. and Forgeng, W.D., "Mechanical Be-   W5/24
         havior of Microcrystals," ACTA Met., $\underline{6}$ pp. 462-469,
         July 1958.

529.   Webb, W.W. and Stern, M., "Effect of Surface Films   W2c/2
         on the Strength of Metal Whiskers," J. Appl. Phys.,
         $\underline{30}$ (9) p. 1471, 1959.

530.   Wechsler, A.E. and Glaser, P.E., "Investigation of   A2b/2
         the Thermal Properties of High Temperature Insula-   CF1/12
         tion Materials," A.D. Little, Inc., Cambridge, Mass.,   CF2/21
         AFML, WPAFB, AF 33(657)-9172, ASD-TDR-63-
         574, July 1963.

531.   Weik, H., "Investigation of Mechanical and Physical   CF2/22
         Properties of Silicon Dioxide and Zirconium Dioxide
         Fibers," WADC, WPAFB, Dayton, Ohio, WCLT TM
         58-72, May 1958.

532.   Weik, H., "Whisker Structure and Tensile Strength,"   W2b/15
         J. Appl. Phys., $\underline{30}$ (5) pp. 791-792, 1959.   W2e/3

533. Weil, N.A., et al, "Investigation of Glass Fiber   F2a/13
      Strength Enhancement Through Bundle Drawing,"   F2d/9
      IITRI, Chicago, Ill., BuWeps, NOw-61-0259-c, 1st   F3a/23
      Qtrly Rpt., May 1961.   F5/18
                                                                C2a/47

534. Weisbart, H., "Investigations on the Strength of   F1/59
      Alkali-Containing Glass Fibers," SPE Trans., $\underline{2}$ (4)   F2b/4
      pp. 339-353, October 1962.   F2e/8

535. Werren, F. and Heebink, B.G., "Interlaminar Shear   C2a/48
      Strength of Glass-Fiber-Reinforced Plastic Lam-   C3/8
      inates," For. Prdts. Lab., Madison, Wisc., Rpt. No.
      1848, Sept. 1955.

536. Westbrook, J.H., "The Source of Strength and Brit-   R6a/8
      tleness in Intermetallic Compounds," G.E. Co., Re-
      search Laboratory, Schenectady, N.Y., Rpt. No. 64-
      RL-3785M, September 1964.

537. Whitehurst, H.B., "Investigation of Glass-Metal Com-   C2c/9
      posite Materials," Owens-Corning Fiberglas Corp.,
      U.S.N. cont. NOrd 15764, 3rd Qtrly, Rpt., Dec.
      1955.

538. Whitehurst, H.B. and Ailes, H.B., "Investigation of   C2c/10
      Glass-Metal Composite Materials," Owens-Corning
      Fiberglas Corp., U.S.N. cont. NOrd 15764, 5th
      Qtrly. Rpt., Sept. 1956.

539. Whitney, J.M., "Applying Engineering Mechanics   F4a/14
      Solutions to Deflections of Statically Loaded Coplanor
      Fibers," AFML, WPAFB, ML TDR 64-118, May
      1964.

540. Witucki, R.M., "Boron Filament Composite Materials   C2c/11
      for Space Structures," Astro Research Corp., Santa   CF3/22
      Barbara, Calif., NASw-652, Rpt. ARC-R-168, Nov.
      1964.

541. Witucki, R.M., "Boron Filaments," Astro Research   CF3/21
      Corp., Santa Barbara, Calif., NASw-652, Rpt. NASA
      CR-96, Sept. 1964.

542. Wolff, E.G., "Growth of α-Tungsten Oxide Whiskers   W1/58
      by Vapor Deposition," private communication, Oct-   W3a/27
      ober 1964.

543. Wolff, E.G. and Coskren, T.D., "Growth and Mor-   W1/59
      phology of MgO Whiskers," AVCO Corp., private   W3a/28
      communication (paper submitted for publication in J.
      Am. Cer. Soc.), October 1964.

544. Wolff, F. and Siuta, T., "Factors Affecting the Per-　C2a/49
formance and Aging of Filament Wound Fiberglass
Structures," ARS Journal, 32 (6) pp. 948-950, June
1962.

545. Wolters, H.B.M. and Schapink, F.W., "Tensile Test-　W4/29
ing Machine for Whiskers," J. Sci. Inst., 38 (6) pp.
250-252, June 1961.

546. Yates, P.C. and Trebilcock, J.W., "The Chemistry　C2a/50
of Chromium Complexes Used as Coupling Agents in
Fiberglass Resin Laminates," SPE Trans., 1 (4) pp.
199-213, Oct. 1961.

547. Yerkovitch, L. and Kirchner, H.P., "Growth and　W1/60
Mechanical Properties of Filamentary Silicon Carbide
Crystals," Cornell Aeronautical Lab. Inc., Buffalo,
N.Y., AF 33(616)-7005, WADD TR 61-252, June 1962.

548. Zeilberger, E.J., "Plastics," Space Aeronautics R&D　R4/3
Handbook, 42 (4) pp. 159-162, Sept. 1964.

549. Zimmerman, F.J., "Fiber Metals for Meteoroid Pro-　A2e/1
tection," Proc, 5th Annual Structures and Materials
Conf., AIAA, Palm Springs, Calif., April 1964.

550. Zisman, W.A., "Surface Chemistry of Glass-Fiber　C2a/51
Reinforced Plastics," Naval Res. Lab., Washington,
D.C., NRL Rpt. 6083, Mar. 1964.

C. Categorized section arranged by subject

W- WHISKERS

<u>W1- Measured Properties</u>

W1/1   Aleksandrov, L.N. and Kogan, A.N., "Investigation of the
        Strength of Needle-Like Tungsten Crystals," Fiz. Tverd. Tela,
        <u>6</u> (1) p. 307, 1964. (in Russian)

W1/2   Anon., "Growing Whiskers for Strength," The Thiokol
        Magazine, <u>3</u> (1) pp. 28-29, 1964.

W1/3   Brenner, S., "Mechanical Behavior of Sapphire Whiskers at
        Elevated Temperature," J. App. Phys., <u>33</u> (1) pp. 33-39, 1962.

W1/4   Bryant, P.J., "Mechanism of Lubrication of Graphite Single
        Crystals," Midwest Res. Inst., Kansas City, Mo., ASD,
        WPAFB, AF 33(616)-6277, WADD TR 60-529, Part I, October
        1960.

W1/5   Ibid., WADD TR 60-529, Part II, June 1962.

W1/6   Bushong, R.M., et al, "Research and Development on Advanced
        Graphite Materials, Vol. XLII, Summary Tech. Rpt., "Union
        Carbide Corp., Lawrenceburg, Tenn., AFML, WPAFB,
        AF 33(615)-6915, WADD TR 61-72, Vol. XLII, pp. 76, 89-104,
        174-184, January 1964.

W1/7   Cabrera, N. and Price, P.B., <u>Growth and Perfection of</u>
        <u>Crystals</u>, edited by R.H. Doremus, B.W. Roberts, and
        D. Turnbull, John Wiley and Sons, Inc., New York, p. 212, 1958.

W1/8   Calow, C.A. and Begley, B., "A Study of the Handling, the
        Mechanical Properties and Crystalline Structure of Silicon
        Nitride Whiskers," Atomic Energy Auth., Aldermaston,
        Berkshire, Rpt. AWRE 0-70/64, September 1964.

W1/9   Campbell, W.B., "Feasibility of Forming Refractory Fibers by
        a Continuous Process," Lexington Labs. Inc., Cambridge,
        Mass., AMRA Cont. DA-19-020-AMC-0068 (X), Qtrly. Prgr.
        Rpt. 1, June 1963.

W1/10   Ibid., Prgr. Rpt. 3, November 1963.

W1/11   Ibid., Prgr. Rpt. 4, April 1964.

W1/12   Davies, G.J., "On the Strength and Fracture Characteristics of
         Intermetallic Fibers, " Phil. Mag., 9 (102) pp. 953-964, June
         1964.

W1/13   Dikina, L.S. and Shpunt, A.A., "Strength of Cleavage
         Whiskers," Soviet Physics-Solid State, 4, pp. 405-406, 1962.
         (in English)

W1/14   Fridman, V. Ya. and Shpunt, A.A., "Investigation of the
         Strength of Lithium Fluoride Crystal Fragments," Soviet
         Physics-Solid State, 5, (3) pp. 575-579, 1963. (in English)

W1/15   Gatti, A. et al, "The Synthesis of B_4C Filaments, " G.E. Co.
         Missile and Space Div., King of Prussia, Pa., NASA-Hqs.,
         NASw-670, 1st Qtrly. Rpt., October 1963.

W1/16   Ibid., 2nd Qtrly. Rpt., January 1964.

W1/17   Ibid., 3rd Qtrly. Rpt., April 1964.

W1/18   Ibid., Final Rpt., September 1964.

W1/19   Gordon, J.E., Growth and Perfection of Crystals, edited by
         R.H. Doremus, B.W. Roberts, and D. Turnbull, John Wiley and
         Sons, Inc., New York, p. 219, 1958.

W1/20   Gordon, J.E. and Menter, J.W., "Mechanical Properties of
         Whiskers and Thin Films," Nature, 182 (4631) pp. 296-299, 1958.

W1/21   Guzeev, V.V. and Malinski, Iv.M., "Device for Measuring
         Stress Relaxation in Fibers," Industrial Laboratory, 29 (3)
         pp. 1529-1530, April 1964. (in English)

W1/22   Hasselman, D.P.H. and Batha, H.D., "Strength of Single
         Crystal Silicon Carbides, " Appl. Phys. Letters, 2 (6) pp. 111-
         113, March 1963.

W1/23   Herring, C. and Galt, J.K., "Elastic and Plastic Properties of
         Very Small Metal Specimens," Phys. Rev., 85 (6) pp. 1060-1061,
         March 1952.

W1/24   Hulse, C.O., "Formation and Strength of MgO Whiskers,"
         J. Am. Cer. Soc., 44 (11) pp. 572-575, 1961.

W1/25   Jeszenski, B. and Hartmann, E., "Growth and Mechanical
         Properties of NaC1 Whiskers, " Soviet Physics-Crystallography,

<u>1</u> (3) pp. 341-344, 1962. (in English)

W1/26   Johnson, R. C., et al, "Synthesis and Some Properties of
         Fibrous Silicon Nitrides," Bureau of Mines, Norris, Tenn.
         Rpt. BM-RI-6467, 1964.

W1/27   Kirchner, H. P. and Knoll, P., "Silicon Carbide Whiskers,"
         J. Am. Cer. Soc., <u>46</u> (6) pp. 299-300, 1963.

W1/28   Kliman, M. I., "Formation of Alumina Whiskers," Watertown
         Arsenal Laboratories, Watertown, Mass., D/A Proj. 59332007,
         Rpt. No. WAL TR 371/52, November 1962.

W1/29   Kliman, M. I., "Formation and Properties of Alumina Fiber,"
         Watertown Arsenal Laboratories, Watertown, Mass., D/A Proj.
         59332007, Rpt. No. WAL TR 371/50, November 1962.

W1/30   Kostyuk, V. E., et al, "Strength of Metal Whisker Crystals Con-
         taining Impurities," Soviet Physics-Solid State, <u>5</u> (11) pp. 2241-
         2245, May 1964. (in English)

W1/31   Luborsky, F. E., and Morelock, C. R., "Magnetization Reversal
         in Cobalt Whiskers," G. E. Co., Research Laboratory,
         Schenectady, N. Y., Rpt. 64-RL-3787M, October 1964.

W1/32   Luborsky, F. E., et al, "Crystallographic Orientation and
         Oxidation of Sub-Micron Whiskers of Iron, Iron Cobalt, and
         Cobalt," J. Appl. Phys., <u>34</u> (9) pp. 2905-2909, 1963.

W1/33   Markali, J., <u>Mechanical Properties of Engineering Ceramics,</u>
         edited by W. W. Kriegel and H. J. Palmour, III, Interscience
         Publishers, New York, p. 93, 1961.

W1/34   Mehan, R. L., Feingold, E., and Gatti, A., "Evaluation of
         Sapphire Wool and Its Incorporation into Composites of High
         Strength," G. E. Co., Missile and Space Div., King of Prussia,
         Pa., AFML, WPAFB, AF 33(615)-1696, 3rd Qtrly. Rpt.,
         February 1965.

W1/35   Papalegis, F. E. and Bourdeau, R. G., "Pyrolytic Reinforce-
         ments for Ablative Plastic Composites," High Temp. Metls.,
         Inc., Boston, Mass., AFML WPAFB, Cont. AF 33(657)-11094,
         ML TDR 64-201, July 1964.

W1/36   Papalegis, F. E. and Bourdeau, R. G., "Research on New and
         Improved Thermally Protective Fiber Reinforced Composite

Materials," High Temp. Mtls. Inc., Boston, Mass., ASD WPAFB, AF 33(657)-8651, August 1962.

W1/37   Ibid., Qtrly. Prgr. Rpt., November 1962.

W1/38   Ibid., Qtrly. Prgr. Rpt., February 1963.

W1/39   Popper, P. and Ruddlesden, S.N., "The Preparation, Properties and Structure of Silicon Nitride," Trans. Brit. Cer. Soc., 60 (9) pp. 615-619, 1961.

W1/40   Price, P.B., "The Mechanical Properties of Whiskers," Ph.D. Thesis, U. of Virginia, 1958.

W1/41   Price, P.B., et al, "On the Growth and Properties of Electrolytic Whiskers," ACTA Met., 6, pp. 524-531, August 1958.

W1/42   Regester, R.F., "The Strength and Perfection of Aluminum Oxide Whiskers," Masters Thesis, U. of Penna., 1963.

W1/43   Sandulova, A.V., et al, " Preparation and Some Properties of Whisker and Needle-Shaped Single Crystals of Germanium, Silicon and Their Solid Solutions," Soviet Physics-Solid State, 5 (9) pp. 1883-1888, March 1964. (in English)

W1/44   Shaffer, P.T.B. and Batha, H.D., "A Study of Fundamental Mechanical Properties of Ceramic Single Crystals," Carborundum Co., Niagara Falls, N.Y., BuWeps, N600 (19) 59749, Bi-Monthly Prgr. Rpt. No. 3, July 1963.

W1/45   Sharkitt, R.L. et al, "High Temperature Resistant Beryllia Fiber-Reinforced Structural Composites," National Beryllia Corp., AFML, WPAFB, AF 33(616)-8066, ASD-TDR-62-632, Part II, June 1963.

W1/46   Sherby, O.D., "Cambridge Conference on the Strength of Metal Whiskers and Thin Films," Office of Naval Res., Washington, D.C., ONRL Rpt. 34-58, April 1958.

W1/47   Soltis, P.J., "Anisotropic Mechanical Behavior in Sapphire (Al_2O_3) Whiskers," Naval Air Eng. Center, Phila., Pa., AML Rpt. 1831, April 1964.

W1/48   Sutton, W.H., et al, "Development of Composite Structural Materials for High Temperature Applications," G.E. Co. Missile and Space Div., King of Prussia, Pa., Contr. NOw-60-

0465-d, U. S. Navy, BuWeps, Prgr. Rpt. No. 10, November 1962.

W1/49   Ibid., Prgr. Rpt. 11, February 1963.

W1/50   Ibid., Prgr. Rpt. 12, May 1963

W1/51   Ibid., Prgr. Rpt. 13, August 1963.

W1/52   Ibid., Prgr. Rpt. 14, November 1963.

W1/53   Ibid., Prgr. Rpt. 15, February 1964.

W1/54   Ibid., Prgr. Rpt. 16, May 1964.

W1/55   Ibid., Prgr. Rpt. 17, September 1964.

W1/56   Treuting, R. G., "Torsional Strain and the Screw Dislocation in Whisker Crystals," Acta Met. , 5, pp. 173-175, Mary 1957.

W1/57   Webb, W. W., "Dislocation Structure and the Formation and Strength of Sodium Chloride Whiskers," J. Appl. Phys., 31 (1) pp. 194-206, 1960.

W1/58   Wolff, E. G., "Growth of α -Tungsten Oxide Whiskers by Vapor Deposition," private communication, October 1964.

W1/59   Wolff, E. G. and Coskren, T. D., "Growth and Morphology of MgO Whiskers," AVCO Corp., private communication (paper submitted for publication in J. Am. Cer. Soc.), October 1964.

W1/60   Yerkovitch, L. and Kirchner, H. P., "Growth and Mechanical Properties of Filamentary Silicon Carbide Crystals," Cornell Aeronautical Lab. Inc., Buffalo, N. Y., AF 33(616)-7005, WADD TR 61-252, June 1962.

W2- Factors Affecting Measured Properties

W2/1   Bacon, R., "Growth, Structure and Properties of Graphite Whiskers," J. Appl. Phys. 31 (2) pp. 284-290, 1960.

W2/2   Brenner, S. S., "Factors Influencing the Strength of Whiskers," presented to ASM Seminar on Fiber Composite Materials, Phila., Pa., October 1964.

W2/3   Brenner, S.S. (see W1/3).

W2/4   Bushong, R.M., et al (see W1/6).

W2/5   Cabrera, N. and Price, P.B. (see W1/7).

W2/6   Calow, C.A. and Begley, B. (see W1/8).

W2/7   Gordon, J.E. (see W1/19 and W1/20).

W2/8   Hasselman, D.P.H. and Batha, H.D. (see W1/22).

W2/9   Herring, C. and Galt, J.K. (see W1/23).

W2/10   Hulse, C.O. (see W1/24).

W2/11   Kliman, M.I., "Transverse Rupture Strength of Alumina Fiber-
        Ceramic Composites," Watertown Arsenal Laboratories, .
        Watertown, Mass., D.A. Proj. 59332007, Rpt. WAL TR 371/53,
        November 1962.

W2/12   Shaffer, P.T.B. and Batha, H.D. (see W1/44).

W2/13   Sharkitt, R.L., et al (see W1/45).

W2/14   Sherby, O.D. (see W1/46).

W2/15   Soltis, P.J. (see W1/47).

W2a- Effect of Size and/or Shape

W2a/1   Amelinckx, S., "On Whisker Growth Shapes," Phil. Mag.,
        pp. 425-428, May 3, 1958.

W2a/2   Baker, G.S., "Angular Bends in Whiskers," ACTA Met 5
        pp. 353-357, July 1957.

W2a/3   Bokshtein, S.Z. and Svetlov, I.L., "Determination of the Shape
        and Dimensions of Transverse Cross Sections of Threadlike
        Crystals," Translated from Zavodskaya Laboratoriya, 28 (5)
        pp. 595-596, May 1962.

W2a/4   Edwards, P.L. and Svager, A., "Spiral Copper Whiskers,"
        J. Appl. Phys. 35 (2) pp. 421, 1964.

W2a/5   Papkov, V.S. and Berezhkova, G.V., "Production of Fila-
         mentary Al_2O_3 Crystals," Kristallografiya 9 (3) pp. 442-443,
         1963. (in Russian)

W2a/6   Strelkov, P.G. and Shpunt, A.A., "Dependence of the Strength
         of Cleavage Whiskers on Their Dimensions," Soviet Physics-
         Solid State 4 (8), February 1963. (in English)

W2b- Effect of Structural Perfection and/or Composition

W2b/1   Amelinckx, S., "Dislocations in Alkali Halide Whiskers," Growth
         and Perfection of Crystals, edited by R. H. Doremus, B.W.
         Roberts and D. Turnbull, John Wiley and Sons, Inc., New York,
         p. 139, 1958.

W2b/2   Austerman, S.B., "Role of Si, Al, and Other Impurities in BeO
         Crystal Growth," Atomic International, Canoga Park, California,
         Atomic Energy Commission Contract AT (11-1)-GEN-8, Report
         Number NAA-SR-8235, July 15, 1963.

W2b/3   Bryant, P.J., et al, "On Graphite Whiskers," J. Appl. Phys.,
         30 (11) p. 1839, 1959.

W2b/4   Coleman, R.V., "Observations of Dislocation in Iron Whiskers,"
         J. Appl. Phys. 29 (10) pp. 1487-1492, 1958.

W2b/5   Drum, C.M. and Mitchell, J.W., "Electron Microscopic
         Examination of Role of Axial Dislocations in Growth of AlN O
         Whiskers," Appl. Phys. Letters, 4 (9) pp. 164-165, May 1964.

W2b/6   Edwards, P.L. and Happel, R.J., Jr., "Beryllium Oxide
         Whiskers and Platelets," J. Appl. Phys., 33 (3) pp. 943-948,
         1962.

W2b/7   Ibid., "Alumina Whisker Growth on a Single Crystal Alumina
         Substrate," pp. 826-827.

W2b/8   Hamilton, D.R., "Interferometric Determination of Twist and
         Polytype in Silicon Carbide Whiskers," J. Appl. Phys., 31 (1)
         pp. 112-116, 1960.

W2b/9   Roberts, E.C. and Lawrence, S.C., "Whisker-on-Whisker
         Growth," Acta Met. , 5 p. 335, June 1957.

W2b/10 Sears, G.W., "Effect of Poisons on Crystal Growth," J. Chem. Phys., 29 (5) pp. 1045-1048, 1958.

W2b/11 Vahldiek, F.W., et al, "Determination of Impurities in Aluminum Oxide Whiskers," Anal. Chem., 34 (12) pp. 1667-1668, November 1962.

W2b/12 Webb, W.W. (see W1/57).

W2b/13 Webb, W.W., et al, "Dislocations in Whiskers," Phys. Rev., 108 (2) pp. 498-499, October 15, 1957.

W2b/14 Webb, W.W. and Forgeng, W.D., "Growth and Defect Structure of Sapphire Microcrystals," J. Appl. Phys., 28 (12) pp. 1449-1454, 1957.

W2b/15 Weik, H., "Whisker Structure and Tensile Strength," J. Appl. Phys., 30 (5) pp. 791-792, 1959.

W2c- Effect of Surface Treatment and/or Coatings

W2c/1 Bushong, R.M., et al, "Pyrolytic Coating of Carbon Filaments," Union Carbide Corp., Lawrenceburg, Tenn., AFML, WPAFB, Contract AF 33(657)-11297, Qtrly. Prgr. Rpt., November 10, 1963.

W2c/2 Webb, W.W. and Stern, M., "Effect of Surface Films on the Strength of Metal Whiskers," J. Appl. Phys., 30 (9) pp. 1471, 1959.

W2c/3 Weik, H. (see W2b/15).

W2d- Effect of Environment

W2d/1 Anderson, J.A., "Metal Whiskers," NASA - Goddard Space Flight Center, Greenbelt, Maryland, Goddard Summer Workshop, Program in Measurements and Simulation of the Space Environment, Summer 1963.

W2d/2 Rollins, F.R., Jr., "Use of Graphite Whiskers in a Study of the Atmosphere Dependence of Graphite Friction," J. Appl. Phys., 32 (4) pp. 1454-1458, 1961.

W3- Growth Techniques

W3/1   Anderson, J.A. (see W2d/1).

W3/2   Bacon, R. (see W2/1).

W3/3   Brenner, S.S., "Growth and Properties of Whiskers," Science
          128 (3324), pp. 569-575, 1958.

W3/4   Bushong, R.M., et al (see W1/6).

W3/5   Cabrera, N. and Price, P.B. (see W1/7).

W3/6   Canter, J.M. and Heller, R.B., "Experimental Method for
          Observing Growth of Graphite Whiskers," NASA Goddard Space
          Flight Center, Greenbelt, Md., Goddard Summer Workshop
          Program in Measurement and Simulation of the Space Environ-
          ment, N 64-28219, 1963.

W3/7   Cech, R.E., "On Whisker-Forming Reactions," G.E. Co.,
          Research Laboratory, Schenectady, N. Y., Rpt. 59-RL-2316M,
          November 1959.

W3/8   Clausen, E.M. and Reutter, J.W., "Preparation Techniques for
          Growth of Single Crystals of Non-Metallic Materials," G.E. Co.,
          Research Laboratory, Schenectady, N. Y., AFML, WPAFB,
          AF 49(638)-1247, Final Rpt., June 1964.

W3/9   Coleman, R.V., "The Growth and Properties of Whiskers,"
          Metallurgical Reviews, 9 (35) pp. 261-304, 1964.

W3/10   Cunningham, A.L., "Mechanism of Growth and Physical
          Properties of Refractory Oxide Fibers," Horizons Inc.,
          Cleveland, Ohio, Office Naval Rsch., NOw-2619(00), Final Rpt.,
          April 1960.

W3/11   Lynch, C.T., et al, "Growth and Analysis of Alumina Whiskers,"
          ASD, WPAFB, Dayton, Ohio, Proj. 7350, ASD-TDR-62-272,
          May 1962.

W3/12   Markali, J. (see W1/33).

W3/13   Nabarro, F.R.N. and Jackson, P.J., Growth and Perfection of
          Crystals, edited by R.H. Doremus, B.W. Roberts and
          D. Turnbull, John Wiley and Sons, Inc., p. 13, 1958.

W3/14   Nannichi, Y., "Formation of Silicon Whiskers on a Subliminating Surface," Appl. Phys. Letters, 3 (8) pp. 139-142, October 1963.

W3/15   Nieto, M.M. and Russell, A.M., "Growth of Whiskers Due to Solid-to-Solid Phase Transformation in Zr," J. Appl. Phys., 35 (2) p. 461, 1964.

W3/16   Papalegis, F.E. and Bourdeau, R.G., "Pyrolytic Reinforcing Agents for Ablative Erosion-Resistant Composites," High Temp. Mtls. Inc., Boston, Mass., ASD, WPAFB, Cont. AF 33(657)-8651, ASD-TDR 63-403, May 1963.

W3/17   Papalegis, F.E. and Bourdeau, R.G. (see W1/35-W1/38).

W3/18   Pierce, C.M., "The Growth of Metal and Metallic Oxide Whiskers: An Annotated Bibliography," Lockheed Missile and Space Co., Sunnyvale Calif., SB 63-68, May 1964.

W3/19   Popper, P. and Ruddleson, S.N. (see W1/39).

W3/20   Ruth, V. and Hirth, J.P., "The Kinetics of Diffusion Controlled Whisker Growth," Met. Eng. Dept., Ohio State Univ., Columbus, Ohio, Office of Naval Res., Contr. Nonr 495(26), Tech. Rpt. 4, March 1964.

W3/21   Sandulova, A.V., et al (see W1/43).

W3/22   Sears, G.W., "Mercury Whiskers," Acta Met., 1 (8) pp. 457-458, 1953.

W3/23   Sears, G.W., "A Mechanism of Whisker Growth," Acta Met., 3 (7) pp. 367-369, 1955.

W3/24   Shaffer, P.T.B. and Batha, H.D. (see W1/44).

W3/25   Sharkitt, R.L., et al (see W1/45).

W3- Growth Techniques

W3a- Vapor Deposition

W3a/1   Austerman, S.B., "Growth of BeO Single Crystals," J. Am. Cer. Soc., 46 (1) pp. 6-10, 1963.

W3a/2   Austerman, S.B. (see W2b/2).

W3a/3   Bigot, J., "Preparation and Study of Iron Oxide Whiskers," Mem. Sci. Rev. Met., LX (7-8) pp. 541-550, 1963. (in French)

W3a/4   Bigot, J. and Talbot-Besnard S., "Origin and Growth of Iron Sesquioxide Filaments and Platelets," Compt. Rand, 255, pp. 1927-1929, October 15, 1962. (in French)

W3a/5   Campbell, W. B. (see W1/9).

W3a/6   Coleman, R. V. and Sears, G. W., "Growth of Zinc Whiskers," Acta Met. , 5, pp. 131-136, March 1957.

W3a/7   Davies, T. J. and Evans, P. E., "Aluminum Nitride Whiskers," Nature, 197 (4867) p. 587, 1963.

W3a/8   Edwards, P. L. and Happel, R. J., Jr., (see W2b/6-W2b/7).

W3a/9   Gatti, A. et al (see W1/15 and W1/18).

W3a/10   Hargreaves, C. M., "On the Growth of Sapphire Microcrystals," J. Appl. Phys., 32 (5) pp. 936-938, 1961.

W3a/11   Hulse, C. O. (see W1/24).

W3a/12   Jaccodine, R. J. and Kline, R. K., "Whisker Growth from Quartz," Nature, 189 (2) p. 298, 1961.

W3a/13   Johnson, E. R. and Amick, J. A., "Formation of Single Crystal Silicon Fibers," J. Appl. Phys., 25 (9) pp. 1204-1205, 1954.

W3a/14   Johnson, R. C., et al (see W1/26).

W3a/15   Kliman, M. I. (see W1/28-W1/29).

W3a/16   Knippenberg, W. F., et al, "Crystal Growth of Silicon Carbide," Philips Tech. Rev., 24 (6) pp. 181-183, 1962/1963.

W3a/17   Luborsky, F. E. and Morelock, C. R. (see W1/31).

W3a/18   Luborsky, F. E. et al (See W1/32).

W3a/19   Morelock, C. R., "Sub-Micron Whiskers by Vapor Deposition," Acta Met., 10 pp. 161-167, 1962.

W3a/20   Morelock, C. R. and Sears, G. W., "Growth of Whiskers by Chemical Reaction," J. Chem. Phys., 34 pp. 1008-1009, 1961.

W3a/21   Papkov, V.S. and Berezhkova, G.V. (see W2a/5).

W3a/22   Roberts, R.W. and McElligott, P.E., "Formation of Silicon
         Whiskers," G.E. Co., Research Lab., Schenectady, N. Y.,
         Rpt. 63-RL-3427C, September 1963.

W3a/23   Sears, G.W., "Growth of Anthracene Whiskers by Vapor
         Deposition," J. Chem. Phys., 39 (11) pp.2846-2847, 1963.

W3a/24   Sears, G.W. and DeVries, R.C., "Growth of Alumina Crystals
         by Vapor Deposition," G.E. Co., Research Lab., Schenectady,
         N. Y., Rpt. 60-RL-2377M, March 1960.

W3a/25   Wagner, R.S., et al, "Study of the Filamentary Growth of Silicon
         Crystals from the Vapor," J. Appl. Phys., 35 (10) pp.2993-3000,
         1964.

W3a/26   Webb, W.W. and Forgeng, W.D. (see W2b/14).

W3a/27   Wolff, E.G. (see W1/58).

W3a/28   Wolff, E.G. and Coskren, T.D. (see W1/59).

W3b- High Temperature Reduction of Metallic Salts

W3b/1   Herring, C. and Galt, J.K. (see W1/23).

W3c- Hydrogen Reduction of Halides

W3c/1   Coleman, R.V. and Sears, G.W. (see W3a/6).

W3c/2   Dixon, T.E. and Kane, R.A., "Growth of Manganeous Oxide
         Whiskers," J. Appl. Phys., 34 (9) pp.2774-75, 1963.

W3c/3   Kirchner, H.P. and Knoll, P. (see W1/27).

W3c/4   Kostyuk, V.E., et al (see W1/30).

W3c/5   Luborsky, F.E. and Morelock, C.R. (See W1/31).

W3c/6   Mazur, J. and Rafalowicz, J., "Hypothesis on the Ionic
         Mechanism of the Growth of Whiskers Obtained by the Reduction
         of Metal Halides," Brit. J. Appl. Phys., 12 pp.569-571,
         October 1961.

W3d- Electrolysis

W3d/1   Price, P.B., et al (see W1/41).

W3e- Controlled Solidification of Eutectic Alloys

W3e/1   Davies, G.J. (see W1/12).

W3e/2   Ford, J.A., et al, "Analytical and Experimental Investigations
        of Fracture Mechanisms of Controlled Polyphase Alloys," United
        Aircraft Corp., U.S. Navy Cont. N600(19)-59361, Final Rpt.,
        November 1963.

W3e/3   Hertzberg, R.W., "Effect of Growth Rate and Cr Whisker
        Orientation on Mechanical Behavior of a Undirectional by
        Solidified Cu-Cr Eutectic Alloy," Trans. ASM, 57 (2) pp. 434-
        441, 1964.

We3/4   Villers, R., "New Muscle for Metals," United Aircraft Corp.,
        Bee Hive Qtrly., 34 (2) pp. 30-32, Spring 1964.

W3f- From Solution

W3f/1   Jezenski, B. and Hartmann, E. (see W1/25).

W3f/2   Newkirk, J.B. and Sears, G.W., "Growth of Potassium Halide
        Crystals from Aqueous Solution," ACTA Met., 3 (1) pp. 110-111,
        1955.

W3f/3   Webb, W.W. and Bertolone, N.P., "Novel Mechanism for Mass
        Transport During Whisker Growth of Cesium Chloride from
        Aqueous Films," J. Appl. Phys., 31 (1) pp. 207-209, 1960.

W3g- Spontaneous

W3g/1   Arnold, S.M. and Koonce, S.E., "Filamentary Growths on
        Metals at Elevated Temperatures," J. Appl. Phys., 27 (8)
        pp. 962-963, 1956.

W3g/2   Norris, L.F. and Parravano, G., "Sintering of Zinc Oxide,"
        J. Am. Cer. Soc., 46 (9) pp. 449-452, 1963.

W3g/3   Sines, G., "Filamentary Crystals Grown from the Solid Metals,"
        J. Phys. Soc. Japan, 15 (10) pp. 1199-1210, 1960.

W3h- Proper

W3h/1   Cabrera, N. and Price, P.B. (see W1/7).

W3i-Vapor-Solid-Liquid

W3i/1   Wagner, R.S. and Ellis, W.C., "Vapor-Liquid-Solid Mechanism
         of Single Crystal Growth," Appl. Phys. Letters, 4 (5) pp. 89-90,
         1964.

W3i/2   Wagner, R.S. and Ellis, W.C., "The Vapor-Liquid-Solid
         Mechanism of Crystal Growth and Its Application to Silicon," to
         be published in Trans. AIME, 233, 1965.

W3j- Pressure

W3j/1   Franks, J., "Growth of Whiskers in the Solid Phase," Acta
         Met., 6 pp. 103-109, February 1958.

W3j/2   Henning, R.G., "Growth Parameters of Tin Metal Whiskers
         Produced by Pressure," Master's Thesis, Univ. of Utah,
         August 1963.

W3j/3   Pitt, C.H. and Henning, R.G., "Pressure Induced Growth of
         Metal Whiskers," J. Appl. Phys., 35 (2) pp. 459-460, 1964.

W3k- Cleavage

W3k/1   Venables, J.D., "Cleavage Whiskers," J. Appl. Phys., 31 (8)
         pp. 1503-1504, 1960.

W4- Test Methods and Equipment

W4/1   Barber, D.J., "Electron Microscopy and Diffraction of Al_2O_3
         Whiskers," Phil. Mag., 10 (103) pp. 75-94, July 1964.

W4/2   Bokshtein, S.Z., et al, "Tensile Testing of Filamentary Crystals
         of Copper Nickel and Cobalt to Failure," Soviet Physics-Solid
         State, 4 (7) pp. 1272-1277, 1963. (in English)

W4/3   Bokshtein, S.Z. and Svetlov, I.L. (see W2a/3).

W4/4   Brenner, S.S., "Tensile Strength of Whiskers," J. Appl. Phys.,
         27 (12) pp. 1484-1491, 1956.

W4/5   Bushong, R.M., et al (see W1/6).

W4/6   Cabrera, N. and Price, P.B. (see W1/7).

W4/7   Fridman, V.Ya. and Shpunt, A.A., "Tensile Tests on Crystal
         Fragments (Cleavage Whiskers), "Soviet Physics-Solid State, 5
         (3) pp. 570-575, 1963. (in English)

W4/8   Gatti, A., et al (see W1/15 and W1/18).

W4/9   Gordon, J.E. (see W1/19).

W4/10   Guzeev, V.V. and Malinski, Iv. M. (see W1/21).

W4/11   Lemkey, F.D. and Kraft, R.W., "Tensile Testing Technique for
         Submicron Specimens," Rev. Sci. Inst., 33 (8) pp. 846-849, 1962.

W4/12   Marsh, D.M., "Tensile Testing Machine for Microscopic
         Specimens," J. Sci. Inst., 36 (4) pp. 165-169, 1959.

W4/13   Marsh, D.M., "Micro Tensile Testing Machine," J. Sci. Inst.,
         38 (6) pp. 229-234, 1961.

W4/14   Mehan, R.L., et al (see W1/34).

W4/15   Morris, W.G. and Ogilvie, R.E., "Kossel Studies of Iron
         Whiskers," MIT, Cambridge, Mass., AFML, WPAFB, Cont.
         AF 33(657)-8906, RTD-TDR-63-4198, January 1964.

W4/16   Muller, E.W. and Nishikawa, O., "Field Ion Microscopy of Iron
         Whiskers," Penn. State U., Univ. Park, Pa., AFML, WPAFB,
         AF 33(616)-6397, Final Rpt., November 1963.

W4/17   Papalegis, F.E. and Bourdeau, R.G. (see W1/35).

W4/18   Pearson, G.L., et al, "Deformation and Fracture of Small
         Silicon Crystals," Acta Met., 5 (4) pp. 181-191, 1957.

W4/19   Rambauske, W. and Gruenzel, R.R., "Micro-X-Ray Diffraction
         of Whiskers," Dayton U., Ohio Res. Inst., AFML, WPAFB,
         AF 33(616)-8416, July 1963.

W4/20   Revere, A., "Replication Techniques for Electron Microscopic
         Studies of Iron Whisker Surface," Althea Revere, Vineyard Haven,
         Mass., AFML, WPAFB, AF 33(616)-6604, WADD, TR 60-287,
         March 1961.

W4/21   Revere, A., "Electron Microscope Studies of the Surface
        Structure of Whiskers Grown from Iron Chloride," Althea Revere,
        Vineyard Haven, Mass., AFML, WPAFB, AF 33(657)-8554,
        ASD TDR 63-4136, November 1963.

W4/22   Schulman, S. and Epting, J., "A Survey of Test Procedures for
        Evaluating Short Brittle Fibers," ASD, WPAFB, Dayton, Ohio,
        ASD-TDR-63-92, May 1963.

W4/23   Sears, G.W., "Ultimate Strength of Crystals," J. Chem. Phys.,
        36 (3) p. 862, February 11, 1962.

W4/24   Shaffer, P.T.B. and Batha, H.D. (see W1/44).

W4/25   Smith, H.G. and Rundle, R.E., "X-Ray Investigation of Per-
        fection in Tin Whiskers," J. Appl. Phys., 29 (4) pp. 679-683,
        1958.

W4/26   Soltis, P.J. (see W1/47).

W4/27   Strelkov, P.G. and Shpunt, A.A. (see W2a/6).

W4/28   Treuting, R.G. (see W1/56).

W4/29   Wolters, H.B.M. and Schapink, F.W., "Tensile Testing Machine
        for Whiskers," J. Sci. Inst., 38 (6) pp. 250-252, June 1961.

W5- General Review and/or Theoretical Strength

W5/1   Accountius, O.E., "Whiskered Metals Reach for 1,000,000 psi,"
        Machine Design, 35 (11) pp. 194-199, May 9, 1963.

W5/2   Baker, W.S. and Kaswell, E.R., "Handbook of Fibrous
        Materials," McGraw-Hill Tech. Writing Service and Fabric Res.
        Labs. Inc., Dedham, Massachusetts, ASD, WPAFB, AF 33(616)-
        7504, WADD TR 60-584 Pt. II, October 1961.

W5/3   Brenner, S.S., "Metal Whiskers," Sci. American, 203 (7)
        pp. 65-72, 1960.

W5/4   Brenner, S.S. (see W2/2).

W5/5   Cunningham, A.L. (see W3/10).

W5/6   Gordon, J.E., "Whiskers," Endeavour, 23 (88) pp. 8-12, 1964.

W5/7   Gorsuch, P.D., "On the Crystal Perfection of Iron Whiskers,"
        G.E. Co., Research Laboratory, Schenectady, N. Y., 58-RL-
        2113, 1958.

W5/8   Growth and Perfection of Crystals, edited by R.H. Doremus,
        B.W. Roberts, and D. Turnbull, John Wiley & Sons, Inc.,
        New York, pp. 275, 1958.

W5/9   Hoffman, G.A., "The Exploitation of the Strength of Whiskers,"
        The Rand Corp., Santa Monica, Calif., Rpt. P-1294, March
        1958.

W5/10   Kress, E., "Whiskers--- But Not on Your Face," Space Digest,
        pp. 70-72, 1960.

W5/11   Krock, R.H. and Kelsey, R.H., "Whiskers-Their Promise and
        Problems," Ind. Res., 1 (2) pp. 46-57, 1965.

W5/12   Lynch, C.T., et al (see W3/11).

W5/13   Nabarro, F.R.N. and Jackson, P.J. (see W3/13).

W5/14   Nadgornyi, E.M., "The Properties of Whiskers," Soviet
        Physics-Uspekhi, 5 (3) pp. 462-477, 1962. (in English)

W5/15   Nadgornyi, E.M., et al, "Filamentary Crystals with Almost the
        Theoretical Strength of Perfect Crystals," Soviet Physics-
        Uspekhi, 67 (2) pp. 282-304, 1959. (in English)

W5/16   Nadgornyi, E.M., "Perfection and Strength of Crystals," FTD
        Rpt., Contemp. Probl. of Phys. Met., N64-16606, July 1963.
        (in English)

W5/17   Pierce, C.M. (see W3/18).

W5/18   Poremba, F.J., comp., "Mechanical Properties of Whiskers
        (Filamentary Crystals) 1957-1964," U.S. Steel Corp.,
        Monroeville, Pa., Metals/Materials Div., Bibliography No. 64-4,
        October 1964.

W5/19   Price, P.B. (see W1/40).

W5/20   Regester, R.F. (see W1/42).

W5/21   Salkind, M. J., "Whiskers and Whisker Strengthened Composite Materials," Watervliet Arsenal, Watervliet, N. Y., Rpt. WVT-RR-6315, October 1963.

W5/22   Salkind, M. J., "Candidate Materials for Whisker Composites," Watervliet Arsenal, Watervliet, N. Y., Rpt. WVT-RR-6411, May 1964.

W5/23   Stambler, I., "Whisker Research Promises Ultra-Strong Materials," Space/Aeronautics, 34 (5) pp. 48-52, 1960.

W5/24   Webb, W. W. and Forgeng, W. D., "Mechanical Behavior of Microcrystals," Acta Met. , 6 pp. 462-469, July 1958.

F- FUSED SILICA AND GLASS FIBERS

F1- Measured Properties

F1/1   Anderegg, F. O., "Strength of Glass Fiber," Ind. & Eng. Chem., 31 (3) pp. 290-298, 1939.

F1/2   Anon., "Glass Gets Tough," Chem. Week, 94 (15) pp. 59-60, April 11, 1964.

F1/3   Bartenev, G. M., "Flawless Glass Fibres," presented to Symposium on Physics of Noncrystalline Solids, sponsored by Int. Union Pure and Applied Phys., Delft, Holland, July 6-10, 1964.

F1/4   Bevis, R. E. and Thomas, G. L., "Glass Fiber Bundles, Theoretical vs. Actual Tensile Strengths," Proc. 19th Annual Mtg. Soc. Plastics Ind., 17-D, pp. 1-6, February 1964.

F1/5   Brossy, J. F. and Provance, J. D., "Development of High Modulus Fibers from Heat Resistant Materials," Houze Glass Co., Point Marion, Pa., AF 33(657)-8904, WADC TR-58-285, Part II, March 1960.

F1/6   Butler, J. B., "An Investigation of Glass Tensile Strength Variations Caused by Induced Mechanical Damage," G. E. Co., Missile & Space Division, King.of Prussia, Pa., Report 63 SD 291, August 1963.

F1/7   Butler, J. B., "A Statistical Evaluation Technique Applied to Glass Filament Yarn Tensile Strength," G. E. Co., Missile &

Space Division, King of Prussia, Pa., Report 62SD798, October 1962.

F1/8   Cameron, N.G., "The Effect of Heating on the Room Temperature Strength of Pristine E-Glass Rods," Univ. of Ill., Urbana, Ill., Naval Res. Lab., Washington, D.C., NONR 2947 (02) (X), T&A. M. Rpt. 252, November 1963.

F1/9   Capps, W. and Blackburn, D.H., "The Development of Glass Fibers Having High Young's Modulus of Elasticity," National Bureau of Standards, Washington, D.C., NBS Rpt. 5188, April 1957.

F1/10   Chase, G.H. and Phillips, C.J., "Elastic Modulii and Tensile Strength of Glasses and Other Ceramic Materials," Rutgers Univ., New Brunswick, N. J., Progress Rpt. 4, June 1964.

F1/11   Cobb, E.S., Jr., "High Temperature Resistant Flexible Fibrous Structural Materials of New Glass Composition," Owens-Corning Fiberglas Corp., Ashton, R.I., AFML, WPAFB, AF 33(616)-7441, ML TDR 64-10, December 1963.

F1/12   Fiedler, W.S., et al, "Quartz Fibers in High Temperature Resistant Materials," Bjorksten Res. Labs. Inc., Madison, Wis., AFOSR, Wash. D.C. Contr. AF 18(603)-42, Rpt. TR 58-91, May 1958.

F1/13   Gates, L.E., et al, "Development of Ceramic Fibers for Reinforcement in Composite Materials," Hughes Aircraft Co., Culver City, Calif., NASA Contract NAS 8-50, NASA Rpt. CR-50793, April 1961.

F1/14   Girard, E.H., "Continuous Filament Ceramic Fibers, Part II," Carborundum Co., Niagara Falls, N.Y., ASD, WPAFB, AF 33(616)-6246, WADD TR-60-244, Part II, February 1961.

F1/15   Hollinger, D.L., "Influence of Stress Corrosion on Strength of Glass Fibers," G.E. Co., AETD, Evendale, Ohio, Naval Res. Lab. Contract NOnr 4486(00)(X), 1st Bi-monthly Prgr. Rpt., May 1964.

F1/16   Hollinger, D.L., et al, "Influence of Stress Corrosion on Strength of Glass Fibers," G.E. Co., AETD, Evendale, Ohio, Naval Res. Lab. Contract NOnr 4486(00)(X), 2nd Bi-monthly Prgr. Rpt., July 1964.

F1/17   Ibid, 3rd Bi-monthly Prgr. Rpt., September 1964.

F1/18   Ibid, 4th Bi-monthly Prgr. Rpt., November 1964.

F1/19   Hollinger, D. L., et al, "High Strength Glass Fibers Develop-
          ment Program," G. E. Co., FPLD, Evendale, Ohio, Dept. of
          Navy Contract NOw 61-0641-c (FBM), Final Rpt., May 1963.

F1/20   Holloway, D. G., "The Strength of Glass Fibres," Phil. Mag., 4
          pp. 1101-1106, 1959.

F1/21   Holloway, D. G. and Schlapp, D. M., "The Strength of Glass
          Fibres Drawn in Vacuum," Trans. Brit. Cer. Soc., 59 pp. 424-
          431, 1960.

F1/22   Kies, J.A., "The Strength of Glass Fibers and the Failure of
          Filament Wound Pressure Vessels," U.S. N. Res. Lab.,
          Washington, D. C., NRL Rpt. 6034, February 1964.

F1/23   Kroenke, W.J., "Linear Structured Glass Fibers," B. F.
          Goodrich Co., Akron, Ohio, AFML, WPAFB, Contract
          AF 33(657)-8905, Qtrly. Prog. Rpt., February 1964.

F1/24   Ibid., Qtrly. Prog. Rpt., December 1964.

F1/25   Kroenke, W.J., et al, "Structured Glasses Patterned After
          Asbestos," The B. F. Goodrich Co., Akron, Ohio, AFML,
          WPAFB, Contract AF 33(657)-8905, RTD-TDR-63-4043, Part I,
          April 1963.

F1/26   Kroenke, W.J., "Flexible Glass Fibers," The B. F. Goodrich
          Co., Akron, Ohio, AFML, WPAFB, Contract AF 33(657)-8905,
          ML-TDR 64-119, March 1964.

F1/27   Lambertson, W.A., et al, "Continuous Filament Ceramic
          Fibers," Carborundum Co., Niagara Falls, N. Y., ASD WPAFB,
          AF 33(616)-6246, WADD TR 60-244, June 1960.

F1/28   Lasday, A.H., "Development of High Modulus Fibers from Heat
          Resistant Materials," Houze Glass Co., Point Marion, Pa.,
          ASD, WPAFB, AF 33(616)-5263, WADC TR 58-285, Oct. 1958.

F1/29   Lewis, A., et al, "Research to Obtain High Strength Continuous
          Filaments from Type 29-A Glass Formulations," Aerojet-
          General Corp., Azuza, Calif., AFML, WPAFB, AF 33(657)-
          8904, RTD-TDR-63-4241, Dec. 1963.

F1/30   Lindsay, E. M. and Rood, J. C., "Glass Reinforcements for Fila-
         ment Wound Composites," Owens-Corning Fiberglas Corp.,
         Toledo, Ohio, AFML, WPAFB, AF 33(657)-9623, Final Tech.
         Eng. Rpt. TR-64-8-104, Dec. 1963.

F1/31   Lindsay, E. M., et al, "Glass Reinforcements for Filament
         Wound Composites," Owens-Corning Fiberglas Corp., Toledo,
         Ohio, AFML, WPAFB, Contract AF 33(657)-9623, Int. Eng.
         Prgr. Rpt., Aug. 1963.

F1/32   Machlan, G. R., "The Development of Fibrous Glasses Having
         High Elastic Moduli," Owens-Corning Fiberglas Corp., Newark,
         Ohio, WADC, WAPAFB, AF 33(616)-2422, WADC TR 55-290,
         Nov. 1955.

F1/33   Mallinder, F. P. and Proctor, B. A., "Elastic Constants of Fused
         Silica as a Function of Large Tensile Strain," Phys. & Chem.
         Glasses, $\underline{5}$ (4) pp. 91-103, Aug. 1964.

F1/34   Metcalfe, A. G. and Schmitz, G. K., "Effect of Length on the
         Strength of Glass Fibres," ASTM Preprint No. 87, presented at
         67th Annual Mtg., June 1964.

F1/35   Morley, J. G., "Reinforcing Metals with Fibers," New Scientist,
         (322) pp. 122-125, Jan. 17, 1963.

F1/36   Morley, J. G., "Strength of Glass Fibres," Nature, $\underline{184}$ (4698)
         p. 1560, Nov. 1959.

F1/37   Morley, J. G., et al, "Strength of Fused Silica," Phys. and Chem.
         Glasses, $\underline{5}$ (1) pp. 1-10, Feb. 1964.

F1/38   Morrison, A. R., et al, "High Modulus, High Temperature
         Laminates with Fibers and Flakes," Owens-Corning Fiberglas
         Co., Granville, Ohio, AFML, WPAFB, AF 33(616)-5802, WADD
         TR-60-24, Suppl. 4, Feb. 1962.

F1/39   Mould, R. E., "Cross-Bending Tests of Glass Fibers and the
         Limiting Strength of Glass," J. App. Phys., $\underline{29}$ (11) pp. 1263-64,
         1958.

F1/40   Murgatroyd, J. B., "The Delayed Elastic Effect in Glass Fibers
         and the Constitution of Glass in Fiber Form," J. Soc. Glass
         Tech., $\underline{37}$ pp. 291-300, 1953.

F1/41   Murgatroyd, J.B., "The Strength of Glass Fibers. Part I:
         Elastic Properties," J. Soc. Glass Tech., 28 (130) pp. 368-387,
         1944.

F1/42   Ibid., "The Strength of Glass Fibers. Part II: The Effect of
         Heat Treatment on Strength," pp. 388-405T.

F1/43   Otto, W.H., "Compaction Effects in Glass Fibers," J. Am. Cer.
         Soc., 44 (2) pp. 68-72, 1961.

F1/44   Otto, W.H., "Relationship of Tensile Strength of Glass Fibers to
         Diameter," J. Am. Cer. Soc., 38 (3) pp. 122-124, 1955.

F1/45   Otto, W.H., et al, "Potential of Filament Wound Composites,"
         Narmco Res. & Dev., San Diego, Calif., BuWeps Contr. NOw-
         61-0623-C (FBM), Qtrly. Rot., May 1962.

F1/46   Otto, W.H. and Vidanoff, R.B., "Silica Fiber Forming and Core
         Sheath Composite Fiber Development," Narmco Ind. Inc.,
         San Diego, Calif., BuWeps, Cont. N600(19)59607, Qtrly. Prgr.
         Rpt. 1, April 1963.

F1/47   Ibid., Final Summary Rpt., Jan. 1964.

F1/48   Petker, I., et al, "Development of High Strength Impregnated
         Roving for Filament Winding," Aerojet-General Corp.,
         Sacramento, Calif., U.S. Navy Spl. Proj. Contr. NOw-63-0627-
         C (FBM), Final Rpt., July 1964.

F1/49   Provance, J.D., "Stronger Glass Fibers Lighten Aerospace
         Vehicles," Cer. Ind., 82 (4) pp. 109-111, 132, 142, 1964.

F1/50   Provance, J.D., "Glass Fiber Properties of a Magnesia-
         Alumina-Silica System" presented at Fall Mtg., Glass Div., Am.
         Cer. Soc., Bedford Springs, Oct. 1963.

F1/51   Provance, J. and Kelly, L.W., "Optimizing Properties of Houze
         29-A Glass Fibers," Houze Glass Corp., Point Marion, Pa.,
         AFML, WPAFB, AF 33(657)-8904, 1st Qtrly. Prgr. Rpt.,
         Sept. 1962.

F1/52   Ibid., 2nd Qtrly. Rpt., Dec. 1962.

F1/53   Schmitz, G.K., "Exploration and Evaluation of New Glasses in
         Fiber Form" Int'l. Harvester Corp., Solar Div., San Diego,

Calif., U.S. Naval Res. Lab. Contract NOnr-3654(00)(X) A2, Bi-monthly Prog. Rpt. 2, May 1964.

F1/54   Thomas, W.F., "An Investigation of the Factors Likely to Affect the Strengths and Properties of Glass Fibers," Phys. and Chem. Glasses, 1 (1) pp. 4-18, Feb. 1960.

F1/55   Tighe, N.J., "Fused Quartz Fibers: A Survey of Properties, Applications and Production Methods," National Bureau of Standards, Washington, D.C., NBS Circular 569, Jan. 1956.

F1/56   Tomashot, R.C., "AF-994: A Superior Glass Fiber Reinforcement for Structural Composites," AFML, WPAFB, Dayton, Ohio, ASD-TDR-63-81, March 1963.

F1/57   Veazie, F.M. and Brady, W.C., "New High Performance Vitreous Reinforcements," Conference on Structural Plastic Adhesives and Filament Wound Composites, ASD-TDR-63-396, pp. 310-314, April 1963.

F1/58   Waugh, J.A., et al, "The Development of Fibrous Glasses Having High Elastic Modulii," Owens-Corning Fiberglas Corp., Granville, Ohio, AFML, WPAFB, AF 33(616)-2422, WADC TR 55-290, Part II, May 1958.

F1/59   Weisbart, H., "Investigations on the Strength of Alkali-Containing Glass Fibers," SPE Trans., 2 (4) pp. 339-353, October 1962.

F2- Factors Affecting Measured Properties

F2/1   Anderegg, F.O. (see F1/1).

F2/2   Barish, L., et al, "Mechanical and Thermal Degradation Mechanisms of Quartz Fibers and the Development of Experimental Quartz Fabrics with Improved Finishes," Fabrics Research Labs., Inc., Dedham, Mass., AFML, WPAFB, Cont. AF 33(616)-7557, ASD-TDR-63-802, September 1963.

F2/3   Bateson, S., "Critical Study of the Optical and Mechanical Properties of Glass Fibers," J. Appl. Phys., 29 (1) pp. 13-21, 1958.

F2/4   Bernal, J.D., et al, "A Discussion of New Materials," Proc. Royal Soc. A., 282 pp. 1-154, 1964.

F2/5   Cameron, N.G., "An Introduction to the Factors Influencing the Strength of Glass," Univ. of Illinois, Urbana, Illinois, Naval Research Lab., Washington, D.C., Proj. 62R05 19A, Tech. Memo 158, March 1961.

F2/6   Capps, W. and Blackburn, D.H. (see F1/9).

F2/7   Cobb, E.S., Jr. (see F1/11).

F2/8   Kies, J.A. (see F1/22).

F2/9   Kroenke, W.J. (see F1/23-F1/26).

F2/10   Lewis, A., et al (see F1/29).

F2/11   Machlan, G.R. (see F1/32).

F2/12   Mallinder, F.P. and Proctor, B.A. (see F1/33).

F2/13   Morley, J.G., et al (see F1/35 and F1/37).

F2/14   Morrison, A.R., et al (see F1/38).

F2/15   Parratt, N.J., "Defects in Glass Fibers and Their Effect on the Strength of Plastic Mouldings," Rubber and Plastics Age, 41 (3) pp. 263-266, March 1960.

F2/16   Provance, J.D. (see F1/49-F1/52).

F2/17   Thomas, W.F. (see F1/54).

F2/18   Waugh, J.A., et al (see F1/58).

F2a- Effect of Size and/or Shape

F2a/1   Bell, J.E., "Effect of Glass Fiber Geometry on Composite Material Strength," ARS Journal, 31 (9) pp. 1260-1265, 1961.

F2a/2   Cornish, R.H. and Chaney, R.M., "Glass Fiber Strength Enhancement Through Bundle Drawing Operations," IITRI, Chicago, Ill., BuWeps, Contract N600(19)-58450, Qtrly. Prgr. Rpt. 4, July 1963.

F2a/3   Eakins, W.J. and Humphrey, R.A., "Studies of Hollow Multi-partitioned Ceramic Structures," DeBell & Richardson Inc.,

Hazardville, Conn., NASA-Hqs., Contract NASW-672, NASA CR-142, December 1964.

F2a/4   Grunfest, I.J. and Dow, N.A., "New Shapes for Glass Fibers," Astronautics, 6 (4) pp. 34-35, 1961.

F2a/5   Jobaris, J., "Optimum Filament Diameter," Narmco Res. & Dev., San Diego, Calif., Navy BuShips, Contract NO1s 86347, 4th Qtrly. Prog. Rpt., Jan. 1964.

F2a/6   Metcalfe, A.G. and Schmitz, G.K. (see F1/34).

F2a/7   Otto, W.H. (see F1/44).

F2a/8   Rosen, B.W., "Effect of Glass Form on Strength," G.E. Co., MSD, King of Prussia, Pa., BuWeps, Contr. NOw-63-0674-C, Rpt. R64SD18, Oct. 1963.

F2a/9   Rosen, B.W., "Hollow Glass Fiber-Reinforced Laminates," G.E. Co., MSD, King of Prussia, Pa., BuWeps, Cont. NOw-63-0674-C, Final Rpt., Sept. 1964.

F2a/10 Rosen, B.W. and Ketler, Jr., A.E., "Hollow Glass Fiber Reinforced Plastics," G.E. Co., MSD, King of Prussia, Pa., BuWeps, Contr. NOw-61-0613-d, Final Report, May 1963.

F2a/11 Schmitz, G.K. (see F1/53).

F2a/12 Siefert, R.F., "Hollow Fibers," Reinforced Plastics, 4 (1) pp. 22-23, Jan/Feb. 1965.

F2a/13 Weil, N.A., et al, "Investigation of Glass Fiber Strength Enhancement Through Bundle Drawing," IITRI, Chicago, Ill., BuWeps, NOw-61-0259-c, 1st Qtrly. Rpt., May 1961.

F2b- Effect of Structural Perfection and/or Composition

F2b/1   Bateson, S., "A Note on the Structure of Glass Fibers," J. Soc. Glass Tech., 37 pp. 3-02T-305T, 1953.

F2b/2   Chase, G.H. and Phillips, C.J. (see F1/10).

F2b/3   Murgatroyd, J.B. (see F1/40-F1/42).

F2b/4   Weisbart, H. (see F1/59).

F2c- Effect of Surface Treatment and/or Coatings

F2c-1   Abbott, R. L., "Evaluation of National Research Corporation
        Glass Monofilaments," Naval Air Engineering Center,
        Philadelphia, Pennsylvania, Report AML-1884, February 1964.

F2c/2   Ainslie, N.G., et al, "Devitrification Kinetics of Fused Silica,"
        G. E. Co. Research Laboratory, Schenectady, New York,
        Report 61-RL-2640M, (Revised), March 1961.

F2c/3   Arridge, R.G. C., et al, "Metal Coated Fibers and Fiber-
        Reinforced Metals," J. Sci. Instr., 41 pp. 259-261, May 1964.

F2c/4   Barnes, R. E., comp., "Silane Coupling Agents, Bibliography
        for 1953-1962," Armed Services Tech. Inf. Agency, Alexandria,
        Virginia, September 1962.

F2c/5   Baskey, R. H., "Fabrication of Core Materials From Aluminum-
        Coated, Fuel Bearing Fiberglas," Clevite Corporation,
        presented at Winter Meeting, Am. Nuclear Society, Chicago,
        Illinois, November 9, 1961.

F2c/6   Butler, J.B. (see F1/6).

F2c/7   Capps, W. and Blackburn, D.H. (see F1/9).

F2c/8   Eakins, W.J., "The Relative Stability of Cation Impurities and
        Silica Fiber Surfaces," SPE Trans., 2 (4) pp. 354-362, October
        1962.

F2c/9   Eakins, W.J., "The Degree of Bonding Between Coupling Agents
        and Glass Filaments: A Study, "SPE Trans., 1 (4) pp. 234-244,
        Oct. 1961.

F2c/10 Eakins, W.J., "A Study of the Finish Mechanism: Glass to
        Finish, Finish Thickness and Finish to Resin, "SPE Trans., 1
        (4) pp. 214-223, Oct. 1961.

F2c/11 Edmunds, W. M., "Research on Protection of High Temperature
        Glass Fibers Against Flexuring and Abrasion," Owens-Corning
        Fiberglas Corp., Gransville, Ohio, AFML, WPAFB, AF 33(616)-
        7441, Final Rpt., Nov. 1962.

F2c/12 Eischen, G., "Effect of Adsorbed Gases on the Torsional
        Elasticity of Glass Fibers, " Compt. Rend., 248 pp. 3160-3162,
        1959. (in French)

F2c/13   Eischen, G., "The Influence of Adsorbed Gas on the Rupture
         Strength of Glass Fibers," Compt. Rend., 250 pp. 2194-2196,
         1960. (in French)

F2c/14   Erickson, P.W., "Glass Fiber Surface Treatments: Theories
         and Navy Research," Naval Ord. Lab., White Oak, Md., NOL
         TR 63-253, Nov. 1963.

F2c/15   Fahland, F.W., "Survey on Metal Coatings of Glass Filaments,"
         U.S. Naval Res. Lab., Tech. Memo. No. 237, June 1963.

F2c/16   Morelock, G.R., "Pyrolytic Graphite Coated Fused Silica,"
         G.E. Co., Research Lab., Schenectady, N.Y., Rpt. 61-RL-
         2672M, Mar. 1961.

F2c/17   Morley, J.G. (see F1/35-Fi/36).

F2c/18   Outwater, J.O. and Ozaltin, O., "Surface Effects of Various
         Environments and of Thermosetting Resins on the Strength of
         Glass," SPE Trans., 2 (4) pp. 321-325, Oct. 1962.

F2d- Effect of Forming Methods

F2d/1   Anderson, O.L., "Cooling Time of Strong Glass Fibers," J.
         Appl. Phys., 29 (1) pp. 9-12, 1958.

F2d/2   Anon., "Space Age Fiber Goal: Strength at 2000°F," Chem.
         Week, 89 (15) pp. 61-62, April 29, 1961.

F2d/3   Arridge, R.G.C. and Prior, K., "Cooling Time of Silica
         Fibers," Nature, 203 (4943) pp. 386-387, July 1964.

F2d/4   Bartenev, G.M. (see F1/3).

F2d/5   Capps, W. and Blackburn, D.H. (see F1/9).

F2d/6   Cornish, R.H. and Chaney, R.M. (see F2a/2).

F2d/7   Murgatroyd, J.B. (see F1/41).

F2d/8   Stirling, F.J., "Frozen Strains in Glass Fibers," J. Soc.
         Glass Tech., 39 (188) pp. 134-144, June 1955.

F2d/9   Weil, N.A., et al (see F2a/13).

F2e- Effect of Environment

F2e/1   Cameron, N. G. (see F1/8).

F2e/2   Charles, R.J. and Hillig, W.B., "The Kinetics of Glass Failure by Stress Corrosion," G.E. Co., Research Lab., Schenectady, N. Y., Rpt. 61-RL-2790M, July 1961.

F2e/3   Hammond, M.L. and Ravitz, S.F., "Influence of Environment on Brittle Fracture of Silica," J. Am. Cer. Soc., 46 (7) pp. 329-332, 1963.

F2e/4   Hollinger, D.L. (see F1/15).

F2e/5   Hollinger, D.L. et al (see F1/16-F1/18).

F2e/6   Holloway, D.G. (see F1/21).

F2e/7   Murgatroyd, J.B. (see F1/42).

F2e/8   Weisbart, H. (see F1/59).

F3- Manufacturing Processes

F3a- Filaments

F3a/1   Brossy, J. F. and Provance, J. D. (see F1/5).

F3a/2   Capps, W. and Blackburn, D.H. (see F1/9).

F3a/3   Cornish, R.H. and Chaney, R.M. (see F2a/2).

F3a/4   Drummond, W.W. and Cash, B.A., "Development of Textile Type Vitreous Silica Yarns," Bjorksten Res. Labs. Inc., Madison, Wisc., ASD, WPAFB, Contract AF 33(616)-6255, WADC TR 59-699, March 1960.

F3a/5   Edmunds, W.M. (see F2c/10).

F3a/6   Gates, L.E., et al (see F1/13).

F3a/7   Girard, E.H. (see F1/14).

F3a/8   Grove, C.S., et al, "Developments in Fiber Technology," Ind. and Eng. Chem., 54 (8) pp. 55-57, 1962.

F3a/9   Koch, P.A., "Glass Filaments and Textiles," CIBA Review, pp. 19-33, 1963/1965.

F3a/10 Kroenke, W.J. (see F1/23-F1/26).

F3a/11 Lambertson, W.A. (see F1/27).

F3a/12 Lasday, A.H. (see F1/28).

F3a/13 Lewis, A., et al (see F1/29).

F3a/14 Machlan, G.R. (see F1/32).

F3a/15 Morley, J.G., et al (see F1/37).

F3a/16 Otto, W.H. and Vidanoff, R.B. (see F1/46-F1/47).

F3a/17 Petker, I., et al (see F1/48).

F3a/18 Provance, J.D. (see F1/49-F1/52).

F3a/19 Silverman, A., "High Temperature Glass Fiber Production," Trans. Soc. Glass Tech., XL, pp. 413-428, 1956.

F3a/20 Thomas, W.F. (see F1/54).

F3a/21 Tomashot, R.C. (see F1/56).

F3a/22 Vidanoff, R. and Bradford, D., "Silica Fiber/Core Sheath Fibers," Narmco Res. & Dev., San Diego, Calif., BuWeps, N600(19)61810, Qtrly. Prog. Rpt. 1, May 1964.

F3a/23 Weil, N.A., et al (see F2a/13).

F3b- Fabrics, Tapes, Webbings

F3b/1   Coplan, M.J., et al, "Flexible Fibrous Structural Materials," Fabric Res. Labs. Inc., Dedham, Mass., AFML-WPAFB, Contract AF 33(657)-10541, ML-TDR-64-102, Feb. 1964.

F3b/2   Coplan, M.J., "Flexible Low Permeability Metal and Ceramic Cloths," Fabric Res. Labs. Inc., Dedham, Mass., WADD, ARDC, Contract AF 33(616)-7222, Bi-monthly Prog. Rpt. 1, Aug. 1960.

F3b/3   Coplan, M.J., et al, "Conversion of High Modulus Materials into Flexible Fabric Structures," Fabric Res. Labs. Inc., Dedham, Mass., ASD, WPAFB, AF 33(616)-7222, ASD-TDR-62-851, Oct. 1962.

F3b/4   Ehlers, S., "Woven Fabrics: Their Properties and Uses," Mat. in Des. Eng., 59 (5) pp. 88-91, May 1964.

F3b/5   Freeston, W.D. and Gardella, J.W., "Phase II Report on Process Development for Stranding and Plying of Metallic Yarns," Fabric Res. Labs. Inc., Dedham, Mass., AFML, WPAFB, AF 33(657)-11624, Proj. Nr. 8-134, Jan. 1964.

F3b/6   Gates, L.E. and Lent, W.E., "Development of Refractory Fabrics," Hughes Aircraft Co., Culver City, Calif., NASA-Huntsville, NAS8-50, Final Summary Report, April 1963.

F3b/7   Hawthorne, A.T., et al, "Development of Fabric Base Materials for Space Applications," Westinghouse Defense & Space Center, Baltimore, Md., AFML, WPAFB, AF 33(6p6)-8259, ML-TDR-64-20, Jan. 1964.

F3b/8   Koch, P.A. (see F3a/9).

F4- Test Methods and Equipment

F4a- Filaments

F4a/1   Charles, R.J. and Hillig, W.B. (see F2e/2).

F4a/2   Grove, C.S. (see F3a/8).

F4a/3   Machlan, G.R. (see F1/32).

F4a/4   Metcalfe, A. G. and Schmitz, G.K. (see F1/34).

F4a/5   Morley, J.G. et al (see F1/37).

F4a/6   Mould, R.E. (see F1/39).

F4a/7   Provance, J.D. (see F1/49-F1/52).

F4a/8   Schmitz, G.K. (see F1/53).

F4a/9   Schwartz, H., "Tensile Strength of Ceramic Staple Fibers," WADC, WPAFB, Dayton, Ohio, WORT TM-56-104, Feb. 1957.

F4a/10   Statton, W.O. ánd Hoffman, L.C., "Structure in Vitreous
         Silicate Fibers As Shown By Small-Angle Scattering of X-Rays,"
         J. Appl. Phys., 31 (2) pp. 404-409, 1960.

F4a/11   "Tensile Stress-Strain Properties of Fibers," E.I. duPont
         de Nemours, Inc., Wilmington, Delaware, Tech. Information
         Booklet, Bulletin X-82.

F4a/12   Thomas, W.F. (see F1/54).

F4a/13   Tomashot, R.C. (see F1/56).

F4a/14   Whitney, J.M., "Applying Engineering Mechanics Solutions to
         Deflections of Statically Loaded Coplanor Fibers," AFML,
         WPAFB, ML-TDR-64-118, May 1964.

F4b- Fabrics, Tapes, Webbings

F4b/1   Coskren, R.J. and Chu, C.C., "Investigation of the High Speed
         Impact Behavior of Fibrous Materials, Part II: Impact
         Characteristics of Parachute Materials," Fabric Res. Labs.
         Inc., Dedham, Mass., ASD, WPAFB, ASD-TDR-60-511, Part II,
         February 1962.

F4b/2   Davidson, D.A., "The Mechanical Behavior of Fabrics Subjected
         to Biaxial Stress: Canal-Model Theory of the Plain Weave,"
         General American Transportation Company, AFML, WPAFB,
         Contract AF 33(657)-8479, ML-TDR-64-239, July 1964.

F5- General Review and/or Theoretical Strength

F5/1   Anon., "New Sinews for Rugged Service," Chem. Week, 95 (25)
         pp. 33-34, December 19, 1964.

F5/2   Bernal, J.D., et al (see F2/4).

F5/3   Burke, J.E., editor, Progress in Ceramic Science, Volume I,
         Pergamon Press, New York 232pp., 1961.

F5/4   Ibid., Volume III, 262pp., 1963.

F5/5   Cameron, N.G. (see F2/5).

F5/6   Dunn, S.A. and Roth, W.D., "High Viscosity Refractory Fibers,"
         Bjorksten Res. Labs. Inc., Madison, Wisc., BuWeps NOrd-
         19100, Summary Tech. Rpt., February 1962.

F5/7   Kies, J.A. (see F1/22).

F5/8   Koch, P.A. (see F3a/9).

F5/9   Lindsay, E.M. and Rood, J.C. (see F1/30).

F5/10   Mallinder, F.P. and Proctor, B.A. (see F1/33).

F5/11   Mettes, D.G., "Glass Fibers in Aerospace," presented to SPE,
         New York City Meeting, May 1964.

F5/12   Morley, J.G., et al (see F1/37).

F5/13   Provance, J.D. (see F1/49).

F5/14   Slayter, G., "Strength of Glass," Bull. Am. Cer. Soc., 31 (8)
         pp. 276-278, 1952.

F5/15   Slayter, G., "Two Phase Materials," Sci. Am., 206 (1) pp. 124-
         134, 1962.

F5/16   Thomas, W.F. (see F1/54).

F5/17   Tighe, N.J. (see F1/55).

F5/18   Weil, N.A., et al (see F2a/13).

C- COMPOSITES

C1- Whisker Reinforcement

C1/1   Anon., "Whiskers Go Commercial," Chem. Week, 96 (7) pp. 65-
         68, February 13, 1965.

C1/2   Keller, E., "Whiskers as Reinforcement Materials," Ind. and
         Eng. Chem., 56 (4) pp. 9-10, 1964.

C1/3   Kelly, A., "Fiber-Reinforced Metals," Sci. Am., 212 (2) pp. 28-
         37, 1965.

C1/4   Kelly, A., "The Promise of Fiber Strengthened Materials,"
         Discovery, 24 (9) pp. 22-27, 1963.

C1a- Metal Whisker - Metal Matrix

C1a/1   Baskey, R.H., "Fiber Reinforcement of Metallic and Non-

Metallic Composites," Clevite Corp., Cleveland, Ohio, AFML, Contract AF 33(657)-7139, Int. Rpt. Vol. 1, February 1962.

C1a/2   Ibid., Vol. 2, May 1962.

C1a/3   Ibid., Vol. 3, August 1962.

C1a/4   Ibid., Vol. 4, December 1962.

C1a/5   Ibid., Final Rpt., ASD-TDR-63-619, July 1963.

C1a/6   McDaniels, D. L., et al, "Stress-Strain Behavior of Tungsten-Fiber Reinforced Copper Composites," NASA-Lewis Res. Center, Cleveland, Ohio, NASA TN D-1881, October 1963.

C1a/7   Parikh, N.M., "Fiber-Reinforced Metals and Alloys," IITRI, Chicago, Illinois, BuWeps Cont. NOw-64-0066-c, Bi-Monthly Rpt., February 1964.

C1a/8   Ibid., April 1964.

C1a/9   Ibid., June 1964.

C1a/10   Ibid., August 1964.

C1a/11   Parikh, N. M. and Adamski, R., "Fiber Reinforced Metals and Alloys," IITRI, Chicago, Illinois, BuWeps Cont. NOw-64-0066-c, IITRI-B6020-5, October 1964.

C1b- Metal Whisker - Inorganic Matrix

(No entries)

C1c- Metal Whisker - Organic Matrix

(No entries)

C1d- Ceramic Whisker - Metal Whisker

C1d/1   Anon., "Stronger Metals with Silicon Nitride Whiskers," New Sci., 19 (351) p. 291, August 8, 1963.

C1d/2   Brenner, S.S., "The Case for Whisker Reinforced Metals," J. Metals, 14 (11) pp. 809-811, 1962.

C1d/3   Kelsey, R. H. "Reinforcement of Nickel Chromium Alloys with Sapphire Whiskers," Horizons Inc., Cleveland, Ohio, BuWeps Cont. NOw-63-0138-c, Final Report, October 1963.

C1d/4   Mehan, R. L., et al (see W1/34).

C1d/5   Parikh, N. M. (see C1a/7-C1a/10).

C1d/6   Parratt, N. J., "Reinforcing Effects of Silicon Nitride Whiskers in Silver and Resin Matrices," Powder Metallurgy, 7 (14) pp. 152-167, 1964.

C1d/7   Raynes, B. C., "Studies of the Reinforcement of Metals with Ultra High Strength Fibers (Whiskers)," Horizons, Inc., Cleveland, Ohio BuWeps Cont. NOw-61-0207-c, Final Report, November 1961.

C1d/8   Raynes, B. C. and Kelsey, R. H., "Studies of the Reinforcement of Metals with Ultra High Strength Fibers (Whiskers)," Horizons Inc., Cleveland, Ohio, BuWeps Cont. NOw-62-0235-c, Final Report, October 1962.

C1d/9   Ibid., Int. Rpt. No. 2, January 1963.

C1d/10   Sharkitt, R. L., et al (see W1/45).

C1d/11   Sicka, R. W., et al, "Reinforcement of Nickel Chromium Alloys with Sapphire Whiskers," Horizons Inc., Cleveland, Ohio, BuWeps Cont. NOw-64-0125-c, Final Report, October 1964.

C1d/12   Smalley, A. K., Atomics Int'l. Canoga Park, Calif., private communication, February 1965.

C1d/13   Sutton, W. H., et al, "Development of Composite Structural Materials for High Temperature Applications," G. E. Co., Missile and Space Div., King of Prussia, Pa., U.S. Navy BuWeps Cont. NOw-60-0465-d, Prgr. Rpt. No. 1, July 1960.

C1d/14   Ibid., Prgr. Rpt. No. 2, October 1960.

C1d/15   Ibid., Prgr. Rpt. No. 3, January 1961.

C1d/16   Ibid., Prgr. Rpt. No. 4, April 1961.

C1d/17   Ibid., Prgr. Rpt. No. 5, August 1961.

C1d/18   Ibid., Prgr. Rpt. No. 6, November 1961.

C1d/19   Ibid., Prgr. Rpt. No. 7, February 1962.

C1d/20   Ibid., Prgr. Rpt. No. 8, May 1962.

C1d/21   Ibid., Prgr. Rpt. No. 9, August 1962.

C1d/22   Sutton, W. H., et al (see W1/48-W1/55).

C1d/23   Sutton, W. H., "Development of Composite Structural Materials
         for Space Vehicle Applications," ARS Jour., 32 (4) pp. 593-600,
         1962.

C1d/24   Sutton, W. H., "Investigation of Metal-Whisker Composites,"
         G. E. Co., Missile and Space Div., King of Prussia, Pa., Rpt.
         R62SD65, June 1962, presented at the 64th Annual Mtg. Am. Cer.
         Soc., May 1962.

C1d/25   Sutton, W. H. and Chorne, J., "Potential of Oxide-Fiber Rein-
         forced Metals," G. E. Co., Missile and Space Div., King of
         Prussia, Pa., Rpt. R65SD2, January 1965.

C1d/26   Sutton, W. H. and Chorne, J., "Development of High-Strength,
         Heat-Resistant Alloys by Whisker Reinforcement," Metals Eng.
         Qtrly., 3 (1) pp. 44-51, February 1963.

C1d/27   Vukasovich, M. S., "Reinforcement of Nickel Chromium Alloys
         with Sapphire Whiskers," Horizons Inc., Cleveland, Ohio,
         BuWeps Cont. NOw-64-0125-c, Int. Rpt. No. 1, March 1964.

C1d/28   Ibid., Int. Rpt. No. 2, June 1964.

C1e- Ceramic Whisker - Inorganic Matrix

C1e/1   Kliman, M. I., "Impact Strength of Alumina Fiber-Ceramic
         Composites," Watertown Arsenal Labs., Watertown, Mass.,
         D/A Proj. 59332007, Rpt. WAL TR 371/51, November 1962.

C1e/2   Lockhart, R. J., "Experimental Research on Filamentized
         Ceramic Radome Materials," Horizons Inc., Cleveland, Ohio,
         AFML, WPAFB, Cont. AF 33(615)-1167, Int. Eng. Prgr. Rpt.,
         September 1964.

C1e/3   Shoffner, J. E., et al, "Ceramic Matrix-Ceramic Fiber Com-
         posite Systems," General Dynamics/Convair, San Diego, Calif.,

AFML, WPAFB, Cont. AF 33(657)-8926, Rpt. No. 8926-101, October 1959.

C1f- Ceramic Whisker - Organic Matrix

C1f/1   Milewski, J. V. and Shyne, J., "Whiskers - The Ultimate Reinforcement for Plastics," SPE 20th Annual Conf., 10, January 1964.

C1f/2   Milewski, J. V. and Shyne, J., "Whiskers Make Reinforced Plastics Better Than Metals," paper presented Soc. Plastics Industry, Chicago Mtg., February 1965.

C1f/3   Parratt, N. J. (see C1d/6).

C1f/4   Sutton, W. H., et al, "Whisker Reinforced Plastics for Space Applications,"SPE Jour., 20 (11) pp. 1203-1209, 1964.

C2- Fiber Reinforcement

C2a- Glass Reinforced Plastics

C2a/1   Andreyeuskaya, G. D., "Soviet Trends in Oriented Glass Fiber Reinforced Plastics," Aerospace Information Division, Washington, D. C., AID Report No. 62-173, May 1962.

C2a/2   Bershtein, V. A. and Glikman, L. A., "On a Rapid Method for Determining the Fatigue Strength of Fiber Glass Reinforced Plastics," Industrial Laboratory 30, pp. 274-277, Sept. 1964.

C2a/3   Boller, K. H., "Effect of Pre-Cyclic Stresses on Fatigue Life of Plastic Laminates Reinforced with Unwoven Fibers," For. Prdts. Lab., USDA, Madison, Wisc., AFML, WPAFB, Cont. AF 33(657)-63358, ML TDR 64-168, Sept. 1964.

C2a/4   Boller, K. H., "Fatigue Properties of Plastic Laminates Reinforced with Unwoven Glass Fibers," For. Prdts, USDA, Madison, Wisc., AFML, WPAFB, Cont. AF 33(657)-63358 ASD-TDR-62-464, Mar. 1962.

C2a/5   Boller, K. H., "Strength Properties of Reinforced Plastic Laminates at Elevated Temperatures," For. Prdts. Lab., USDA, Madison, Wisc., AFML, WPAFB, Cont. AF 33(657)-63358, ML TDR 64-167, Aug. 1964.

C2a/6   Brookfield, K. J. and Pickthall, D. , "Some Further Studies on the
         Effect of Glass-Resin Type and Cure on the Strength of Lami-
         nates, " SPE Trans. , 2 (4) pp. 332-338, Oct. 1962.

C2a/7   Brown, R. G. , "Fundamentals of Filament Winding, " Missiles
         and Rockets, 13, pp. 27-29, Aug. 12, 1963.

C2a/8   "Conference on Structural Plastics, Adhesives and Filament
         Wound Composites. Vol. I: Organic and Inorganic Resins and
         Adhesives, Vol. II: Filament Wound Composites, Vol. III:
         Reinforcments, " symposium sponsored by Plastics and Com-
         posites Branch, NML, Mat. Central, WPAFB, Dec. 1962.

C2a/9   Cooney, J. J. , "Development of Improved Resin Systems for
         Filament Wound Structures, " Aerojet-General Corp. ,
         Sacramento, California, BuWeps, Cont. NOw-63-0627-c, Task
         VI, Final Rpt. , Aug. 1964.

C2a/10   Cornish, R. L. , et al, "An Investigation of Material Parameters
         Influencing Creep and Fatigue Life in Filament Wound Laminates, "
         IITRI, Chicago, Ill. , BuShips Cont. NOw-86461, Qtrly. Prgr.
         Rpt. 5, May 1963.

C2a/11   Davis, J. W. , "Standards for Filament Wound Plastics, " SPE
         Journal, 20 (7) pp. 601-604, July 1964.

C2a/12   Downey, T. F. , "Development and Evaluation of Improved Resins
         for Filament Wound Plastics, " U. S. Polymeric Chem. Inc. ,
         Santa Ana, Calif. , Proj. 208LR107, Qtrly. Rpt. No. 2, May 1964.

C2a/13   Duft. B. , et al, "Potential of Filament Wound Composites, "
         Narmco Res. & Dev. , San Diego, Calif. , BuWeps Cont. NOw-61-
         0623-c (FBM), Monthly Prgr. Rpt. No. 2, May 1961.

C2a/14   Eakins, W. J. , "Materials Study. High Strength, High Modulus,
         Filament Wound, Deep Submergence Structures, " DeBell and
         Richardson Inc. , Hazardville, Conn. , BuShips Cont. NObs-
         84672, Prgr. Rpt. No. 3, June 1962.

C2a/15   Epstein, G. , "Filament Winding Puts Strength Where Needed, "
         Mat. in Des. Eng. , 59 (2) p. 90, 1964.

C2a/16   Forcht, B. A. and Rudick, M. J. , "Reinforced Carbonaceous
         Materials, " Astronautics, 6 (4) pp. 32-33, 1961.

C2a/17 Glorioso, S.V., "Preliminary Evaluation of Glass Fiber Reinforced Phenolic Laminate," Gen. Dynamics, Ft. Worth, Texas, AFML, WPAFB Cont. AF 33(657)-7248, Rpt. No. FTDM 1968, Feb. 1962.

C2a/18 Goldberg, R.S., et al, "Micromechanics of Fiber-Reinforced Composites," Rocketdyne Corp., Canoga Park, Calif., Cont. AF 33(615)-1627, Qtrly. Rpt. R-5908-1, Oct. 1964.

C2a/19 Grand d'Hauteville, E., "Textile Glass Products for Reinforced Plastic Laminates," CIBA Review, pp. 34-37, 1963/1965.

C2a/20 Griffith, J.R. and Whisenhunt, F.S., Jr., "Filament Winding Plastics, Part I: Molecular Structure and Tensile Properties," Naval Res. Lab., Washington, D.C., NRL Rpt. 6047, Mar. 1964.

C2a/21 Gruntfest, I.J. and Dow. N.F. (see F2a/4).

C2a/22 Howard, J.S., Jr., "When to Design for Filament Winding," Prod. Eng., 35 (22) pp. 102-107, Oct. 26, 1964.

C2a/23 Islinger, J., et al, "Mechanism of Reinforcement of Fiber-Reinforced Structural Plastics and Composites," IITRI, Chicago, Ill., ASD, WPAFB, AF 33(616)-5983, WADC TR 59-600, Part II, June 1960.

C2a/24 McMullen, P., "The Tensile and Flexural Strengths of Some Glass Reinforced Plastics Tested at Various Rates of Loading," Royal Aircraft Est., Farnborough, England, Tech. Note CPM 16, April 1963.

C2a/25 McNeil, D.W., et al, "Mechanism of Water Absorption in Glass Reinforced Plastics," Battelle Mem. Inst., Columbus, Ohio, BuShips Cont. NObs-86871, Qtrly. Prog. Rpt. No. 3, April 1963.

C2a/26 Marco, D.M., "New and Improved Materials for Expandable Structures," Goodyear Aircraft Corp., Akron, Ohio, AFML, WPAFB, AF 33(616)-7865, ASD-TDR-62-542, Part I, June 1962.

C2a/27 Matta, J.A. and Outwater, J.O., "The Nature, Origin and Effects of Internal Stresses in Reinforced Plastic Laminates," SPE Trans., 2 (4) pp. 314-320, Oct. 1962.

C2a/28 Merolo, S.A., "Properties and Characteristics of Ablative Plastic Chars. Part I: Organic-Fiber Reinforced Plastics," AFML, WPAFB, Dayton, Ohio, ASD-TDR-62-1028, Feb. 1963.

C2a/29 Mettes, D.G. (see F5/11).

C2a/30 Morrison, A.R., et al (see F1/38).

C2a/31 Myers, N.C., et al, "Investigation of Structural Problems with
        Filament Wound Deep Submersibles," H.I. Thompson Fiber
        Glass Co., Gardena, Calif., BuShips Cont. NObs-88351, 3rd
        Qtrly. Rpt., Oct. 1963.

C2a/32 Otto, W.H., et al (see F1/45).

C2a/33 Parratt, N.J. (see F2/15).

C2a/34 Peterson, G.P., "Optimum Filament Wound Composites," ASD,
        WPAFB, Dayton, Ohio, WADD TN 61-50, June 1961.

C2a/35 Petker, I., et al (see F1/48).

C2a/36 Pladek, O.J., "Reinforced Polycarbonates," Reinforced Plastics,
        4 (1) pp. 28-29, Jan./Feb. 1965.

C2a/37 Prosen, S.P., et al, "Interlaminar Shear Properties of Filament
        Wound Composite Materials for Deep Submergence," Naval Ord.
        Lab., White Oak, Md., Proc. 19th Annual Mtg. Soc. Plastics
        Ind., 9-D, pp. 1-8, Feb. 1964.

C2a/38 Riley, M.W., et al, "Filament Wound Reinforced Plastics: State
        of the Art," Mat. in Des. Eng., 52 (2) pp. 127-146, Aug. 1960.

C2a/39 Riley, M.W., "Reinforced Plastics Get Stronger," Mat. in Des.
        Eng., 56 (2) pp. 99-103, Aug. 1962.

C2a/40 Romstad, K., "Investigation of Methods for Evaluating Unwoven
        Glass-Fiber-Reinforced Plastic Laminates in Flexure," For.
        Prdts. Lab., Madison, Wisc., FPL Rpt. 024, Feb. 1964.

C2a/41 Romstad, K., "Methods for Evaluating Tensile and Compressive
        Properties of Plastic Laminates Reinforced with Unwoven Glass
        Fibers," For. Prdts. Lab., USDA, Madison, Wisc., Rpt. No.
        FPL-052, Aug. 1964.

C2a/42 Rosato, D.V., "Non-Woven Fibers in Reinforcing Plastics,"
        Ind. & Eng. Chem., 54 (8) pp. 30-38, 1962.

C2a/43 Rosen, B.W. (see F2a/8-F2a/10).

C2a/44 Roskos, T.G., et al, "Interfacial Properties of an Amine-
Modified Silane and Its Preformed Polymer at E-Glass Surfaces,
SPE Trans., $\underline{2}$ (4) pp. 326-331, Oct. 1962.

C2a/45 Saunders, R.D., "Filament Winding of Thick Laminates for Deep
Submergence," Proc. 19th Annual Meeting, Soc. Plastics Ind.,
Sect. 9-B, pp. 1-5, Feb. 1964.

C2a/46 Stevens, D.W., et al, "Potential of Filament Wound Composites,"
Narmco Res. & Dev., San Diego, Calif. BuWeps Cont. NOw-61-
0623-3 (FBN), Final Summary Rpt., May 1963.

C2a/47 Weil, N.A., et al (see F2a/13).

C2a/48 Werren, F. and Heebink, B.G., "Interlaminar Shear Strength of
Glass-Fiber-Reinforced Plastic Laminates," For. Prdts. Lab.,
Madison, Wisc., Rpt. No. 1848, Sept. 1955.

C2a/49 Wolff, F. and Siuta, T., "Factors Affecting the Performance and
Aging of Filament Wound Fiberglass Structures," ARS Journal,
$\underline{32}$ (6) pp. 948-950, June 1962.

C2a/50 Yates, P.C. and Trebilcock, J.W., "The Chemistry of
Chromium Complexes Used as Coupling Agents in Fiberglass
Resin Laminates," SPE Trans., $\underline{1}$ (4) pp. 199-213, Oct. 1961.

C2a/51 Zisman, W.A., "Surface Chemistry of Glass-Fiber Reinforced
Plastics," Naval Res. Lab., Washington, D.C., NRL Rpt. 6083,
Mar. 1964.

C2b- Carbon/Graphite Reinforced Plastics

C2b/1 Forcht, B.A. and Rudick, M.J. (see C2a/16).

C2b/2 Prosen, S.P., "Some Properties of NOL Rings of Carbon-Base
Fibers and Epoxy Resin," Naval Ordnance Lab., White Oak, Md.,
NOL TR 64-183, Dec. 4, 1964.

C2b/3 Schmidt. D.L. and Hawkins, H.T., "Pyrolyzed Rayon Fiber
Reinforced Plastics," AFML, WPAFB, Dayton, Ohio, ML TDR
64-47, AD 438-892, April 1964.

C2b/4 Schmidt, D.L. and Jones, W.C., "Carbon-Base Fiber Reinforced
Plastics," AFML, WPAFB, Dayton, Ohio, ASD-TDR-62-635,
Aug. 1962.

C2b/5   Schmidt, D. L. and Jones, W. C., "Carbon-Base Fiber-Reinforced
         Plastics," ASD, AFSC, WPAFB, Chem. Eng. Prog., 58 (10) pp.
         42-50, 1962.

C2b/6   Schmidt, D. L. and Jones W. C., "Carbon Fiber Plastic Com-
         posites," SPE Journal, 20 (2) pp. 162-169, Feb. 1964.

C2c- Fiber Reinforced Metal

C2c/1   Arridge, R. G. C., et al (see F2c/3).

C2c/2   Clausen, E. M., et al, "Synthesis of Fiber Reinforced Inorganic
         Laminates," Illinois Univ., Urbana, Ill. AFML, WPAFB,
         AF 33(616)-6283, WADD TR 60-299, Pt. 2, Final Rpt., June
         1962.

C2c/3   Cratchley, D. and Baker, A. A., "The Tensile Strength of a
         Silica Fibre Reinforced Aluminum Alloy," Metallurgia, 69,
         pp. 153-159, April 1964.

C2c/4   Cratchley, D., "Factors Affecting the UTS of a Metal/Metal-
         Fibre Reinforced System," Powd. Met., 11, pp. 59-72, 1963.

C2c/5   Davis, R. M. and Milewski, J. C., "High Temperature Composite
         Structure," Martin Company, Baltimore, Md., ASD-TDR-62-418,
         Sept. 1962.

C2c/6   Harris, R. S., "Investigation of Glass-Metal Composite
         Materials," Owens-Corning Fiberglas Corp., Newark, Ohio,
         Naval Res. Lab. Cont. NOrd 15764, Qtrly. Rpt., July 1960.

C2c/7   Morley, J. G. (see F1/35).

C2c/8   Schuerch, H., "Boron Filament Composite Materials for Space
         Structures, Part I: Compressive Strength of Boron-Metal
         Composites," Astro Research Corp., Santa Barbara, Calif.,
         Cont. NASw-652, Rpt. ARC-R-168, Nov. 1964.

C2c/9   Whitehurst, H. B., "Investigation of Glass-Metal Composite
         Materials," Owens-Corning Fiberglas Corp., U.S.N. Cont.
         NOrd 15764, 3rd Qtrly. Rpt., Dec. 1955.

C2c/10   Whitehurst, H. B. and Ailes, H. B., "Investigation of Glass-Metal
         Composite Materials," Owens-Corning Fiberglas Corp., U.S.N.
         Cont. NOrd 15764, 5th Qtrly. Rpt., Sept. 1956.

C2c/11   Witucki, R.M., "Boron Filament Composite Materials for Space
         Structures," Astro Research Corp., Santa Barbara, Calif.,
         NASw-652, Rpt. ARC-R-168, Nov. 1964.

C2d- Fiber Reinforced Inorganics

C2d/1   Lauchner, J.H., et al, "High Temperature Inorganic Structural
         Composite Materials," Miss. State Univ., State College, Miss.,
         AFML, WPAFB, AF 33(616)-7765, ASD-TDR-62-202, Pt. I.
         Nov. 1962.

C2d/2   Ibid., ASD-TDR-62-202, Part II, Jan. 1963.

C2d/3   Margo, G., "Exploratory Investigation of Inorganic Fiber Rein-
         forced Inorganic Laminates," Goodyear Aircraft Corp., Akron,
         Ohio, ASD, WPAFB, AF 33(616)-5251, WADC TR 58-298,
         Part I, Oct. 1958.

C2d/4   Ibid., Part II, June 1959.

C2d/5   Swica, J., et al, "Metal Fiber Reinforced Ceramics," Alfred
         Univ., Alfred, N.Y., AFML, WPAFB, AF 33(616)-5298, WADC
         TR 58-452, Pt. II, Jan. 1960.

C2d/6   Vondracek, C.H., et al, "Fiber Reinforced Boron Phosphate
         Structural Composites," Westinghouse Electric Corp.,
         Pittsburgh, Pa., AFML, WPAFB, Dayton, Ohio, AF 33(657)-
         7587, ML TDR 64-60, Jan. 1964.

C2d/7   Vondracek, C.H., et al, "Synthesis and Formulation of Inorganic
         Bonded-Inorganic Fiber Reinforced Non-Metallic Structural
         Materials," Westinghouse Electric Corp., Pittsburgh, Pa.,
         AFML, WPAFB, Dayton, Ohio, AF 33(657)-7587, Qtrly. Prog.
         Rpt., June 1963.

C3- Test Methods

C3/1   Bershtein, V. A. and Glikman, L.A. (see C2a/2).

C3/2   Boller, K. H. (see C2a/2-C2a/5).

C3/3   Cratchley, D. (see C2c/3-C2c/4).

C3/4   Glorioso, S.V. (see C2a/17).

C3/5   McDaniels, D.L., et al (see C1a/6).

C3/6   McMullen, P. (see C2a/24).

C3/7   Romstad, K. (see C2a/40-C2a/41).

C3/8   Werren, F. and Heebink, B.G. (see C2a/48).

C4- Interfacial Studies

C4a- Whisker Composites

C4a/1   Schuster, D.M. and Scala, E., "The Mechanical Interaction of
        Sapphire Whiskers with a Birefrigent Matrix," Trans. AIME,
        230, pp. 1635-1640, Dec. 1964.

C4a/2   Sutton, W.H., "Investigation of Bonding in Oxide-Fiber (Whisker)
        Reinforced Metals," G.E. Co., Missile and Space Div., King of
        Prussia, Pa., U.S. Army Materials Research Agency, AMRA
        CR 6301/1, November 1962.

C4a/3   Ibid., AMRA CR 6301/2, January 1963.

C4a/4   Ibid., AMRA CR 6301/3, March 1963.

C4a/5   Ibid., AMRA CR 6301/4, June 1963.

C4a/6   Ibid., AMRA CR 6301/5, September 1963.

C4a/7   Ibid., AMRA CR 6301/6, December 1963.

C4a/8   Ibid., AMRA CR 6301/7, March 1964.

C4a/9   Ibid., AMRA CR 6301/8, June 1964.

C4a/10 Sutton, W.H. et al (see C1d/13-21).

C4a/11 Ibid. (see W1/48-55).

C4b - Fiber Composites

C4b/1   Eakins, W.J., "Glass/Resin Interface: Patent Survey, Patent
        List and General Bibliography," DeBell & Richardson Inc.,
        Hazardville, Conn., Plastics Tech. Eval. Cntr., Picatinny
        Arsenal, Dover, N.J., Plastic Rpt. 18, Sept. 1964.

C4b/2   Eakins, W.J., "The Interfaces: The Critical Region of a Com-
        posite Material," DeBell & Richardson Inc., Hazardville, Conn.,

USAMC, Picatinny Arsenal, Cont. DA-28-017-AMC-343(A), A State-of-the-Art Evaluation, Nov. 1963.

C4b/3   Erickson, P.W. and Volpe, A.A., "Chemical and Physical Effects of Glass Surfaces Upon an Epoxide Polymer System," Naval Ord. Lab., White Oak, Md., NOL TR 63-10, Mar. 1963.

C4b/4   Erickson, P.W., et al, "Chemical and Physical Effects of Glass Surfaces Upon Laminating Resins," Naval Ord. Lab., White Oak, Md., NOL TR 63-267, Jan. 1964.

C4b/5   Gutfreund, K. and Weber, H.S., "Interaction of Organic Monomers and Water with Fiberglass," SPE Trans., 1 (4) pp. 191-198, Oct. 1961.

C4b/6   Laird, J.A., et al, "Glass Surface Chemistry for Glass Fiber Reinforced Plastics," A.O. Smith Corp., Milwaukee, Wisc., BuWeps Cont. NOw-62-0679-c (FBM), Final Summary Report, June 1963.

C4b/7   McNeil, D.W., et al (see C2a/25).

C4b/8   Patrick, R.L. and Port, E.A., "Interfacial Interaction in Composite Structures," Alpha Res. & Dev. Inc., Blue Island, Ill., BuWeps Cont. NOw-64-0413-c, Qtrly. Prgr. Rpt. 2, Sept. 1964.

C4b/9   Patrick, R.L., "Physical Techniques Used in Studying Interfacial Phenomena," SPE Trans., 1 (4) pp. 181-190, Oct. 1961.

C4b/10   Roskos, T.G., et al (see C2a/44).

C4b/11   Sterman, S. and Bradley, H.B., "A New Interpretation of the Glass-Coupling Agent Surface Through Use of Electron Microscopy," SPE Trans., 1 (4) pp. 224-233, October 1961.

C4b/12   Vanderbilt, B.M. and Jarvzelski, J.J., "The Bonding of Fillers to Thermosetting Resins," SPE Trans., 2 (4) pp. 363-366, October 1962.

C5- General Review and/or Strength Theory

C5/1   Abbott, H.M., "Composite Materials: An Annotated Bibliography," Lockheed Missiles and Space Co., Sunnyvale, Calif., SB-62-58, February 1963.

C5/2   Ibid., "Supplement No. 1," SB-63-49, June 1963.

C5/3   Accountius, O.E. (see W5/1).

C5/4   Anon., "The Promise of Composites," Mat. in Des. Eng., 58 (3)
         pp.89-126, 1963.

C5/5   Bradstreet, S.E., "Principles Affecting High Strength to Density
         Composites with Fibers or Flakes," AFML, WPAFB, Dayton,
         Ohio, ML TDR 64-85, May 1964.

C5/6   Cratchley, D. and Baker, A.A. (see C2c/3).

C5/7   Dietz, A.G.H., "Fibrous Composite Materials," Int. Sci. and
         Tech., (32) pp.58-62, 64, 66-69, August 1964.

C5/8   Dow, N.F., "Study of Stresses Near a Discontinuity in a Fila-
         ment-Reinforced Composite Metal," G.E. Co., Missile and
         Space Div., King of Prussia, Pa., U.S. Navy BuWeps Cont.
         NOw-60-0465-d, August 1963.

C5/9   Glasser, J.H. and Sump, C.H., "Report on Evaluation of Manu-
         facturing Methods for Fibrous Composite Materials," survey
         conducted for MPB, Mfg, Tech. Div., AFML, WPAFB,
         July 10, 1964.

C5/10   Goldberg, R.S., et al (see C2a/18).

C5/11   Grand d'Hauteville, E. (see C2a/19).

C5/12   Kelly, A. (see C1/3-C1/4).

C5/13   King, H.A., "Reinforced Plastics in Aerospace and Hydro-
         space in the Year 2000," Proc. Soc. Plastic Ind., 19th Annual
         Mtg., 16-A, pp.1-12, Feb. 1964.

C5/14   King, H.A., "Reinforced Plastics for Aerospace and Hydro-
         space," Research and Dev., 15 (12) pp.17-20, 1964.

C5/15   Kiselev, B.A., "Glass Fiber Reinforced Plastics," U.S. Dept.
         Comm., Washington, D.C., Joint Publications Res. Ser. Rpt.
         19389, May 1963. (in English)

C5/16   Krock, R.H. and Kelsey, R.H. (see W5/11).

C5/17   Machlin, E.S., "Non-Metallic Fibrous Reinforced Metal Com-
         posites," Mat. Res. Corp., Yonkers, N.Y., BuWeps NOw-61-
         0209-c, Status Report, Sept. 1961.

C5/18   Pladek, O.J. (see C2a/36).

C5/19   Riley, M.W., et al (see C2a/38).

C5/20   Ruane, A.G., comp., "Fiber Reinforcement in Metals and
         Materials, A Literature Search," G.E. Co., Res. Lab.,
         Schenectady, N.Y., Aug. 1964.

C5/21   Sadowsky, M.A. and Hussain, M.A., "Thermal Stress Discon-
         tinuities in Microfibers," Watervliet Arsenal, Watervliet, N.Y.,
         WVT RR 6401, AD 601-057, April 1964.

C5/22   Sadowsky, M.A. and Hussain, M.A., "Alpha-Classification of
         Microfilms, Part IV," Watervliet Arnsenal, Watervliet, N.Y.,
         Rpt. WVT RR 6317, October 1963.

C5/23   Sadowsky, M.W., et al, "Effect of Couple Stresses on Force
         Transfer Between Embedded Microfibers," Watervliet Arnsenal,
         Watervliet, N.Y., Rpt. WVT RR 6407, June 1964.

C5/24   Salkind, M.J. (see W5/21-W5/22).

C5/25   Scala, E., "The Design and Performance of Fibers and Com-
         posites," Cornell University, Materials Science Center, Ithaca,
         New York, Rpt. 286, November 1964.

C5/26   Stowell, E.Z. and Liu, Tien-Shih, "Parametric Studies of Metal
         Fiber Reinforced Ceramic Composite Materials," Southwest Res.
         Inst., San Antonio, Texas, BuWeps Cont. NOas-60-6077-c, Int.
         Rpt. No. 2, May 1960.

C5/27   Sutton, W.H. and Chorne, J. (see C1d/25).

C5/28   Tyson, W.R. and Davies, G.J., "A Photoelastic Study of the
         Shear Stresses Associated with the Transfer of Stress During
         Fibre Reinforcement," Brit. J. Appl. Phys., 16 (3) pp. 199-206,
         1965.

C5/29   Wagner, H.J., "Review of Recent Developments: Fiber-
         Reinforced Metals," Battelle Mem. Inst., Columbus, Ohio,
         DMIC Bulletin, March 6, 1964.

C5/30   Ibid., December 11, 1964.

A- APPLICATIONS

A1- Whiskers

A1/1   Arledter, H. F. and Knowles, S. E., <u>Synthetic Fibers in Paper-making</u>, edited by O. A. Battista, Interscience Publishers, New York, pp. 185-243, 1964.

A1/2   Carroll-Porczynski, C. Z., <u>Advanced Materials,</u> Chemical Publishing Co., 286 pp., 1962.

A1a- Whisker Reinforcement

(No entries)

A1b- Thermal Insulation

(No entries)

A1c- Fillers

(No entries)

A1d- Papers

(No entries)

A1e- Other Applications

A1e/1   Murphy, W. K. and Sears, G. W., "Iron Whisker Resonators," Appl. Phys. Letters, <u>3</u> (4) pp. 55-57, Aug. 1963.

A2- Fibers

A2/1   Arledter, H. F. and Knowles, S. E. (see A1/1).

A2/2   Carroll-Porczynski, C. Z., (see A1/2).

A2/3   Johnson, D. E. and Newton, E. H., "Metal Filaments for High Temperature Fabrics," A. D. Little, Inc., Cambridge, Mass., AFML, WPAFB, AF 33(657)-10539, ASD-TDR-62-180, Part II, Feb. 1963.

A2/4   Johnson, D.E., et al, "Candidate Materials for High Temperature Fabrics," A.D. Little Inc., Cambridge, Mass., WADC, WPAFB, AF 33(616)-5880, WADC TR 59-155, Sept. 1959.

A2/5   Tighe, N.J. (see F1/55).

A2a- Fiber Reinforcement

A2a/1   Carroll, M.T. and Pritchard, D.J., "Evaluation of Metal Fiber Composite," General Dynamics, Ft. Worth, Texas, AFML, WPAFB, AF 33(657)-7248, Rpt. No. ERR-FW-121, Dec. 1961.

A2a/2   Otto, W.H., et al (see F1/48).

A2a/3   Ross, J.H., "Needed: Textiles for the Space Age," Mat. on Des. Eng., 51 (6) pp. 138-140, June 1960.

A2a/4   Ross, J.H. and Little, C.O., "Thermal Stability of Flexible Fibrous Materials," AFML, WPAFB, Dayton, Ohio, Rpt. RTD-TDR 63-4031, Oct. 1964.

A2a/5   Schmidt, D.L. and Hawkins, H.T. (see C2b/3).

A2a/6   Schmidt, D.L. and Jones, W.C. (see C2b/4-C2b/6).

A2a/7   Sollenberger, N.J., "Investigation of Glass Fibers as Reinforcement for Prestressing Concrete," Dept. Civil Eng., Princeton U., Dept. Navy Cont. NBy 8996, Final Rpt., July 1963.

A2b- Thermal Insulation

A2b/1   Ross, J.H. and Little, C.O. (see A2a/4).

A2b/2   Wechsler, A.E. and Glaser, P.E., "Investigation of the Thermal Properties of High Temperature Insulation Materials," A.D. Little, Inc., Cambridge, Mass., AFML, WPAFB, AF 33(657)-9172, ASD-TDR-63-574, July 1963.

A2c- Fillers

A2c/1   Schmidt, D.L. and Jones, W.C. (see C2b/4-C2b/6).

A2d- Fabrics and Papers

A2d/1   Barish, L., et al (see F2/2).

A2d/2   Cobb, E.S., Jr., "High Strength Glass Fiber Webbings, Tapes
         and Ribbons for High Temperature Pressure Packaged
         Decelerators," Owens-Corning Fiberglas Corp., Ashton, R.I.,
         AFML, WPAFB, AF 33(616)-7441, ASD-TDR-62-518, July 1962.

A2d/3   Epting, J.L., Jr., "Silicone Finished Glass Fiber Tapes and
         Webbing for Decelerators," ASD, WPAFB, ASD-TDR-63-808,
         Sept. 1963.

A2d/4   Ross, J.H. and Little, C.O. (see A2a/4).

A2e- Other Applications

A2e/1   Zimmerman, F.J., "Fiber Metals for Meteoroid Protection,"
         Proc. 5th Annual Structures and Materials Conf., AIAA, Palm
         Springs, Calif., April 1964.

CF - COMPOSITE AND POLYCRYSTALLINE FIBERS

CF1- Carbon or Graphite Fibers; Properties, Used and Forming Methods

CF1/1   Bacon, R. and Tang, M.M., "Carbonization of Cellulose Fibers
         - II. Physical Property Study," Carbon, 2 (3) pp.221-225,
         December 1964.

CF1/2   Dawn, F.S. and Ross, J.H., "Investigation of the Thermal
         Behavior of Graphite and Carbon-Based Fibrous Materials,"
         ASD, WPAFB, Dayton, Ohio, ASD-TDR-62-782, Oct. 1962.

CF1/3   Hough, R.L., "Continuous Pyrolytic Graphite Composite Fila-
         ments," Proc. 5th Annual Structures and Materials Conference
         AIAA, pp. 471-479, 1964.

CF1/4   Nikolayov, V. Yu and Kasatochkin, V.I., "Fibrous Carbon,"
         AFSC, WPAFB, FTD Rpt. TT63-330, May 1963, (in English)

CF1/5   Papalegis, F.E. and Bourdeau, R.G. (see W1/35-W1/38).

CF1/6   Schmidt, D.L. and Jones, W.C. (see C2b/4-C2b/6).

CF1/7   Schulman, S., "Elevated Temperature Behavior of Fibers," ASD,
         WPAFB, Dayton, Ohio, Rpt. WADD TN 60-298, August 1961.

CF1/8   Ibid., Rpt. ASD-TDR-63-62, April 1963.

CF1/9 Shindo, A., "Studies on Graphite Fibers," Japanese Gov't Ind. Res. Inst., Osaka, Japan, Rpt. #317, December 1961.

CF1/10 Stover, E.R., "Structure of Whiskers in Pyrolytic Graphite," G.E. Co., Res. Lab., Schenectady, N.Y., Rpt. 64-RL-3609M, March 1964.

CF1/11 Tang, M.M. and Bacon, R., "Carbonization of Cellulose Fibers - I. Low Temperature Pyrolysis," Carbon, $\underline{2}$ (3) pp.211-220, December 1964.

CF1/12 Wechsler, A.E. and Glaser, P.E. (see A2b/2).

CF2 - Polycrystalline Fibers; Properties, Uses and Forming Methods

CF2/1 Anon., "Structural Ceramic Fibers of Virtually Any Oxide," Mat. in Des. Eng., $\underline{52}$ (1) p.5, July 1960.

CF2/2 Ellis, R.B., "Investigation of Techniques and Materials for the Formation of High Temperature (1500^0) Inorganic Fibers," Southern Res. Inst., Birmingham, Ala., ASD, WPAFB, AF 33(616)-6955, ASD-TDR-61-134, June 1961.

CF2/3 Farmer, R.W., "Air Aging of Inorganic Fibrous Insulations," AFML, WPAFB, Dayton, Ohio, ASD-TDR-64-137, June 1964.

CF2/4 Fellows, B.T., "Polycrystalline Ceramic Fiber Reinforcements for High Temperature Structural Composites," H.I. Thompson Fiber Glass Co., Gardena, Calif., ASD, WPAFB, AF 33(616)-8080, Qtrly. Prgr. Rpt. 7, January 1963.

CF2/5 Ibid., ASD-TDR-62-1108, May 1963.

CF2/6 Ibid., Qtrly. Prgr. Rpt. 9, June 1963.

CF2/7 Ibid., Qtrly. Prgr. Rpt. 11, December 1963.

CF2/8 Ibid., Qtrly. Prgr. Rpt. 12, March 1964.

CF2/9 Florentine, R.A., et al, "Magnesium Oxide in Fiber Form by Extrusion Methods," Penn-Salt Chemicals Corp., Phila., Pa., Naval Res. Lab. Cont. Nonr 2687(00), Qtrly. Prgr. Rpt., August 1960.

CF2/10 Hough, R.L., "Refractory Reinforcements for Ablative Plastics. Part I: Synthesis and Reaction Mechanism of Fibrous Zirconium

Nitride," AFML, WPAFB, Dayton, Ohio, Proj. 7340, ASD-TDR-62-260, Pt. I, June 1962.

CF2/11   Ibid., "Part II: Synthesis of Zirconium Nitride and ZrO_2 Flakes," October 1963.

CF2/12   Ibid., "Part III: Pyrolytic Boride Reinforcing Agents," August 1964.

CF2/13   Ibid., "Part IV: Synthesis Apparatus for Continuous Filamentous Reinforcements," December 1963.

CF2/14   Lachman, W. L. and Sterry, J. P., "Ceramic Fibers," Chem. Eng. Prog., 58 (10) pp. 37-41, 1962.

CF2/15   Mazdiyasni, K. S. and Lynch,· C. T., "An Approach to the Preparation of Powders, Fibers and Films of Ultra-High Purity Ceramics," AFML, WPAFB, Dayton, Ohio, ML TDR 64-269, September 1964.

CF2/16   Schulman, S. (see CF1/6).

CF2/17   Soltis, P. J., "Investigation of Horizons Inc. Zirconia Fibers," Naval Air Eng. Cntr., Phila., Pa., AML Rpt. 1549, AD 404 358L, November 1962.

CF2/18   Spencer, S. B. and Jackson, W. O., "A New Fiber for Reinforced Plastic Aerospace Applications," presented at Regional Meeting of SPE, Garden City, Long Island, May 13-14, 1964.

CF2/19   Sterry, J. P., "New Refractory for Ablation, Thermal Insulation," Mat. in Des. Eng., 56 (4) pp. 12-13, 1962.

CF2/20   Stites, J. and Sterry, J. P., "Non-Vitreous, Non-Metallic Fiber Synthesis," H. I. Thompson Fiber Glass Co., Gardena, Calif., ASD, WPAFB, AF 33(616)-8080, 6th Qtrly. Rpt., October 1962.

CF2/21   Wechsler, A. E. and Glaser, P. E. (see A2b/2).

CF2/22   Weik, H., "Investigation of Mechanical and Physical Properties of Silicon Dioxide and Zirconium Dioxide Fibers," WADC, WPAFB, Dayton, Ohio, WCLT TM 58-72, May 1958.

CF3 - Composite Fibers; Properties, Uses, and Forming Methods

CF3/1   Adams, J.J. and Sterry, J.P., "High Temperature Fibrous
        Insulation," H.I. Thompson Fiber Glass Co., Gardena, Calif.,
        AFML, WPAFB, ML TDR 64-156, October 1964.

CF3/2   Ibid., "Zirconia Fibrous Insulations," ASD-TDR-63-725,
        December 1963.

CF3/3   Cox, J.E., et al, "Exploratory Investigation of Glass-Metal
        Composite Fibers," United Aircraft Corp., East Hartford,
        Conn., BuWeps Cont. NOw-64-0389-f, Bi-Monthly Prgr. Rpt.
        No. 1, July 13, 1964.

CF3/4   Ibid., Prgr. Rpt. No. 2, September 1964.

CF3/5   Ibid., Prgr. Rpt. No. 3, November 1964.

CF3/6   Davies, L.G. and Withers, J.C., "A Study of High Modulus,
        High Strength Filament Materials by Deposition Techniques,"
        General Tech. Corp., Alexandria, Va., BuWeps Cont. NOw-
        64-0176-c, 1st Bi-Monthly Rpt., January 1964.

CF3/7   Ibid., 2nd Bi-Monthly Rpt., March 1964.

CF3/8   Ibid., 3rd Bi-Monthly Rpt., May 1964.

CF3/9   Ibid., 4th Bi-Monthly Rpt., July 1964.

CF3/10 Ibid., 5th Bi-Monthly Rpt., September 1964.

CF3/11 Davies, L.G., et al, "A Study of High Modulus, High Strength
        Filament Materials by Deposition Techniques," General Tech.
        Corp., Alexandria, Va., BuWeps Cont. NOw-64-0176-c, Final
        Report, January 1965.

CF3/12 Fechek, F. and Hennessey, M., "Boron Fiber Reinforced
        Structural Composites," paper presented AIAA Launch and Space
        Vehicle Shell Structure Conference, Palm Springs, Calif.,
        April 1-3, 1963.

CF3/13 Gatti, A. and Higgins, J., "Research on High Strength, High
        Modulus, Low-Density Continuous Filaments of Boron Carbide,"
        G.E. Co., Missile and Space Div., King of Prussia, Pa., Cont.
        AF 33(615)-1644, 2nd Qtrly. Prgr. Rpt., February 1965.

CF3/14   Hough, R. L. (see CF1/3).

CF3/15   Otto, W. H. and Vidanoff, R. B. (see F1/46-F1/47).

CF3/16   Schuerch, H. (see C2c/8).

CF3/17   Souttar, H. S., "Demonstration of a Method of Making Capillary
         Filaments," Phys. Soc. Proc., London, 24, pp. 166-167, 1912.

CF3/18   Talley, C. P., et al, "Boron Reinforcements for Structural
         Composites," Texaco Exp. Inc., Richmond, Va., AFML,
         WPAFB, AF 33(616)-8067, Summary Rpt. 1, May 1962.

CF3/19   Ibid., ASD-TDR-62-257, Part II, April 1963.

CF3/20   Vidanoff, R. and Bradford, D. (see F3a/22).

CF3/21   Witucki, R. M., "Boron Filaments," Astro Research Corp.,
         Santa Barbara, Calif., NASw-652, Rpt. NASA CR-96, Sept. 1964.

CF3/22   Witucki, R. M. (see C2c/11).

CF4 - Metal Filaments

CF4/1   Arledter, H. F., Synthetic Fibers in Papermaking, edited by
         O. A. Battista, Interscience Publishers, New York, pp. 118-184,
         1964.

CF4/2   Geminov, U. N. and Kop'yev, I. M., "The Strength of Fine
         Metallic Threads," Joint Publications Res. Service, OTS,
         U. S. Dept. Commerce, TPRS 16567, Dec. 1962.

CF4/3   Gorton, C. A. and McMahon, C. C., "Ultra-Fine, High Tem-
         perature, High Strength Metallic Fibers," Hoskins Mfg. Co.,
         Detroit, Mich., AFML, WPAFB, Cont. AF 33(616)-8366, ML
         TDR 64-48, Feb. 1964.

CF4/4   Johnson, D. E. and Newton, E. H. (see A2/3).

CF4/5   Johnson, D. E., et al (see A2/4).

CF4/6   Murphy, E. A. and O'Rourke, R. G., "Fabrication of Ultrafine
         Beryllium Wire," The Brush Beryllium Co., Cleveland, Ohio,
         BuWeps Cont. NOw-63-0137-c, TR 318-240, Aug. 1963.

CF4/7   Newton, E. H. and Johnson, D. E., "Fine Metal Filaments for
         High Temperature Applications," A. D. Little, Inc., Cambridge,
         Mass., AFML, WPAFB AF 33(657)-10539, ML TDR 64-92,
         Feb. 1964.

CF4/8   Roberts, D. A., "Physical and Mechanical Properties of Some
         High-Strength Fine Wires," Battelle Memorial Institute, DMIC
         Memo 80, January 1961.

CF4/9   Soltis, P. J., "Study of Size Effect in Fine Beryllium Wires,"
         Naval Air Eng. Cntr., Phila., AML Rpt. 1909, AD 438 478,
         March 1964.

R- RELATED STUDIES WITH BULK MATERIALS AND/OR
   THEORETICAL REVIEWS

R1- Graphite

R1/1   Amelinckx, S. and Delavignette, P., "Electron Optical Study of
         Basal Dislocations in Graphite," J. Appl. Phys., $\underline{31}$ (12)
         pp. 2126-2135, 1960.

R1/2   Bushong, R. M., et al (see W1/6).

R1/3   Pullam, G. R., "Ceramics," Space Aeronautics R&D Handbook,
         $\underline{42}$ (4) pp. 156-158, Sept. 1964.

R2- Sapphire

R2/1   Conrad, H., "Mechanical Behavior of Sapphire," Aerospace
         Corporation, El Segundo, Calif., Rpt. No. ATN-64 (9236)-14,
         Mar. 1964.

R2/2   Pullam, G. R. (see R1/3).

R2/3   Stephens, D. L. and Alford, W. J., "Dislocation Structures in
         Single Crystal Alumina," J. Am. Cer. Soc., $\underline{47}$ (2) pp. 81-84,
         1964.

R3- Glass

R3/1   Douglas, R. W., et al, "A Study of Configuration Changes in
         Glass by Means of Density Measurements," J. Soc. Glass Tech.,
         $\underline{32}$ pp. 309-339, 1948.

R3/2   Ernsberger, F. M., Progress in Ceramic Science, Vol. III,
          edited by J. H. Burke, The Macmillan Co., New York, p. 57, 1963.

R3/3   Gordon, J. E., et al, "On the Strength and Structure of Glass,"
          Proc. Royal Soc., London, A249, pp. 65-72, 1959.

R3/4   Hillig, W. B., "The Strength of Bulk Fused Quartz," G. E. Co.,
          Research Laboratory, Schenectady, N. Y., Rpt. No. 60-RL-
          2449M, Aug. 1960.

R3/5   Hillig, W. B., "The Factors Affecting the Strength of Bulk Fused
          Silica," G. E. Co., Research Laboratory, Schenectady, N. Y.,
          Rpt. 61-RL-2777M, July 1961.

R3/6   Hillig, W. B. and Charles, R. J., "Concerning Dislocations in
          Glass," G. E. Co., Research Laboratory, Schenectady, N. Y.,
          Rpt. 60-RL-2535M, Oct. 1960.

R3/7   Holland, L, The Properties of Glass Surfaces, John Wiley &
          Sons Inc., New York, 546pp., 1964.

R3/8   Kerper, M. J., et al, "Properties of Glasses at Elevated Tem-
          peratures," National Bureau of Standards, Washington, D. C.,
          ASD, WPAFB, AF 33(657)-62-362, WADC TR 56-645, Part VIII,
          Mar. 1963.

R3/9   Kies, J. A., "The Strength of Glass," U.S. N. Res. Lab.,
          Washington, D. C., NRL Rpt. 5098, April 1958.

R3/10   Loewenstein, K. L., "Studies in the Composition and Structure
          of Glasses Possessing High Young's Modulus Part 1: The Com-
          position of High Young's Modulus Glasses and the Function of
          Individual Ions in the Glass Structure," Phys. and Chem.
          Glasses, 2 (3) pp. 69-82, June 1961.

R3/11   Ibid., "Part 2: The Effects of Changes in the Configuration
          Temperature," Phys. and Chem. Glasses, 2 (4) pp. 119-125,
          Aug. 1961.

R3/12   Phillips, C. J., Glass: Its Industrial Applications, Reinhold
          Publishing Corp., New York, 252pp., 1960.

R3/13   "Properties of Selected Commercial Glasses," Bulletin B-83,
          Corning Glass Works, Corning, New York.

R3/14   Pustovalov, V. V., "Change in Thermal Conductivity of Quartz
        Glass in the Process of Crystallization," Steklo i Keramika, 5
        pp. 28-30, May 1960, trans. by A. J. Peat, G. E. Co., Res. Lab.,
        Schenectady, N.Y., Rpt. 60-RL-2525M, Sept. 1960.

R3/15   Shand, E. B., "Experimental Study of Fracture of Glass: I, The
        Fracture Process," J. Am. Cer. Soc., 37 (2) pp. 52-60, 1954.

R3/16   Ibid., "II, Experimental Data," J. Am. Cer. Soc., 37 (12)
        pp. 559-572, 1954.

R4- Plastics

R4/1   Feltzin, J., "Research and Development of High Temperature
        Stable Inorganic Resins and Elastomers," Aerojet-General
        Corp., Azuza, Calif., Dept. of the Army Cont. DA-04-495-ORD-
        3075, Final Report 2365, Aug. 1962.

R4/2   Frazer, A. H. and Reed, T. A., "Research on Aromatic Polymers
        for Thermally-Stable Fibers and Films," E. I. duPont De
        & Co., Textile Fibers Dept., ASD, WPAFB, AF 33(616)-8253,
        Qtrly. Prgr. Rpt., April 1963.

R4/3   Zeilberger, E. J., "Plastics," Space Aeronautics R&D Handbook,
        42 (4) pp. 159-162, Sept. 1964.

R5- Metals

R5/1   Dix, E. H., Jr., "Aluminum Alloys for Elevated Temperature
        Service," Symp. on Structures for Thermal Flight, ASME
        Aviation Conf., Los Angeles, Calif., paper 56-AV-8, March 14-
        16, 1956.

R5/2   Spretnak, J. W., "A Summary of the Theory of Fracture in
        Metals," Battelle Mem. Inst., DMIC Rpt. 157, August 1961.

R6- Theoretical Reviews

R6/1   Fracture, edited by B. L. Averbach, D. K. Felbeck, G. T. Hahn,
        and D. A. Thomas, Technology Press and John Wiley and Sons
        Inc., New York, 646pp., 1959.

R6/2   Gilman, J. J., "High Strength Materials of the Future," G. E.
        Co., Research Laboratory, Schenectady, N.Y., Rpt. No. 60-
        RL-2579M, December 1960.

R6/3   Hillig, W. B. and Charles, R. J., "Surface, Stress-Dependent
          Reactions, and Strength," G. E. Co., Research Laboratory,
          Schenectady, N. Y., Rpt. No. 64-RL-3756M, October 1964.

R6/4   Hillig, W. B., "The Sources of Weakness and the Ultimate
          Strength of Brittle Amorphous Solids," G. E. Co., Research
          Laboratory, Schenectady, N. Y., Rpt. No. 62-RL-2896M,
          February 1962.

R6a- Crystals

R6a/1   Cottrell, A. H., Dislocation and Plastic Flow in Crystals,
          Oxford Press, New York, 223pp., 1958.

R6a/2   Garber, R. I. and Gindin, I. A., "Physics of the Strength of
          Crystalline Materials," Soviet Physics-Uspekhi, $\underline{3}$ (70) pp. 41-77,
          1960. (in English)

R6a/3   Griffith, A. A., "Phenomena of Rupture and Flow in Solids,"
          Trans. Royal Soc., London, A221, pp. 163-169, 1920.

R6a/4   Hull, E. H. and Malloy, G. T., "The Strength of Diamond," G. E.
          Co., Research Laboratory, Schenectady, N. Y., Rpt. No. 64-
          RL-3824C, November 1964.

R6a/5   Klassen-Neklyudova, M. V. and Rozhanskii, V. N., "Funda-
          mental Problems in the Physics of the Strength and Plasticity of
          Crystals," Soviet Physics-Crystallography, $\underline{1}$ (4) pp. 403-409,
          1963. (in English)

R6a/6   Polanyi, Von M., "The Nature of Fracture," Z. Phys., $\underline{7}$ pp.
          323-327, 1921.

R6a/7   Vuillard, G., "Boron Carbide Crystals," Compt. Rend., $\underline{257}$
          (25) pp. 3927-3929, 1963. (in French)

R6a/8   Westbrook, J. H., "The Source of Strength and Brittleness in
          Intermetallic Compounds," G. E. Co., Research Laboratory,
          Schenectady, N. Y., Rpt. No. 64-RL-3785M, September 1964.

R6b- Filaments

R6b/1   Coleman, B. D., "Statistics and Time Dependence of Mechanical
          Breakdown in Fibres," J. Appl. Phys., $\underline{29}$ (6) pp. 968-983, 1958.

R6b/2   Coleman, B.D., "On the Strength of Classical Fibres and Fibre
         Bundles," J. Mechs. and Phys. Solids, 7 pp.60-70, 1958.

R6b/3   Cox, H.L., "The Elasticity and Strength of Paper and Other
         Fibrous Materials," Brit. J. Appl. Phys., 3 p. 72, 1952.

R6b/4   Daniels, H.E., "The Statistical Theory of the Strength of
         Threads," Proc. Royal Soc., London, 183A p. 405, 1945.

R6b/5   Dukes, W.A., "The Endurance of Textile Materials Under Con-
         stant Load, A Review," Expl. Res. and Dev. Est., Waltham
         Abbey, England, ERDE Survey No. 1/5/62, AD 290-678,
         August 1962.

R6b/6   Synthetic Fibers in Papermaking, edited by O.A. Battista,
         Interscience Publishers, New York, 340pp., 1964.

R7 - Crystal Growth

R7/1   Kestigan, M. and Tombs, N.C., "Growth of Single Crystals of
         Electronic Materials," Sperry Eng. Rev., 16 (3) pp.2-9,
         Fall 1963.

R7/2   Laudise, R.A., "Growing Oxide Crystals," Bell Labs. Record,
         40 (7) pp.244-250, 1962.

R7/3   Linares, R.C., "Growth of Refractory Oxide Single Crystals,"
         J. Appl. Phys., 33 (5) pp.1747-1749, 1962.

R7/4   Linz, A., et al, "Growth of Crystals by Flame Fusion," MIT,
         Cambridge, Mass., AFML, WPAFB, AF 33(616)-8353 and
         AF 19(628)-395. Rpt. MIT-TR-185, December 1963.

R7/5   McMurdie, H.F., editor, "Research on Crystal Growth and
         Characterization at the National Bureau of Standards, January to
         June 1964," National Bureau of Standards, Washington, D.C.,
         NBS Tech. Note 251, October 1964.

R7/6   Pieser, H.S., editor, "Research on Crystal Growth and
         Characterization at the National Bureau of Standards, July to
         December 1963," National Bureau of Standards, Washington,
         D.C., NBS Tech. Note 236, April 1964.

NAME INDEX

Note: "T" after page number indicates Table reference.

SUBJECT INDEX